AF567027

NONDESTRUCTIVE EVALUATION IN THE NUCLEAR INDUSTRY

NONDESTRUCTIVE EVALUATION IN THE NUCLEAR INDUSTRY

Proceedings of an International Conference

13-15 February 1978

Salt Lake City, Utah

Sponsored by the
Materials Testing and Quality Control Division
of the
American Society for Metals
in cooperation with
American Society for Testing and Materials,
American Society for Nondestructive Testing,
and
American Nuclear Society

Edited by

R. Natesh

Materials Research, Incorporated

Materials/Metalworking Technology Series
AMERICAN SOCIETY FOR METALS
Metals Park, Ohio 44073

Library of Congress Cataloging in Publication Data

Main entry under title:

Nondestructive evaluation in the nuclear industry.

(Materials/metalworking technology series)
Includes bibliographical references.
1. Non-destructive testing—Congresses. 2. Nuclear engineering—Materials—Testing—Congresses. I. Natesh, R II. American Society for Metals. Materials Testing and Quality Control Division. III. Series.

TA417.2.N66 621.48'3'028 78-25552
ISBN 0-87170-029-8

PRINTED IN THE UNITED STATES OF AMERICA

CONFERENCE ORGANIZING COMMITTEE

RAM NATESH
Conference General Chairman
Materials Research, Incorporated - Centerville, Utah

H. H. ASKWITH
General Physics Corporation - Columbia, Maryland

ELLIOT B. BOARDMAN
American Society for Metals - Metals Park, Ohio

DONALD N. BUGDEN
Magnetic Analysis Corporation - Mt. Vernon, New York

LANCE E. BURGESS
American Society for Testing and Materials
Philadelphia, Pennsylvania

V. (HARY) CHARYULU
Idaho State University - Pocatello, Idaho

ROGER B. CLOUGH
National Bureau of Standards - Washington, D. C.

V. S. GOEL
United States Nuclear Regulatory Commission - Washington, D.C.

HAROLD C. GRABER
Babcock & Wilcox Company, Barberton, Ohio

G. ROBERT KEEPIN
University of California - Los Alamos, New Mexico

K. J. REIMANN
Argonne National Laboratory - Argonne, Illinois

EUGENE R. REINHART
Electric Power Research Institute - Palo Alto, California

GARY M. SANDQUIST
University of Utah - Salt Lake City, Utah

L. E. ZWISSLER
Argonne National Laboratory - Argonne, Illinois

PREFACE

This international conference was held to draw attention to the problems and opportunities in the area of nondestructive evaluation (NDE) in the nuclear industry; particularly concentrating on problem areas and areas of controversy. The conference had been under preparation for over two years and was an outstanding success. This volume summarizes much of the recent progress in this field. Careful study of papers in this book will lead to a better understanding of NDE problems. Several papers, in addition, offer possible solutions in many areas of controversy. Problem areas include nuclear reactor components such as pressure vessels, stainless steel components, steam generators, nozzles, etc. This book contains papers which describe significant new advances in the area of Acoustic Emission, Ultrasonics, Eddy Current Examination, Radiography, Nondestructive Assay and Fuel Materials Quality Control, Spent Fuel Transportation and Storage, etc. Panel discussions were held in the areas of Codes and Standards and 10CFR50. These two areas of controversy were thoroughly discussed with active participation from the audience.

The manuscripts were requested in photo-ready form to make an early publication date possible. They are presented here with minimum editing. It is hoped that the reader will excuse the linguistic and typographical errors that are unavoidable in this process, as these are well outweighed by the value and merit contained in these papers.

The contributions of the keynote speaker, plenary speakers, the invited lectures, contributors, and session chairmen are gratefully acknowledged. The members of the Organizing Committee deserve a special thanks for their efforts that made this conference a great success. Special acknowledgments are due to Drs. Sigvard Eklund, G. Robert Keepin, Roger B. Clough, Gary M. Sandquist, and Eugene R. Reinhart. Thanks are also due to the American Society for Metals and its Managing Director, Allan Ray Putnam, Elliot Boardman, and Patti Gaeta who devoted their expertise and efforts to the conference and contributed much to its success. A sincere expression of gratitude is extended to Dr. V. S. Goel, who actively helped in every possible way during the past two years and contributed much to the success of the conference.

At the conclusion of the conference, it was decided to continue an NDE conference on a regular basis. The Third International Conference on Nondestructive Evaluation in the Nuclear Industry will be held at Salt Lake City, Utah, U.S.A., in mid-February 1980. We are looking forward to meeting with you at that time.

R. Natesh, Ph.D.
Chairman
Conference Organizing Committee

C O N T E N T S

CONFIDENCE IN NUCLEAR SAFETY AND NON-PROLIFERATION OF NUCLEAR WEAPONS: A PREREQUISITE FOR FURTHER DEVELOPMENT OF NUCLEAR POWER

Sigvard Eklund
Director General
International Atomic Energy Agency
Vienna, Austria

It is a privilege to have the opportunity to address this forum, assembled here to discuss the role of Nondestructive Evaluation in the Nuclear Industry The topic is both important and timely.

In my paper, I have outlined the relationship between the development of nuclear power and the confidence of governmental authorities, the nuclear industry, and the general public in the safety of nuclear power plants and in measures to ensure the non-proliferation of nuclear weapons. Such confidence can only be developed through the creation of appropriate regulations and through the planned systematic control of nuclear activities at the national and international levels. For the International Atomic Energy Agency, with its statutory responsibilities relating to the development and implementation of international safeguards and to the development of international nuclear safety standards and recommendations, this relationship is not only considered to be appropriate, it is also considered to be essential for the further development and use of nuclear power on an international scale.

In my presentation, I shall take a broad view of the problem areas involved and shall present the viewpoint of the organization I serve, the IAEA, without reference to specific details but with a recognition of the main problem areas for improving confidence in nuclear power production.

* * * * * * * * * * * * * * * * *

As you know, the IAEA is an organization within the United Nations family. In accordance with its Statute, its objectives are "to accelerate and enlarge the contribution of atomic energy to peace, health and prosperity throughout the world," and to "ensure, so far as it is able, that assistance provided by it or at its request or under its supervision or control is not used in such a way as to further any military purpose." To achieve the first objective, the IAEA is authorized to establish standards of safety for the protection of health and minimization of danger to life and property and to provide for the application of these standards upon the request of the parties to operations under any bilateral or multilateral arrangement.

In respect to the second objective, the IAEA is authorized to apply safeguards at the request of the parties to any bilateral or multilateral arrangement, or at the request of a State to any of that State's activities in the field of atomic energy.

The principal functions of the IAEA have therefore been designed from the outset to ensure a high level of confidence that nuclear energy will be generated in a safe manner and without risk of proliferation of nuclear weapons through transfer of nuclear technology. Behind these functions of the International Atomic Energy Agency there is a strong belief - recently confirmed by two large international conferences held in 1977, the Salzburg Conference on Nuclear Power and its Fuel Cycle, and the World Energy Conference in Instanbul - that with the approach of limits to the world's petroleum resources nuclear energy will be necessary and irreplaceable in relation to the future energy supply. In relatively large units, nuclear power plants are able to compete with fossil-fuelled power plants in both electricity and heat production, and both nuclear and fossil fuels will be required to meet future energy demands.

* * * * * * * * * * * * * * * * *

As you are all aware, however, the last few years in nuclear energy development have been marked by serious doubts and slow-downs. Requests for reassessment of the viability of the nuclear energy concept have been raised by official and unofficial groups all over the world. Several issues concerning nuclear energy have been the objective of controversial discussions and two of them have especially attracted the interest of the general public:

1. There are doubts as regards the safety of nuclear power plants and associated nuclear fuel cycle installations, including those intended for ultimate storage of nuclear waste; and

2. There are fears about the proliferation of nuclear weapons through the transfer of nuclear technology to the non-nuclear weapon states.

Without discussing the controversial aspects of these arguments, one can state that further progress in nuclear power will depend on the development of confidence in the effectiveness of those measures which are, or can be, designed to ensure the safety of nuclear installations and prevent proliferation of nuclear weapons. It is essential that these measures be undertaken internationally, and that they include developing as well as the more advanced countries. All tasks related to the safety and environmental aspects of nuclear power should be performed on the basis of pre-established regulations and accepted standards. Objective evidence should be produced to document that this is indeed the case. Confidence in nuclear power safety and reliability, as well as confidence in the non-proliferation of nuclear weapons, must be tied to:

1. Internationally accepted safety standards and their strict implementation in nuclear activities; and

2. Full commitment to the spirit and intent of international arrangements for non-proliferation of nuclear weapons.

Control and verification at the national, and in some aspects at the international level, are indispensable to assure that safety requirements are met and nuclear materials are adequately safeguarded. Development and improvement of technical methods to verify conformity with established requirements are necessary for this task. These methods must be capable of generating objective evidence of conformity with requirements and must be reliable and effective.

As a great deal has been accomplished, and much is still being done at the national level in respect to safety, I shall limit my comments to the activities of the IAEA, or, as we more familiarly refer to it, the Agency. At present, the Agency is engaged in a programme of development of a comprehensive set of requirements and recommendations, usually referred to as Safety Codes of Practice and Safety Guides, relating to the safety of nuclear power plants using thermal neutron reactors. These documents will be recommended to the Member States to be used as a standard frame of reference for assuring the safety of their nuclear power installations. The work on development of Safety Codes and Safety Guides is in progress in five main fields, namely, power plant siting, design, operation, quality assurance and governmental organization. The Codes of Practice establish in each of the mentioned fields minimum requirements which must be fulfilled to provide adequate safety for nuclear power plants and their systems and components. All five Codes have been prepared and will be published soon.

The Safety Guides recommend procedures that might be followed to implement the Codes of Practice. In each field, drafts of seven to ten Safety Guides are being prepared and are in the process of review and approval by the representatives of the Agency's Member States.

The objective of this work of the Agency, which is expected to be completed by 1982, is to assist the Member States to create national regulations and institutional mechanisms which will ensure the safety of nuclear installations, provide confidence in plant safety and bring reassurance to a concerned public.

However much assistance may be provided by public bodies, a major share of the responsibility inevitably rests with plant operators, and a quality assurance system established by the plant owner and operator is necessary as a basis for safety and reliability control and verification activities.

The functions of a quality assurance system include planning, management, verification and eventually implementation of corrective action. The objectives of such action should be:

1. The prevention of failures in nuclear installations through a planned and systematic approach to all activities. This includes an assurance of correct performance of all basic functions in building a nuclear power plant such as in design, manufacture, construction and operation; and

2. The timely detection of non-conformity through verification activities such as inspection and testing, and ensuring that subsequent corrective action is taken.

So far as each nuclear power station project is concerned, each area of activity such as design, procurement, construction, installation, commissioning and operation must be covered by quality assurance activities. The primary responsibility for establishing and implementing the quality assurance programme and for providing documented evidence to that effect belongs to the station operator or station owner. This, however, does not mean that the responsibility of plant designers and constructors is reduced. They are responsible for the achievement of quality objectives and their responsibility in respect to the regulatory body is either direct or indirect through contractual arrangements with the plant owner. An efficient and strict quality assurance programme at the level of plant owner and each of the plant constructors is a basic and important step towards acquiring and maintaining confidence in nuclear power.

Additional critical factors in ensuring the safety of nuclear installations are the existence and adequacy of national regulatory organizations. In performing its safety function, the regulatory organization must make decisions about the use of appropriate safety and engineering standards and ensure their implementation in nuclear power activities by means of review and inspections. Here in the U.S.A., these are, of course, the functions of the Nuclear Regulatory Commission (NRC), but in other countries the situation is by no means so clear-cut. The Agency has always insisted on the establishment of regulatory organizations in countries which initiate nuclear power programmes. In developing countries such organizations play an additional specific role in resolving safety problems arising from the importation of non-standardized power plants, interfaces between different suppliers of equipment and services, specific siting conditions, participation of local industry, etc. Perhaps the most difficult problem in establishing such an organization in a developing country is the question of its staffing. Specialists from numerous technical disciplines are required to be able to independently evaluate the work performed by the suppliers as well as by the plant owner and the consultants employed.

The existence of a competent regulatory organization is a pre-condition for a successful nuclear power programme. Efficiency in these regulatory organizations will strengthen and maintain confidence in the safety of nuclear power plants, and significantly reduce the risk of a potential nuclear accident. The key point in implementing requirements both of safety standards and of safeguards agreements is the existence of institutional arrangements on national and international levels to verify compliance with the existing requirements. The complexity of the undertaking and the consequences of inadequacy of established measures, require control of activities and verification to be undertaken at the level of the plant operating organization, at the level of governmental authorities, and also at the international level.

* * * * * * * * * * * * * * * * *

A further critical aspect of building public confidence in nuclear power is the effectiveness of political and technical measures aimed at preventing the proliferation of nuclear weapons. One of the objectives of the Agency's international safeguards system is to provide confidence on a worldwide basis that there are no diversions of nuclear material and equipment from peaceful uses to the production of nuclear weapons or other explosive devices. The basis of this system are the safeguards agreements established between the Member States and the Agency pursuant to the Statute of the IAEA and to the Non-Proliferation Treaty. This Treaty, which entered into force in 1970, now has 100 States as parties in addition to the three nuclear weapons States, USSR, UK and USA, and remains the most important international structure against the non-proliferation of nuclear weapons. It specifies that each State, party to the Treaty, has to conclude an agreement with the IAEA, submitting all its nuclear activities to Agency safeguards. The Agency is responsible for the independent verification of the States' compliance with the provisions of these safeguards agreements.

In implementing its responsibilities under safeguards agreements with individual Member States, the Agency relies on the State's system of accounting for and control of all nuclear material subject to safeguards. It is therefore of importance to establish principles on which this system should operate so as to enable the Agency to evaluate its effectiveness. The Agency's verification efforts under NPT agreements include independent measurements and observation by random sampling and inspections on "strategic points," and take due account of the technical effectiveness of the State's system. The effectiveness of the Agency's safeguards will, however, depend on the availability of effective verification methods. These include the methods of physical verification through accounting, inspections, examinations and checks. There is naturally a preference for instrumental methods of physical verification and the use of nondestructive methods of examination and assay. It is reassuring to note that the value of nondestructive assay to the international safeguards system has been recognized by including this subject in the programme of this conference.

* * * * * * * * * * * * * * * * *

Your conference goal to promote, under the auspices of application of nondestructive evaluation methods, a better understanding of and solution to the problems connected with controversies faced by future nuclear power development is a challenging one. The contribution of these methods and techniques to the resolution of such controversies lies in their extensive use in verification activities for providing objective evidence that nuclear installations are designed, constructed and operated according to established standards of safety, and that nuclear material is not diverted to nuclear weapons or other nuclear explosive devices.

In satisfying the growing needs for efficient verification methods, nondestructive examinations are of special importance because by definition they do not alter the physical characteristic or design of inspected items. The main problem is that they are, at present, limited in scope and applicability.

The IAEA at an early stage recognized the significance of these specialized techniques in relation both to nuclear safety and the application of safeguards. Parallel to the development of its own safety standards and the efforts to strengthen institutional aspects of safety and non-proliferation, the Agency has encouraged the international exchange of information and has promoted research and new development in this field. A Symposium on Nondestructive Testing in Nuclear Technology was held as early as 1965. A panel of experts discussed the same problems again in 1970. There have been many other meetings which addressed these problems such as a Specialist Meeting on In-Service Inspection and Monitoring in 1976 and a Technical Committee on the Use of Nondestructive Testing for In-Service Inspection on Reactor Pressure Components in 1977.

At present, the aspects of the application of nondestructive examination have been included into the scope of permanent activities of the three IAEA International Working Groups, namely, those dealing with Reliability of Reactor Pressure Components, Fuel Technology and Fast Breeder Reactors. Nondestructive assay techniques have been discussed on several occasions, including a Symposium on Safeguards Techniques in 1970.

In the discussions during these meetings, two basic questions are frequently considered, namely:

1. To what extent the nondestructive examination methods can replace direct observation and measurements based on conventional techniques and analyses; and

2. What should be done to improve the existing techniques of verification and develop new ones which might contribute more towards developing confidence in nuclear power?

The demand placed on component and system integrity by the strict safety standards will still call for improvement and development of new examination techniques which will be used in nuclear industry at all stages from nuclear reactor and containment construction to in-service inspection and maintenance.

These problems are considered by the Agency's International Working Group on Reliability of Pressure Components, which meets regularly once a year and organizes several specialists' meetings annually. We may draw a number of observations from recommendations of this working group and from recommendations of several other meetings related to safety and reliability of pressure components, which may be of interest for consideration by this gathering:

1. There is an urgent need for harmonization of national codes and specifications in the field of nuclear testing technology. As a minimum, a better understanding of similarity and differences between various national codes and standards of nondestructive examination and in-service inspection should be developed. Particular attention should be given to required techniques, sensitivity requirements, heat treatment requirements, reference pieces, interpretation of obtained results, acceptance level of defects, etc. Harmonization of national specifications would, of course, facilitate the work of the Agency on further development of international safety standards and of other international organizations responsible for industrial standards development.

2. With the existence of a large number of nondestructive examination techniques and an even larger number of inspection instruments, it would be desirable to have clearly defined sets of criteria for selection of techniques and equipment. At least clear recommendations on different methods of inspection should be made available for potential users, in particular in developing countries.

3. Correct interpretation of the significance of discovered defects is needed. This interpretation should be in some quantitative term, for instance relating the defect to a probability of in-service failure. This requires in the first place a better understanding of the significance of the various discontinuities that are located during nondestructive examination and their effect on the service life of a component and structure. Many of the primary properties (such as embrittlement) which ultimately influence service performance cannot be monitored directly so that one has to find some relationship between these properties and those structural defects which can be discovered either during fabrication or during service life.

4. A most critical activity from the point of view of safety during operation of nuclear power plants is in-service inspection. Recognizing the leading role of ASME (American Society of Mechanical Engineers) Code Section XI in planning and performance of this activity, we at the Agency are now trying to develop a consensus among the Member States in requiring and recommending procedures for that activity. A Safety Guide on In-Service Inspection has been developed and is at present being prepared for final approval. At the same time we are also fostering an exchange of information on present experience and future developments in this area. Better knowledge of the development of flaws and other operational modifications is required in order to identify major differences between in-service inspection results compared with inspections during manufacture of pressure components. The areas of nuclear power plant pressure components which require most critical examination during in-service inspection are those possessing the least tolerance to defects. These are areas where high stresses exist and where the total effect of failures is

large. Unfortunately, these are also areas where good results of nondestructive examination are most difficult to obtain.

5. The concept of continuous monitoring and surveillance of quality and integrity of reactor components, or a "cradle-to-grave" philosophy, has a particular significance for quality achievement and maintenance. Testing should be included in the early stages of design and continuously applied to successive stages of the fabrication process where defects are likely to be introduced. Continuous monitoring of components during in-service life requires the further improvement of inspection equipment, in particular efficient automatic scanning systems, with means of recording and displaying the test results in easy read-out form. This improvement should increase the reliability, reproducibility and resolution of obtained results. Once again, in all these activities a correct interpretation of the significance of defects in respect to the information obtained during examination and monitoring is of predominant importance.

6. The needs of continuous in-service inspection to monitor the appearance and growth of cracks in pressure vessels also require the development of new methods and techniques. The technique of acoustic emission, which has been suggested as most promising for application in both pressure test and on-line usage, seems to present great technical problems because of the many more components involved, such as background noise, high temperature, radiation effects, etc. Investigations into the application of some new techniques and further improvement of the acoustic emission should be encouraged.

7. Nondestructive examination is a rather specialized technology embracing several technical disciplines, and formal university-level training for the personnel involved is rare. The increasing demand of examination, interpretation and evaluation placed on NDE personnel will call for development of educational programmes in all countries embarking on nuclear power programmes. This may include: seminars and classroom-type courses for engineers and technicians supplemented by on-the job training in an established laboratory where a high level of competence is recognized.

The use of nondestructive assay techniques (usually referred to as NDA) for verification and measurement of nuclear materials plays a significant role in safeguards verification. Although measurements of the more classic type cannot be totally replaced by NDA methods, there are many situations in safeguards inspections when NDA measurements are the only alternative. I would like to point out the importance of these methods to safeguards by referring to some of the situations involved:

1. Destructive sampling may be economically unacceptable as, for example, in the case of valuable finished products such as finished fuel elements or assemblies;

2. Destructive sampling may be very difficult, time-consuming or even dangerous as, for example, in the case of plutonium bearing materials, large UF_6 storage cylinders and irradiated fuels; and

3. It can be extremely difficult to make destructive sampling representative as in the case of heterogeneous mixtures which are only too often encountered, e.g., in scrap containers. NDA techniques can, by their nature, offer the possibility of integral measurements of whole containers.

In addition, there can be another distinct advantage of using NDA for safeguards inspections, as they can be much less time-consuming than destructive analyses and can be made on-site with compact portable instrumentation. This enhances the timely acquisition of measurement results which is extremely important in verification activities.

* * * * * * * * * * * * * * * * *

Problem areas still remain in NDA technology to improve the capability of safeguards measurements. I would like to draw your attention to a few:

1. There is a need for smaller, robust and reliable portable instruments, which can be used during on-site inspection and which are still capable of sophisticated and accurate measurements. The requirements cover a broad range from simple identification instruments, capable only of identifying the basic materials present, i.e., enriched uranium and plutonium, to much more sophisticated instruments needed for

verification of element concentrations and isotopic concentrations or of integral assay of non-homogeneous mixtures. It is rewarding to see the progress which is being made in the development of both passive measurement and active interrogation measurement techniques. Work needs to be done also on providing rapid and dependable means of calibration of instruments and standardization of methods of measurement and calibration. A basic problem remaining here, as with the NDE techniques I referred to earlier, is the accumulation of a data base for adequate quantitative assessment and interpretation of measurement results.

2. Better techniques are needed for the assay of finished uranium and plutonium fuel assemblies. For irradiated fuel assemblies NDA techniques are highly desirable but far from perfect. Instruments are required that are capable of measurement in deep storage pools and in high radiation fields and at the same time capable of giving information about burn-up as well as plutonium build-up in individual assemblies. This is certainly a challenge, and may indeed be a goal too difficult to achieve.

These and other problems which may become apparent during this conference should attract our attention in our future work on development and application of nondestructive evaluation techniques and corresponding equipment. I think that it is no exaggeration to say that future development of nuclear energy utilization for peaceful uses may depend, to a great extent, on the adequacy and effectiveness of solutions found in this technical discipline as well as in other activities designed to provide confidence in the safety of nuclear installations and the non-proliferation of nuclear weapons.

* * * * * * * * * * * * * * * * * * *

I would like to close my presentation on the role of controls and verification in building confidence in the safety of nuclear power plants and in the effectiveness of safeguards with a few additional remarks.

The sets of measures which I have referred to and which we, as professional engineers and scientists, establish in order to provide confidence in nuclear power are of a technical and administrative nature. They are designed to be rational and objective. However, the doubts about nuclear energy which exist among the public are of a different nature. They are more often based on psychological, political, moral, religious or other concerns. This is at the heart of the difficulties which we are now encountering in public acceptance of nuclear power. Therefore, a parallel task to the establishment of control and verification measures should be an assurance of their acceptance by the public. Professional societies and organizations should use their authority and technical competence to demonstrate the effectiveness and adequacy of the established measures. Ways must be found to educate the public to understand and accept our standards and verifications. This conference, among its other tasks, might well concern itself with this problem.

EVALUATION OF UT INDICATIONS IN MATERIALS AND WELDS OF LWR COMPONENTS

S. Onodera[1), K. Fujioka[2)],
H. Kobayashi[3)], M. Wataya[4)],
and M. Takeya[5)]

1. INTRODUCTION

The development of heavy section steel technology has played an important role in constructing high quality, ultra large size pressure vessels for light water nuclear steam supply system.

However, even with the highest technique of metallurgy, some defects are unavoidable in the supply of numerous steel products, in which Muroran Plant, The Japan Steel Works, Ltd. (JSW hereafter) has been engaged since 1966.

The nondestructive evaluation of them is therefore the everlasting subject of discussions.

It is the intention of this paper to demonstrate the internal quality of heavy steel forgings and their welds for light water reactor (LWR) components, taking some examples from our recent products and covering use of the nondestructive examinations to the defects from three main points.

1) A statistical data of UT results on large forgings,
2) Cutting examinations through the UT indications in steels to confirm the nature, size and distribution of them, and
3) Cutting examinations of UT indications in welds in the same manner as 2) above, and comparison of the results of UT, RT and visual observation.

---

1) Dr. Eng., General Plant Manager, Muroran Plant, The Japan Steel Works, Ltd. (JSW)
2) General Manager, Atomic Energy Dept., Muroran Plant, JSW
3) Chief NDE Engineer, Inspection Dept., Muroran Plant, JSW
4) Manager, QA Group, Atomic Energy Dept., Muroran Plant, JSW
5) Senior NDE Engineer, Inspection Dept., Muroran Plant, JSW

2. UT INDICATIONS IN LARGE FORGINGS FOR LWR COMPONENTS (1)

2.1 Statisitical Data for UT Examination Results of Large Forgings

A statistical survey was made on the UT results of more than 150 pieces of SA508, CL.2 and CL.3 forgings, covering;

1) Weight of products : 2 to 166 tons
2) Item of products : flange, shell and tube sheet
3) Applicable codes :
   (a) ASME, Boiler and Pressure Vessel Code, Sect. III (ASME Code, Sect. III)
   (b) Japanese Ministry of International Trade and Industry Code (MITI Code)
   (c) European specifications : German requirements as a representative one

As shown in Table 1, no forgings to ASME and MITI Code were rejected and 5 % of forgings to European requirements were rejected.

Table 1 UT Examination Results of Large Forgings

Level of UT results 1)	Applicable Code or Specification			
	ASME Code Sect. III	MITI Code	European Specification 2)	Total
A	37	5	84	126
B	4 3)	0	15	19
C	0	0	5	5

Notes:

1) Grading of UT results per each of Codes and specification
   A : Acceptable with no recordable indication
   B : Acceptable with recordable indication
   C : Unacceptable
2) Evaluated by DGS (AVG) method with acceptable and recordable size of indications as follows (2) ;

UT technique	Size of individual indication (mm EFG)	
	Recordable	Acceptable
Straight beam	6	15
Angle beam	3	9

Table 2 Examination Conditions for Each Forging

		Forgings			
		Cover flange	Shell flange	Nozzle	Tube sheet
Material		SA508, Cl.2	SA508, Cl.3	SA508, Cl.2	SA508, Cl.2
Dimensions at final shape (mm) 1)		$5,750^{OD} x 4,540^{ID} x 1,100^{H}$	$5,750^{OD} x 4,690^{ID} x 2,550^{H}$	$1,200^{OD} x 750^{ID} x 662^{L}$	$3,640^{OD} x 700^{H}$
Weight (kg)		90,000	150,000	1,300	58,000
Examination stage		Before Q & T and final shape	Before Q & T and final shape	UT shape after Q & T	After N & T
Examination Conditions: Equipment		Krautkrämer USIP-11			
Examination Conditions: Scanning method		Straight beam and angle beam			Straight beam
Examination Conditions: Size and frequency of transducer	Straight beam	2 MHz, 24 mm$^{\phi}$			
	Angle beam	2 MHz, 20 x 22 mm			-
Examination Conditions: Calibration method		DGS (AVG) method			

Note : 1) OD : Outside diameter ID : Inside diameter
H : Height L : Length

3) Tube sheet forging examined with additional criteria of customers

2.2 Cutting Examination of the Spots of UT Indications

With the progress of manufacturing process the nature of defects in steel forgings might be changing. Thus these "biopsies" or "autopsies" are of utmost importance.

It is therefore the key point of the present paper to illustrate the cutting examination of spots of UT indications.

2.2.1 Examination conditions for each forging

Examination conditions for each forging are as shown in Table 2.

2.2.2 Cutting results for each forging

a) Cover flange

Many UT indications were detected near the peripheral surface with straight beam technique from both end surfaces, and details of the defects and the typical reflection patterns are as shown in Fig. 1. Small pieces which include the defects were taken from the prolongations for the mechanical test.

These pieces were sliced to determine the nature and the actual size of the defects. The results are as shown in Fig. 2, and nature of the defects are Al_2O_3 inclusions.

b) Shell flange

Similar indications to those of the cover flange were detected for shell flange made from SA508, Class 3 steel material of 400 tons ingot. The examination results are included in Para. 2.3.1.

c) Nozzle

UT indications which are less than 3 mm EFG were found in the exterior zone of the largest diameter to the maximum depth of 50 mm from the exterior surface. Small samples were taken from the exterior and the metallurgical investigations were performed. From Fig. 3, it is concluded that the UT indications are caused by discontinuities which were left after forging operation.

d) Tube sheet

1) UT results

A cluster of indications was revealed by straight beam UT in midthickness approximately 400 mm to 600 mm from primary side surface and some minor indications, in addition, were detected around the cluster indication. Consequently center portion (approximately 1,700 mm in diameter) of the forging was re-forged and again

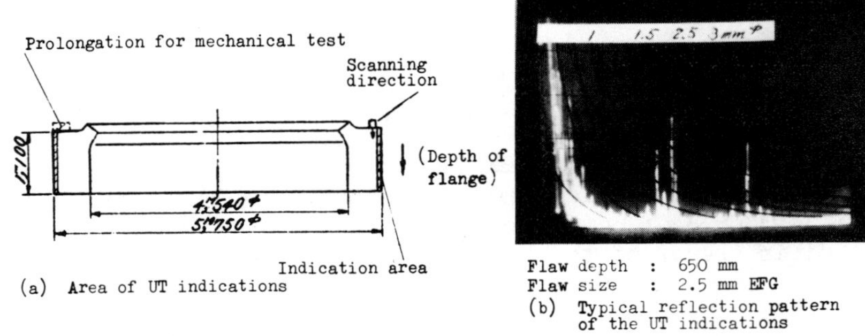

Fig. 1 UT Indications Detected in Cover Flange

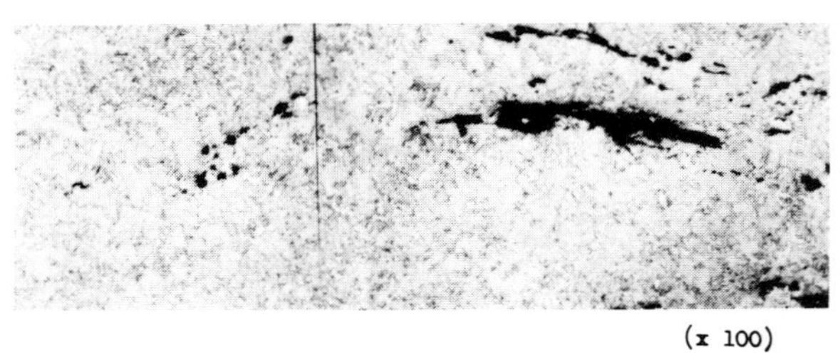

(x 100)

Reduced : x 0.67

Fig. 2 Nature of UT Indications for Cover Flange (Al_2O_3 Inclusions)

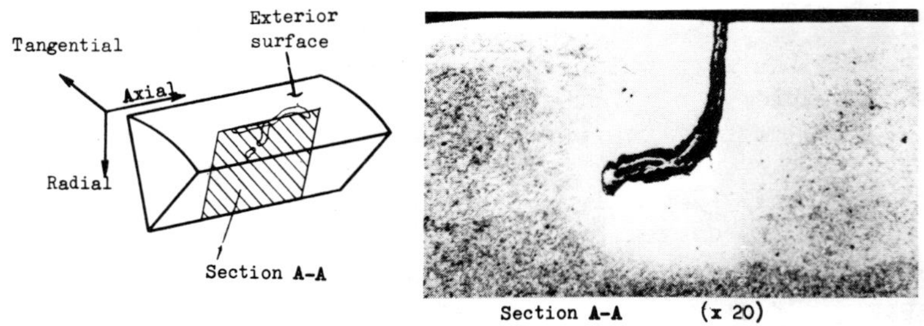

Reduced : x 0.67

Fig. 3 Nature of the UT Defects in Nozzle Forging (Forged Laps)

examined by UT. However, some indications around center got larger as shown in Fig. 4 though center cluster disappeared.

2) Cutting results of the defects

One sample bar, 24 mm dia. by 970 mm length, was trepanned from the forging as shown in Fig. 4, and the sample specimen, 8.5 mm dia., was taken from the bar which locates at a depth of 395 mm from primary surface, corresponding to the depth of defects detected by UT. Fig. 5 shows microphotograph of the defects. It can be clearly understood from this picture that the defects are cracking initiated from loose structural micro cavities and non-metallic inclusions which are aligned in the lamellar bainitic structure, and are propagated thereafter due to subsequent heat treatment cycle.

2.3 Comparison of the Size Estimated by UT with the Actual Defect Size

Normally the sizes of the defects in the forgings are evaluated by the echo height on the screen (cathode ray tube). However, the size estimated by UT is not the same to the actual size because of the nature and orientation of the defects. Here, for the defects of which sizes were determined by DGS method, the cutting and grinding investigation were performed to compare the size estimated by UT with the actual size.

2.3.1 The size estimated by UT

The sizes of the defects were determined with DGS method on the assumption that the defects have a round flat shape.

a) Examination conditions

Equipment	: Krautkrämer USIP-11
Transducer	: Straight beam probe 2MHz, 24 mm dia.
Couplant	: Machine oil

b) Defects investigated

The defects found in the cover flange and the shell flange described in Para. 2.2.2, and those found in near surface of tube sheets, were evaluated by UT and the results are as shown in Table 3.

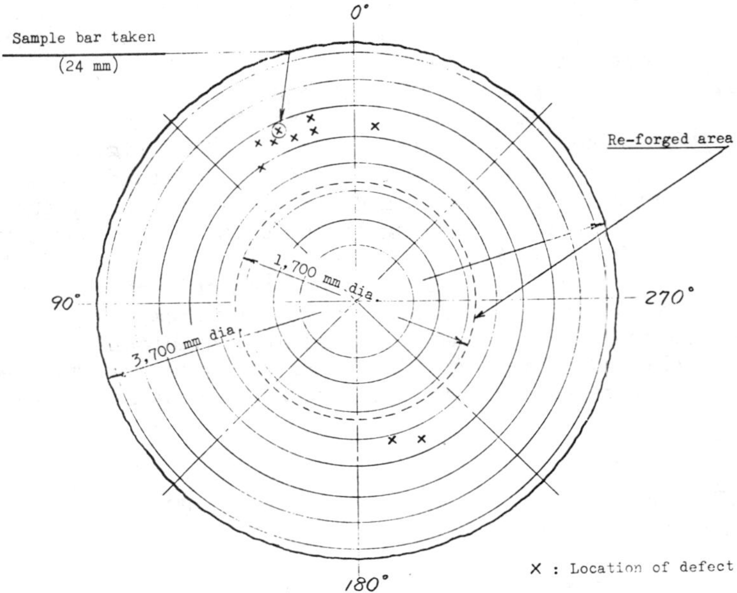

Fig. 4 Results of UT after Re-forging and the Location of the Sample Bar Taken from the Tube Sheet

(x 200)

Reduced : x 0.65

Fig. 5 Micro-observation of the UT Defects in the Tube Sheet

Table 3 The Sizes Estimated by DGS Method for the Defects

Defect No.	Part name	Evaluated size (mm EFG)
1	Cover flange	2
2		1.4
3		1
4	Shell flange	1
5		1
6	Tube sheet	3
7		3

2.3.2 The size determined by the cutting or grinding

a) For the defects of No. 1 to No. 5, small coupons were taken from the prolongations of the forgings, and these piece were sliced by 0.2 mm pitch in the planes parallel to the beam direction of UT to observe the microscopic nature as shown in Fig. 6.

From this observation, the projections of the defects on the plane normal to the UT beam direction were obtained typically as shown in Fig. 7.

b) For the defects of No. 6 and No. 7, the actual sizes were determined by the magnetic particle examination during the excavations of the defects. The results are as shown in Fig. 8. Nature of the defects was non-metallic inclusions (Al_2O_3) which were confirmed by the microscopic observation.

2.3.3 Correlation between the size estimated by UT and the actual size

The correlations between the size estimated by UT and the actual size are as shown in Tables 4 and 5 and in Fig. 9.

From Fig. 9, the estimated diameter of the non-metallic inclusions are about 1.3 to 1.8 times the diameter of those evaluated by UT.

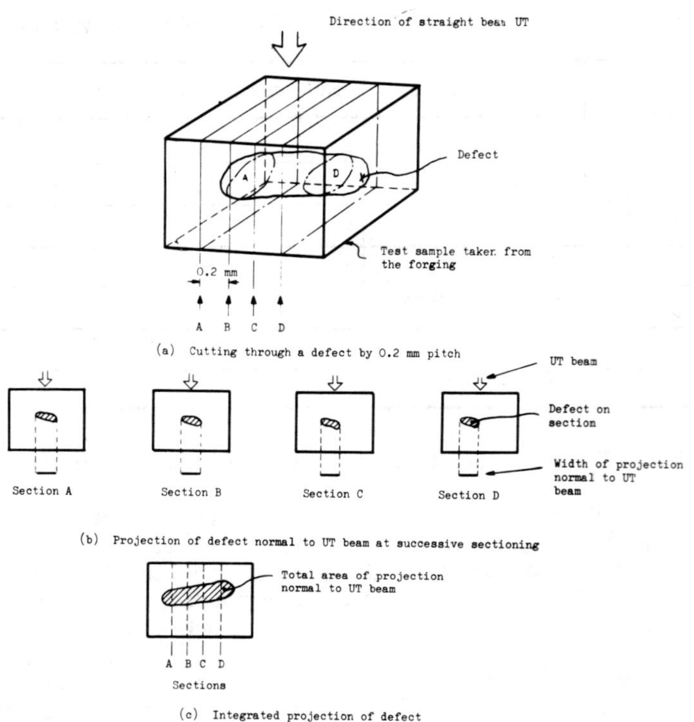

Fig. 6 Cutting through a Defect, and Projected Area of the Defect (a Schematic Explanation)

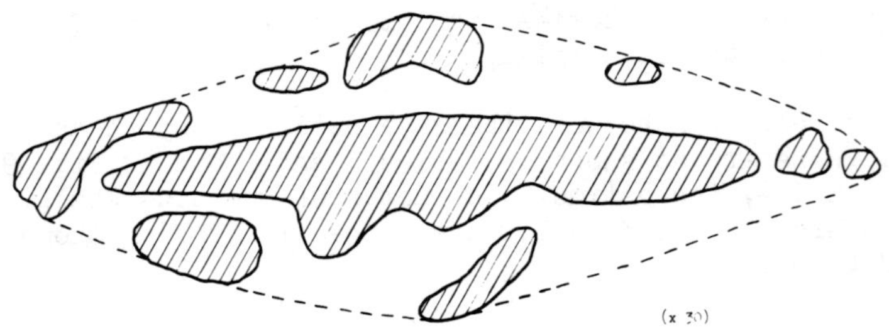

Reduced : x 0.48

Fig. 7 Total Projection of Al_2O_3 Inclusions on the Plane Normal to the UT Straight Beam Direction (Typical)

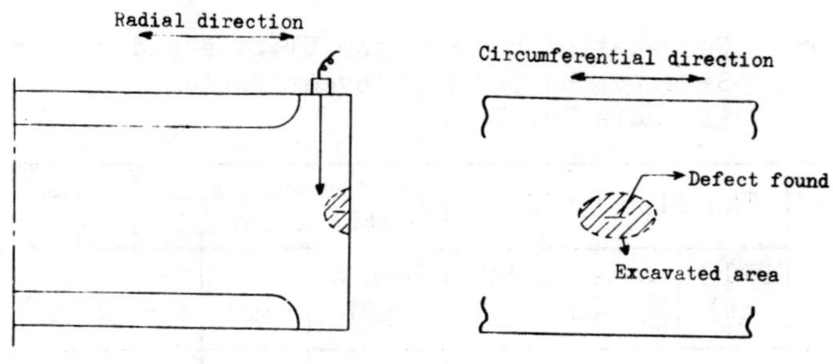

(a) Location of UT indication

No. 6 Indication

Depth from outer surface (mm)	25	26	26.2	27	28.2	29
Dimension of defect (mm)	0	2	2.5	2	1	0.8

No. 7 Indication

Depth from outer surface (mm)	19	20	21	22	23
Dimension of defect (mm)	0	6	6	4	0

(b) Dimensions of the defects during the excavation

Fig. 8 Actual Sizes of the Defect in the Tube Sheet

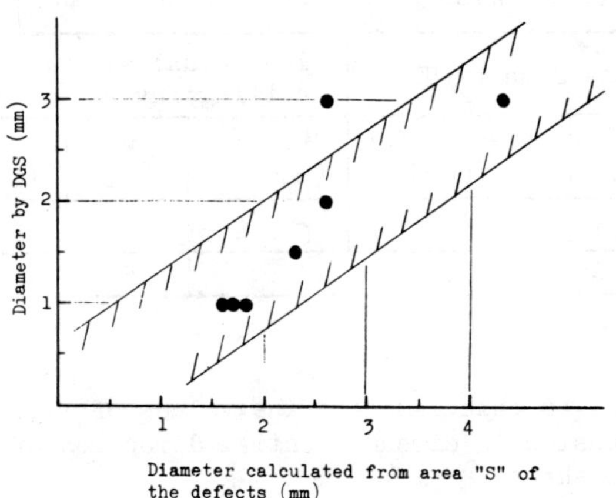

Fig. 9 Correlation between the Size Estimated by UT and the Actual Size Calculated from Cutting Examinations

Table 4 Correlation between the UT Size and the Actual Size by the Cutting Investigation (Defects No. 1 to 5)

Defect No.	The size by UT		The actual size by the cutting investigation 1)			
	EFG (mm)	Area of EFG (mm^2)	Area S (mm^2)	T (mm)	W (mm)	L (mm)
1	2	3.1	5.9	1.7	2.4	7.9
2	1.4	1.5	4.4	1.4	2.1	6.2
3	1	0.8	2.7	0.8	2.3	6.0
4	1	0.8	2.7	0.7	3.0	4.6
5	1	0.8	2.3	1.0	2.5	3.5

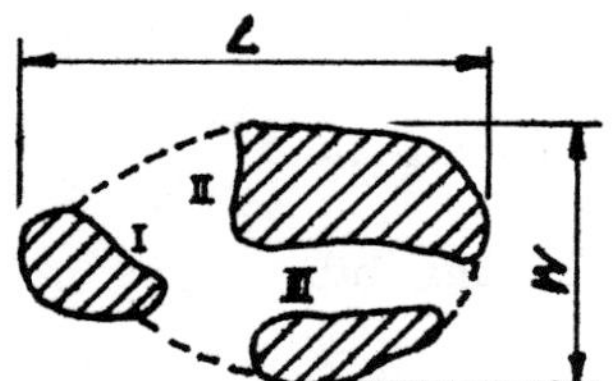

Note : 1)

Area S = Area I + Area II + Area III
T = Dimension parallel to UT beam
W, L = Dimensions normal to UT beam

Table 5 Correlation between the Size by UT and the Actual Size by MT (Defects No. 6 and No. 7)

Defect No.	The size by UT		The actual size by the cutting investigation 1)		
	EFG (mm)	Area of EFG (mm^2)	Area S (mm^2)	W (mm)	L (mm)
6	3	7.0	6	2.5	4
7	3	7.0	15	6	4

Note : 1)

Area S = This was calculated by the values of Fig. 8(b).
W = Dimension in circumferential direction of the tube sheet
L = Dimension in radial direction of the tube sheet

3. UT INDICATIONS IN HEAVY NOZZLE WELDS

As one of the typical examples of heavy component weld, welds of a set-on type, main coolant nozzles are investigated here. Fig. 10 shows the sketch of vessel flange to nozzle weld design being investigated.

3.1 UT Procedure for Set-on Type Nozzle Welds

Most of the nozzle welds to vessel shell are usually examined with RT. However, it is very difficult to apply RT for the so-called set-on type nozzle weld, and to get meaningful results from it. Usefulness of UT rather than RT for complicated shape of set-on type nozzle weld was confirmed by a preliminary test (mock-up test).

Equipment	: Krautkrämer USIP-10W	
Frequency	: 2 MHz	
Transducer	: Straight beam	24 mm dia.
	Angle beam	8 x 9 mm, refraction angle; 35°, 45°, 60°, 70°
Evaluation of indication	: DGS (AVG) diagram	
Couplant	: Machine oil	
Scanning direction	: From inner and outer surfaces (Fig. 11, four (4) different kinds of angle to detect all possible defects oriented in any directions in the welds.	

3.2 Detectability of the Defects with UT

3.2.1 Test block used for examination

The calibration of the equipment was made by using the DGS diagram. Each transducer was calibrated using the IIW type reference block and then the curvature was corrected to fit the curved surface. Prior to actual scanning, the detectabilities were confirmed by using the two test blocks, which were taken from the mock-up weld at 0° and 90° orientation of nozzle.

Four (4) 2 mm dia. holes were drilled near fusion zone of each test block and reflections from each hole were obtained satisfactorily. Typical reflection patterns are as shown in Fig. 12.

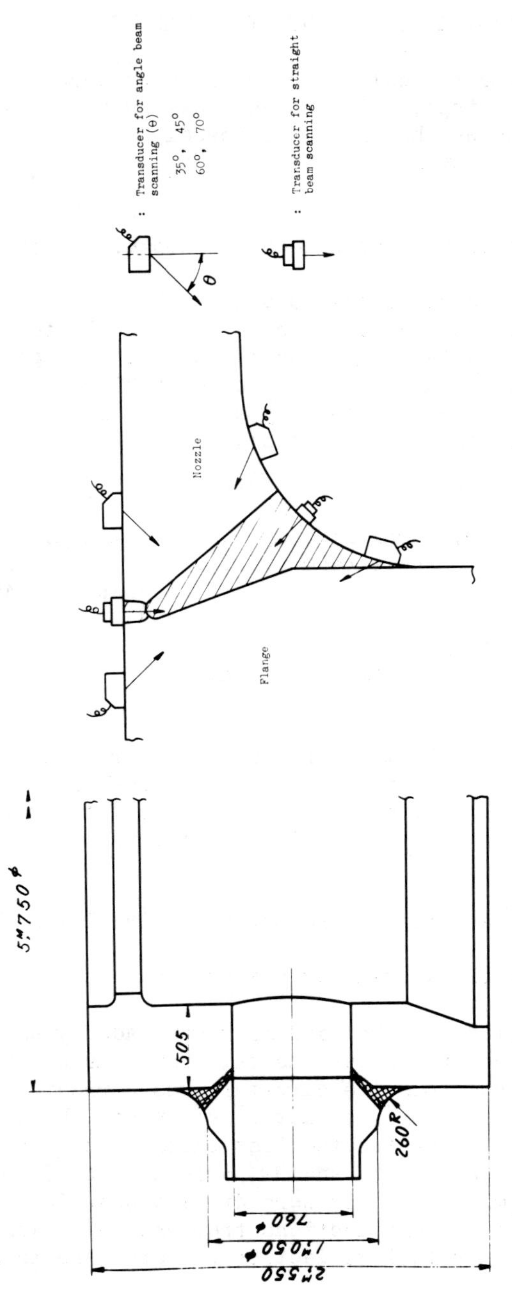

Fig. 10 Set-on Type Flange-to-Nozzle Weld

Fig. 11 Scanning Direction for a Set-on Type Nozzle Weld

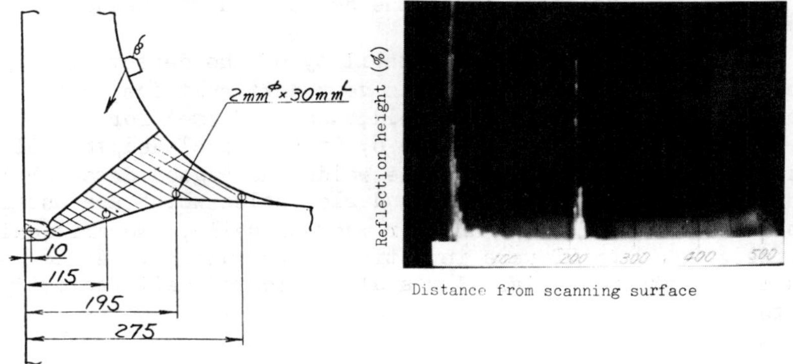

Test block in circumferential section of shell (90°)

Fig. 12 Typical Reflection Pattern from Reference Holes

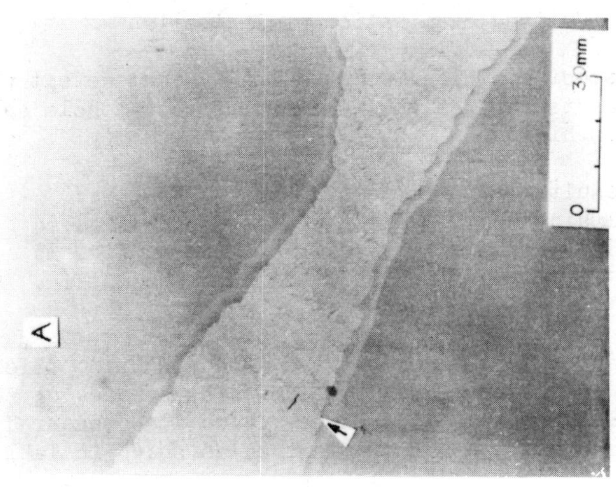

(x 1)

Reduced : x 0.67

Fig. 13 Appearance of Indication on the Cross Section of Weld Joint (Echant : 10 % Nital)

3.2.2 Detectability of the defects in the welds

To demonstrate the detectability of the defects in the welds, RT examination, which is very difficult due to the complicated geometry of the weld, was performed for the mock-up test block. As shown in Table 6, it was concluded that UT examination for the set-on type welds had a good detectability to the defects such as slags and blow holes which are usually insensitive to UT. For another mock-up welds, two critical defects found by UT were investigated by cutting. A lack of fusion as shown in Fig. 13 was also detected satisfactorily with UT.

3.3 Cutting Investigation and Comparison of the Size Estimated by UT with the Actual Defect Size

In order to observe the nature of the defects and to compare the actual sizes with the sizes estimated by UT, the following investigation was made on the indication of actual product.

3.3.1 Cutting examination of indication

Test specimen which includes the typical defect with a length was taken from the weldment between nozzle and flange by machining.

The results of investigation are as follows;

a) UT results of the defects are as shown in Fig. 14
b) RT results of the defects are as shown in Fig. 15
c) Micro-structure of the defect is as shown in Fig. 16

From these observations, it was found that the defect was attributed to slag inclusions accompanied by lack of fusion from Fig. 16. In addition, sizes of the flaw measured by different examinations are summarized as shown in Table 7.

Table 6 Results of UT and RT Examinations for Mock-up Welds

Results of UT examination

Scanning method		Defect size*) (mm dia.)	Number of defects	Nature of defects
Angle beam	45°	1.5	1	-
	70°	1.5	3	-
		2.5	1	3mm x 1mm slag
Straight beam		2.0	1	2ø blow hole
		3.0	9	-
		3.5	1	-
		4.0	2	5mm x 1mm slag
		5.0	1	4mm x 3mm slag

*) Indications exceeding 1.5mm dia. were recorded

Results of RT examination

Examination Conditions :

Equipment ; Linear accelerator (12 MeV).

Film ; Kodak AA (7" x 17")

Screen ; Pb (1mm Thickness)

Penetrameter ; DIN FE 1/7 (Equivalent to ASME penetrameter No. 100)

Examination result : No indication

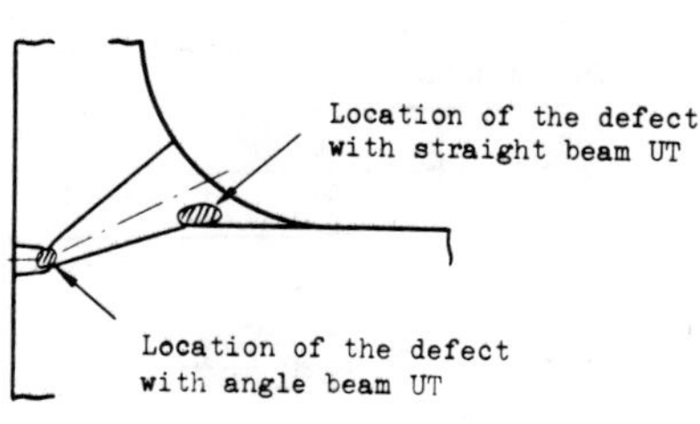

Location of UT indications

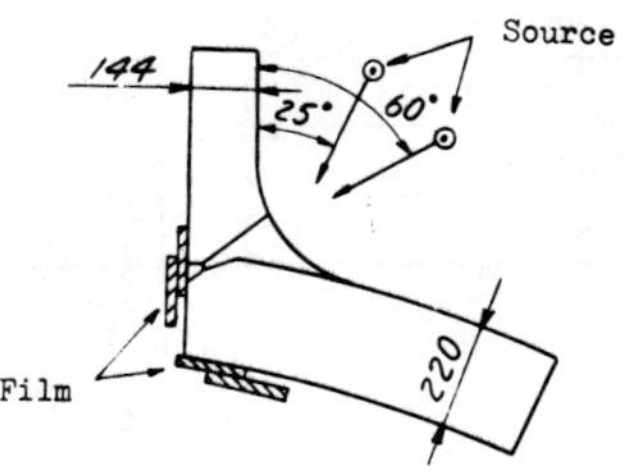

Location of source and film

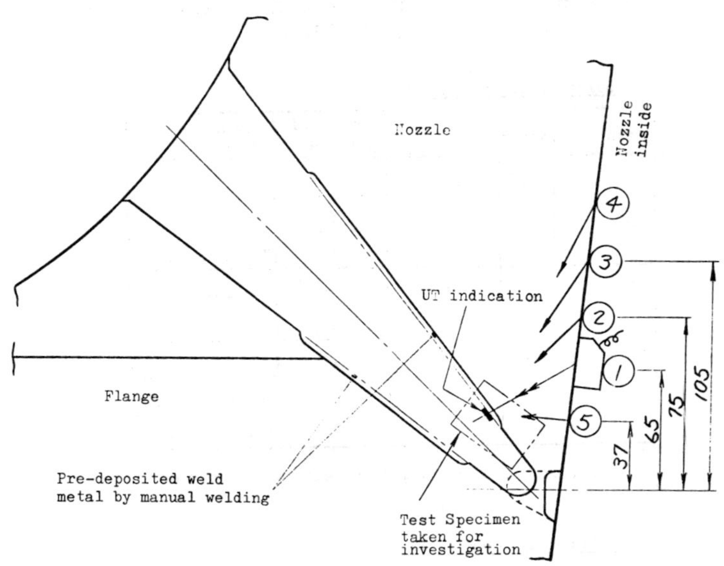

Transducer	Location of transducer	Flaw size (mm EFG)	Length of flaw (mm)	Path Length (mm)	Flaw depth (mm)
WB35-N2	①	5.5	20	45	37
WB45-N2	②	5	20	55	38
WB60-N2	③	2	<10	75	37
WB70-N2	④	<1.5	Point	100	35
MWB35-N4	①	3.5	10	45	37
MWB45-N2	②	3	8	55	38
MWB60-N2	③	1.5	Point	75	37
MWB70-N2	④	No indication			
MSEB4H	⑤	2	12	38	38
MB4S	⑤	1.2	Point	38	38

Fig. 14 UT Examination of Indication

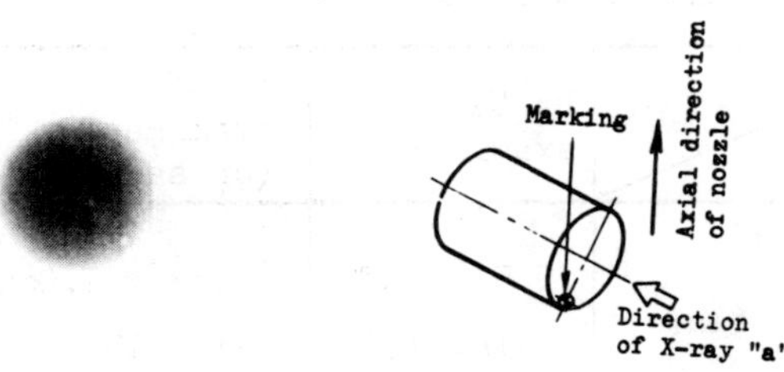

Note : The film meet the requirements of DIN 54109, image quality Class I. (Penetrameter Sensitivity ; 1 %)

Reduced : x 0.67

Fig. 15 Radiographic Examination of Indication

(x 10)

Reduced : x 0.67

Fig. 16 Micro-observation of Indication

Table 7 Size of indications evaluated by different examination methods

	By UT examination	By RT examination (on sample)	By visual observation
Evaluated maximum size of flaw (mm EFG or mm)	5.5 mm EFG (Fig. 14)	4.9mm x 1.0mm (Fig. 15)	4.2mm x 1.0mm (Fig. 16)
Evaluated maximum length of flaw (mm)	20.0	4.9	4.2

3.3.2 Correlation between the size estimated by UT and RT and the actual size

As shown in Table 7, the size of the indication estimated by UT, even though the results from most preferable scanning angle are introduced, is extremely larger than the actual defect size. Size estimated by RT, needless to say, is equal to the actual size since the examination was made on a small sample taken from heavy welds. However, the detectability of RT is quite unfavorable to the present weld of complicated geometry.

4. CONCLUDING REMARKS

Detectability of defects in both material and its weld by ultrasonic examination would decrease with increasing dimensions of the material and weld due to the following reasons.

1) The larger is the thickness, the larger might be the minimum size of the detectable defects.
2) The larger is the material thickness, the more difficult is the control of grain size, cleanliness, overall permeability of ultrasonic wave.

It has been demonstrated so far, through statistical data and several typical examples of investigations on the actual products of JSW, that the present technique of manufacturing heavy walled materials for LWR components assures defects-free products, (3), (4), (5) and the NDE investigations presented here lead to the following conclusions.

1) UT indications found in NSSS heavy steel forgings can be assumed as giving exact information of the defects such as oxide clusters and small cracks in them, although actual size will be larger slightly.
2) UT is sufficiently reliable for the indications in welds of complicated shape, if performed with sufficient number of angles of incidence, except that the linear slags would be evaluated much larger than the image of RT.
3) Simple as they are, this kind of investigation should constantly be done, and accumulation of cases is necessary. Biopsies and autopsies are of utmost concern, and better material and better weld which are the target in our industry will be realized through these investigations.

The authors express their cordial acknowlegdements to the management of JSW, Mr. M. Kashihara, Managing Director and former General Plant Manager of Muroran Plant, for the permission of disclosing data of the company.

REFERENCES

(1) S. Onodera, et al., "UT Indications in Large Forgings for NSSS", B&W-TÜV Nondestructive Examination Conference, Washington, D.C., USA, November, 1976.

(2) Krautkraemer GmbH; Advanced Testing Technique, The Echo, No. 19, 1968, P.195-202.

(3) S. Onodera, et al., "Large Size, Integrated Type Steel Forgings as Intended for Easier I.S.I." IAEA Technical Committee, Kobe, Japan, April, 1977.

(4) S. Onodera, et al., "Mono-Block Vessel Flange Forging for PWRPV 1,000 MEW", The 8th International Forgemasters Meeting, Kyoto, Japan, October, 1977.

(5) S. Onodera, et al., "Advantages in Application of Integrated Flange Forgings for Reactor Vessels", The 3rd MPA-Seminar in Stuttgart, F.R. of Germany, September, 1977.

ULTRASONIC EXAMINATION OF NOZZLE INNER SURFACES IN HEAVY-WALL PRESSURE VESSELS

R. A. Baker
A. S. Greer

Southwest Research Institute

INTRODUCTION

Nuclear power plants with boiling water reactor (BWR) systems provide economical and reliable electric power. Nondestructive testing (NDT) methods are used to help ensure safe and reliable operation. Components of the primary pressure systems are periodically examined during plant operational outages on a regularly scheduled sampling basis. However, known problem areas are removed from the regular schedule listing and placed on an accelerated schedule. Cracking in the feedwater (FW) and the control rod drive (CRD) return line nozzles is such a problem. This paper describes the problem, the NDT methods used to monitor it, and the present stand of the Nuclear Regulatory Commission (NRC). It is

---

basically a statement of a problem for which there are no present solutions.

DISCUSSION OF THE PROBLEM

Cracking in the CRD and FW nozzles has been diagnosed as fatigue resulting from cyclic thermal gradients which occur when cold water is injected into the hot water in the vessel. Differences in the gradient, water flow rate, flow duration, and mechanical design result in somewhat different cracking situations in the two nozzle types and between different power plants.

Feedwater Nozzles

Feedwater is condensate from steam supplied to the turbines. Steam, after being cooled and condensed back to water in the condensers, is stored in the suppression pool. This cooled water is returned to the reactor pressure vessel (RPV) through feedwater piping, ranging in sizes from 10- to 12-inch diameter, and through four feedwater nozzles to be evaporated into steam once again. During low power operations, FW temperature may be 200C° (400F°) below that of the vessel while feedwater heaters may reduce the differential to 100C° (180F°) during high power operations.

FW nozzles contain thermal sleeves which direct the cold water into a sparger or manifold for distribution around the periphery of the RPV. Unfortunately, water may leak around the thermal sleeve and mix with the hot 290°C (550°F) water in the RPV. Mixing occurs, primarily, in the annulus between the outside of the sleeve and the inside nozzle surface (see Figure 1). During mixing, nozzle inner surfaces are subjected to rapidly fluctuating water temperatures causing relatively high, localized stress levels in the surface cladding layer.

Inner surfaces of the RPV and all nozzles into the vessel are clad with welded overlay of Type 304 stainless steel. This clad layer, which ranges from 4.5 mm (.180 inch) to 12 mm (.5 inch), was applied to reduce corrosion products which (1) increase turbidity (thereby decreasing visibility during refueling operations), and (2) become radioactive and must be removed by the reactor water clean-up system. Unfortunately, cladding properties may lead to cracking in the nozzle environment. These properties include:

(1) Lower thermal conductivity than carbon steel;
(2) Higher thermal expansion than carbon steel;
(3) Lower resistance to fatigue cracking;
(4) Lower modules than carbon steel.

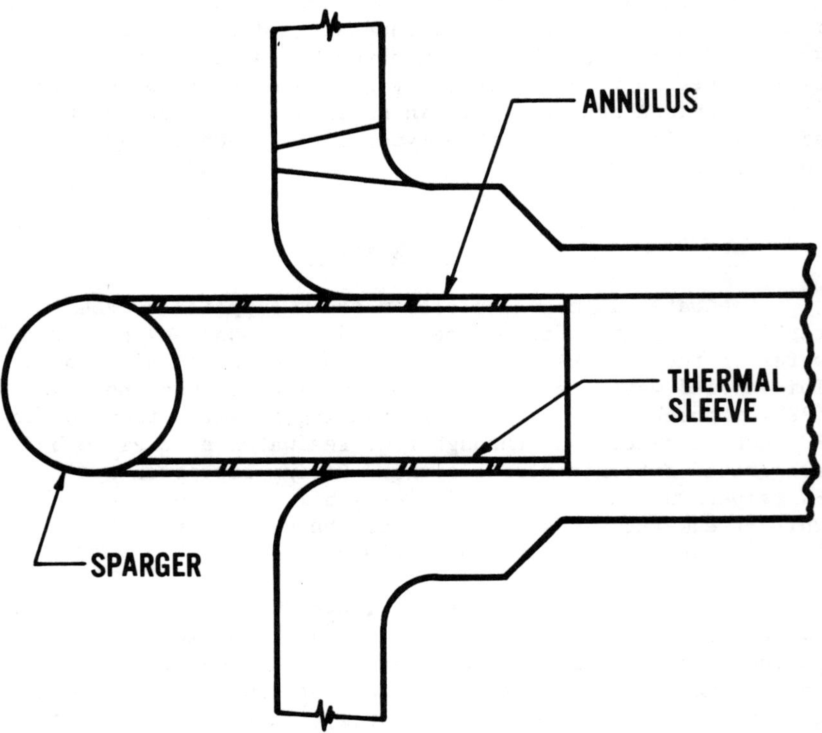

Fig. 1. Schematic of Nozzle With Thermal Sleeve and Sparger

In FW nozzles, cyclic fluctuations at rates up to 100 Hz have been measured. These, combined with large temperature differentials, produce high cycle fatigue conditions leading to multiple crack initiation sites.

Cracks may initiate in any part of the nozzle bore or on the inner corner radius (ICR) (See Figure 2). However, a general tendency toward crack locations at the top and bottom of the nozzles has been observed. Crack orientations, like the locations, are random in nature. Initial crack growth rate is believed high up to crack depths of 6 to 12 mm.

Beyond that depth, growth is slower although the growth rate probably increases at still greater depths.

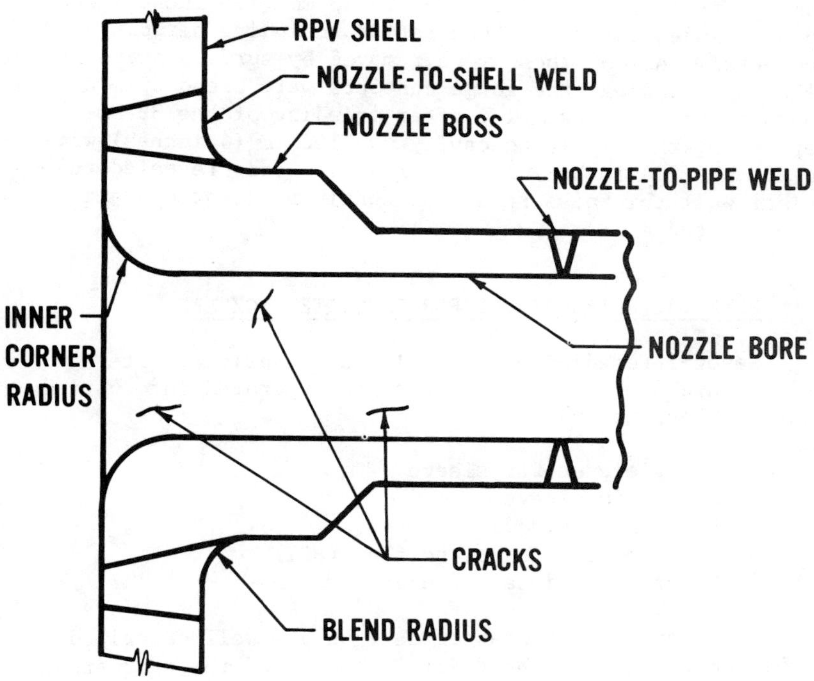

Fig. 2. Nozzle Nomenclature

Because the thermal stresses are generally surface and near-surface phenomena, growth can only continue as a function of mechanical stresses. Mechanical stresses result from expansion of the RPV and the nozzles caused by high operating temperature and pressure (290°C, 6.9 x 10^6Pa) (550°F, 1000 psi). Because these stresses are oriented in specific directions, added crack growth also becomes specifically oriented rather than random. Mechanical stresses are high but the cyclic rate is low, i.e., one cycle per start-up/shut-down operation.

As indicated earlier, the random mixing of hot and cold water can lead to multiple crack initiation sites. Indeed, up to 400 cracks have been found in the four FW nozzles of a single plant. Forty of these extended into the carbon steel base metal. As of June 1977, cracks had been found in all operating BWR plants except Brown's Ferry Unit 2, which had had less than one year operation.

Maximum crack depths in the various plants have ranged from less than 1 mm (0.040 inch) to 38 mm (1.5 inches) where depth includes the total penetration including cladding and base metal. All of these were removed by surface grinding. Following grinding, the gouge contours were blended to a smooth curve. For example, after grinding of the 38-mm-deep crack, the resulting cavity was 100 mm (4 inches) wide and 235 mm (9.4 inches) long. After analysis revealed sufficient wall thickness remained, continued plant operation was permitted.

REMEDIAL ALTERNATIVES FOR FEEDWATER NOZZLES

Several remedial mechanical and operational alternatives have been suggested for or incorporated into various plants. These include:

- Interference-fit sleeve
- Welded-in sleeve
- Removal of cladding
- Close monitoring of nozzles (NDT)
- Other operational changes

Interference-fit sparger designs are well described by the name; that is, the thermal sleeves of the spargers fit very tightly into the nozzles. Tight fits tend to reduce leakage and thereby reduce the thermal stress level. With this approach, at least two problems exist. First, there is heat transfer through the thermal sleeve causing temperature differentials to exist in the annulus between the sleeve and the nozzle. Second, some leakage around the sleeve still may occur. Leakage could increase if system operational cycles, with the resultant nozzle stresses and strains, cause permanent distortion of the sleeve.

Welded spargers have thermal sleeves seal-welded into the nozzle. Unless the weld or sleeve cracks, no leakage can occur. However, heat flow can again occur through the sleeve material and, additionally, two other factors must be considered. Because the sleeve is not easily removable, examination of the nozzle for cracks from the inside surface is not practicable. Of all the viable alternative solutions, welded sleeves are most expensive in initial costs and radiation exposure.

Removal of cladding may be the most attractive alternative. Estimated costs and exposure figures are not available to the authors at this time, so a comparison cannot be made. On the other hand, after removal of cladding, cracking probability is very low because of the improved properties of carbon over stainless steel. Corrosion products resulting from the relatively small exposed areas are not expected to be a problem.

Continued monitoring of nozzles without design changes may be a reasonable alternative, especially on a short-term basis. When confined with operational changes to reduce thermal stresses, plant safety can be maintained.

Of course, operational changes to reduce the thermal stresses are desirable with any of the preceding alternatives. Two of the available changes are to:

- Reduce the duration of periods of low FW temperatures;
- Reduce temperature differential between FW and vessel water.

These two measures alone should reduce the severity of the cracking problem.

Unfortunately, all of these mechanical measures cost money and result in added radiation exposure of personnel. According to W. J. Zarella (1), General Electric Product Service, costs and exposure were estimated to be:

	Interference Fit Sparger Design	Welded Sparger	Grinding Out of Cracks
Total Cost* Replacement/ Repair	$455/$680	$2512/$2850	$250
Man-Rem	425/530	2000	390

CONTROL ROD DRIVE RETURN LINE NOZZLES

Excess water from the control rod drives (CRD) is returned to the RPV through a 4-inch nozzle located 1.5 to 2.5 meters above the reactor core. Not only are there cyclic thermal loads from the 10-38°C (50-100°F) water, but there is also a large temperature gradient of 280C° (500F°) from the top to the bottom of the nozzle. Cracks have been found in the nozzle as well as horizontal cracks in the RPV wall 200 to 250 mm below the nozzle. Thermal sleeves in some plants have reduced but not eliminated the problem.

NONDESTRUCTIVE TEST METHODS

Two NDT methods are being used to detect the cracks--ultrasonics (UT) and liquid penetrants (PT). These are supplemented by visual techniques (VT) where appropriate.

Liquid Penetrants

PT, one of the most widely known and easily used NDT methods, is sensitive to very small surface cracks. However, two limitations of the method are of concern in nozzle examinations. First, oxide deposits may build over and in the thermal cracks, reducing test sensitivity. Usually, flapper wheel grinding will remove surface deposits, resulting in satisfactory sensitivity in spite of possible internal deposits. Unfortunately, this grinding operation increases total radiation exposure and costs.

Second, PT requires access to the affected surface. This means the RPV closure head and some internal devices must be removed from the vessel, an expensive and time-consuming operation. Radiation exposure also increases considerably whenever operations are carried on inside the vessel.

Ultrasonic Testing

UT from the outside surface of the RPV and the nozzle offers many advantages. First, it is relatively fast. Second, radiation exposure can be quite low. Third, and most important of all, the RPV does not have to be opened unless crack-like indications are found. At least this is true in theory. Unfortunately, the NRC does not see it that way. NUREG-0312 contains the statement:

> "We conclude that the UT methods, when applied to nozzle geometry, have not demonstrated a level of reliability that would allow UT to be used as a sole basis for a decision to permit continued reactor vessel operation."

If higher reliability in the UT method in detecting nozzle cracks is shown in the future, this stand could be revised.

Two basic problems are encountered in UT examinations of nozzles. Since the physical configuration is that of the intersection of two thick-walled, hollow cylinders with the inner and outer surface junctions blended to form a smooth surface, the geometrical relationship between inner and outer surface features changes with distance around the nozzle. Similar extreme conditions occur at the top and bottom (0° and 180°) as well as at the 90° and 270° positions. These changing geometrical relationships lead to difficulties in choosing the surface from which to perform the examination, the preferred direction of the sound beam, and location of reflectors. Three choices of examination surfaces are available, i.e., (1) nozzle boss, (2) blend radius, and (3) RPV shell. Examinations with the sound beam directed from the boss to the ICR result in an undesirable angle at the ICR. Examinations from the blend radius are constrained by the nonuniform blend radius (usually manually ground) and the continually changing direction of the sound beam to intersect the ICR. Examinations from the RPV shell are less affected by geometrical considerations but are affected by long metal paths to the ICR.

Cladding echoes present a second severe problem to nozzle examinations. When a sound beam in the base metal is incident on the clad-base metal interface, several things can happen to the sound beam:

- Direct reflection to the search unit;
- Refracted sound penetrates the clad and is reflected from clad surface irregularities;
- Mode converted sound is refracted to the clad surface where reflection occurs.

All of these echoes or reflections may combine to form what is called "clad ripple" or "clad roll." This appears on the UT instrument cathode ray tube as a cluster of individual echoes which "walk" through the cluster as the search unit is moved. Amplitudes often exceed recording levels. Echoes from surface cracks would appear in the same cluster. Detection and identification of flaw echoes is, therefore, extremely difficult.

At least three basic techniques are used for inspection of the nozzle ICR. Simply stated, these are:

(1) Shear wave from the RPV shell

   (a) 1 MHz, 60-70° refracted angle, 45° intercept angle at the ICR;
   (b) 2 shoes for clockwise (cw) and counterclockwise (ccw) scans;

(2) Shear wave from the blend radius (See Figure 3)

   (a) 1 MHz, intercept tangent to ICR;
   (b) 2 shoes for cw and ccw scans;

(3) Longitudinal wave from the blend radius (See Figure 4)

   (a) 2.25 MHz, 45° intercept at the ICR;
   (b) 1 shoe, reversed for cw and ccw scans.

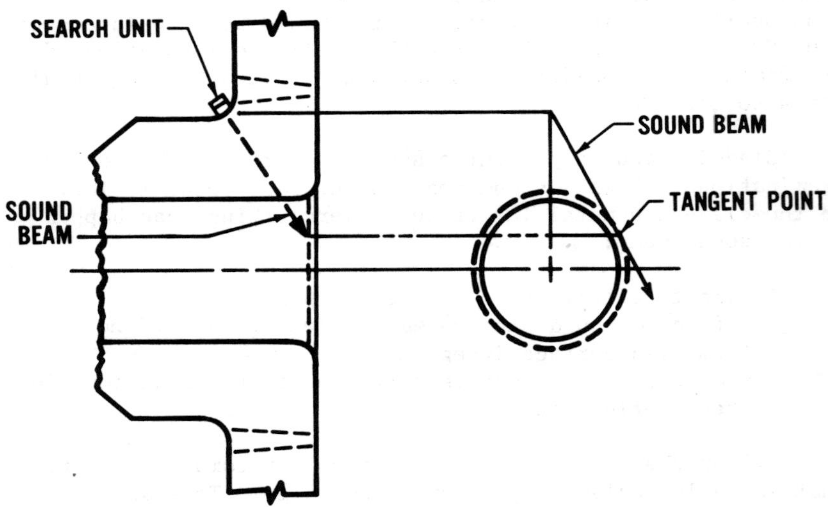

Fig. 3. Sound Beam Direction for Inner Corner Radius Examination Shear Wave

On the other hand, nozzle bore examinations are conducted from the blend radius and nozzle boss much the same as thick wall cylinder examinations (See Figure 5). That is, sound beams are directed in cw or ccw directions to intercept the inside surface at roughly 45° to a tangent at the point of intercept.

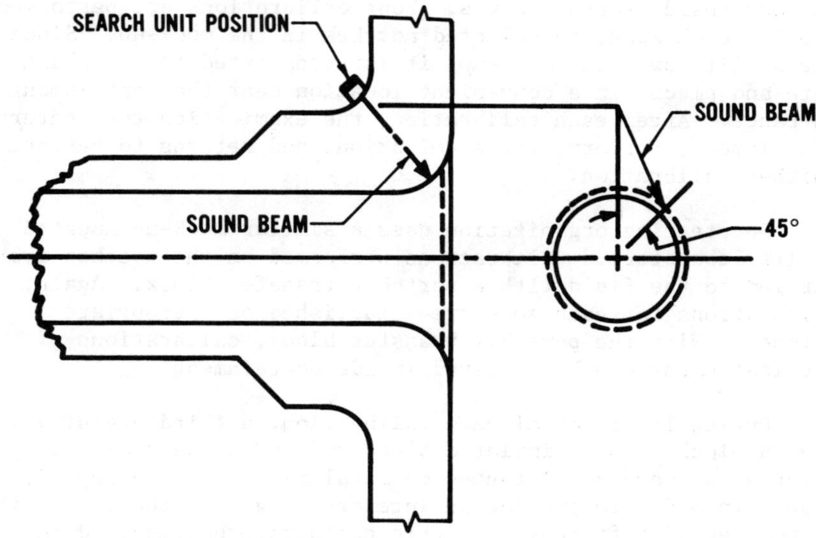

Fig. 4. Sound Beam Direction for Inner Corner Radius Examination Longitudinal Wave

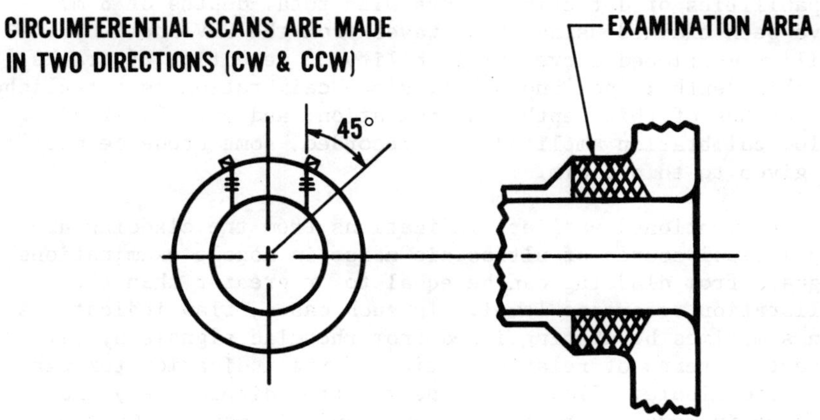

Fig. 5. Nozzle Examination Area

Calibration methods also vary. In one case, a utility uses a full-scale mock-up of a FW nozzle mounted in a segment of RPV shell material. Notches are machined to various depths in four inside surface zones. Four calibrations are performed, one for each zone, on selected notches in the mock-up. Since the utility owns the mock-up, it is transported to the plant site and placed in a convenient location near the containment entrance. After each calibration, the examination team enters containment, performs the examination, and returns to perform another calibration.

One testing organization uses a similar mock-up located at its laboratory. Calibrations performed on the notches are carried to the field with a portable transfer block. Again, calibrations for each zone are established on appropriate notches. With the portable transfer block, calibrations of the instrument can be achieved inside containment.

During longitudinal wave calibration, a third organization uses a block with a simulated blend radius on one side and notches, at varying distances or metal paths, on the opposite side. In order to provide an intercept angle at the ICR of 45° in nozzles of different geometric contours, the required refracted entry angle is estimated by use of a computer program. Shoes are designed for each specific nozzle configuration.

All of these techniques have been successful in locating cracks on the nozzle inner surface with penetrations into the base metal of 6.0 mm or more (total depth of 10 to 16 mm). Capabilities of detecting cracks with total depths of 8 mm have been claimed using shear waves from the RPV shell by the utility mentioned above. Reliability of detection of cracks of this depth is not known but, since calibration is established on notches of this depth and indication, and signals 12 dB below calibration amplitude are recorded, some credence must be given to this claim.

As mentioned earlier, indications from the cladding are a principal source of ultrasonic noise in nozzle examinations. Signals from cladding can be equal to or greater than the calibration reference level. In such cases, flaw indications can sometimes be distinguished from the clad signals by different patterns of relative motion of the indication through the clad ripple. Clad ripple peaks often display very small changes in metal path as the search unit is moved. On the other hand, flaw signals may show large changes in metal path with similar motion of the search unit. Again, reliability of this parameter for identification has not been established. It appears that considerable study is required in this area.

NRC REQUIREMENTS

The NRC has established several basic ground rules for future plant operation and inservice inspections of FW and CRD nozzles. Specific ground rules are applied on a case-by-case basis. For FW nozzles, important operational actions include:

- Monitor and record FW temperature and flow rate;
- Minimize rapid changes in FW flow and temperature;
- Minimize duration of cold FW flow;
- Reduce the temperature differential between the FW and vessel by heating the FW.

Improved repair procedures should be developed to reduce exposure. Repair personnel should be trained and special tools should be developed. Localized radiation levels should be reduced by shielding and decontamination methods such as hydrolasing.

NDT examinations should be performed on vessels without welded spargers as follows:

- UT of nozzle inside surfaces from nozzle and vessel outside surfaces;
- Remove 1 sparger;
- Remove surface oxides by flapper wheel grinding;
- PT examine the one nozzle and the accessible portions of the remaining nozzles;
- If any cracks are found, remove all spargers, flapper wheel grind, and PT;
- Remove all cracks by grinding and record all with depths over 6 mm;
- Report results to the NRC.

As a final interim measure, improved sparger designs should be incorporated.

Vessels with welded spargers will be treated on a case-by-case basis.

CRD nozzles and the surrounding area of the vessel inner surface will be flapper wheel ground and examined with PT. All cracks will be repaired. Other measures are reduction in flow, increasing the pressure to force the water through CRD seals, and reroute return water to another line.

On a long-term basis, the utility should consider cutting the line and rerouting through other lines. The nozzle would be capped.

SUMMARY

In summary, the following items should be noted.

- First, serious cracking problems have been found in FW and CRD nozzles. All of these have been repaired by shallow surface grinding.
- UT methods from the exterior are not recognized as adequate for the sole determination of the presence of cracks.
- PT examinations of inside surfaces are reliable.
- Operational changes and sparger redesign can reduce the cracking problem as can clad removal.
- Additional development of UT techniques are encouraged to improve confidence and reliability, thus increasing plant safety and reducing inspection costs.

BIBLIOGRAPHY

Hannah, K. J. "Test Procedure for Feedwater Nozzle Inner Radius Zone 1 Ultrasonic Examination," General Electric Co., June 21, 1977.

Hannah, K. J. "Test Procedure for Feedwater Nozzle Inner Radius Zone 2 Ultrasonic Examination," General Electric Co., June 21, 1977.

Pickett, A. G. "Discussion of the Flaw Depth Detection Capability Goal for Ultrasonic Inspection of Light Water Reactor Nozzles," Southwest Research Institute.

U.S. Nuclear Regulatory Commission. NUREG 0312, "Interim Technical Report on BWR Feedwater and Control Rod Drive Return Line Nozzle Cracking," July 1, 1977.

U.S. Nuclear Regulatory Commission. "Report to Congress on Abnormal Occurrences," October-December 1976.

U.S. Nuclear Regulatory Commission. "Summary of Meeting Regarding BWR Feedwater Nozzle Cracking," March 24, 1976.

REFERENCES

(1) U.S. Nuclear Regulatory Commission. NUREG 0312, "Interim Technical Report on BWR Feedwater and Control Rod Drive Return Line Nozzle Cracking," July 1, 1977.

THE DIFFICULTIES OF APPLYING NONDESTRUCTIVE EVALUATION TO STAINLESS STEEL COMPONENTS IN THE NUCLEAR INDUSTRY

Allen R. Whiting, Amos S. Greer,
and Jerry T. McElroy
Southwest Research Institute
and
Ake Junghem
AB Sveriges Tekniska Kontrollinstitut

ABSTRACT

This paper discusses the application of ultrasonic examination technology to assess the integrity of stainless steel components in the nuclear industry. These examinations are routinely required during the preservice and ongoing inservice examinations of each nuclear power plant. Selected areas of the systems, such as piping and pipe components, are subject to repetitive examination throughout the life of the plant. Examinations reflecting preservice of "fabrication-type" reflectors, as well as inservice or "service-induced" reflectors, are highlighted; and correlations and comparisons of information regarding these respective conditions are in-

---

A. R. Whiting, Director, Department of Engineering Services, Quality Assurance Systems and Engineering Division, Southwest Research Institute, San Antonio, Texas.

Ake Junghem, Technical Director, Sveriges Tekniska Kontrollinstitut, Stockholm, Sweden.

Amos S. Greer, Assistant Director, Department of Engineering Services, Quality Assurance Systems and Engineering Division, Southwest Research Institute, San Antonio, Texas.

Jerry T. McElroy, Senior Electro-Mechanical Designer, Department of Engineering Services, Quality Assurance Systems and Engineering Division, Southwest Research Institute, San Antonio, Texas.

cluded as they relate to the current American Society of Mechanical Engineers (ASME) Code requirements. Construction practices, metallurgical conditions, accessibility problems, and design considerations that cause difficulty in applying conventional inspection technology are discussed. Data generated by examinations of the required areas are discussed as related to the ASME Section XI Code requirements currently used for establishing the acceptability of these components. Data are shown regarding the metallurgical, geometrical, recording threshold, and data gathering process effects of examinations currently being applied to these areas on an international basis. Discussions regarding advancing technology and its application to these regions of inspection concern are included within this presentation.

INTRODUCTION

The difficulties of applying nondestructive evaluation to stainless steel components in the nuclear industry are manyfold. Stainless steel materials exhibit metallurgical properties that, in many cases, preclude complete examination using conventional ultrasonic techniques. The application of ultrasonic examination technology is critical in the nuclear industry in order to assess the integrity of components used in nuclear power plants. These examinations are required by regulatory agencies and are routinely performed during the preservice and ongoing inservice examination of each plant. Selected areas of systems, such as piping and pipe components, are subject to repetitive examination throughout the life of the plant. As such, it is critical that viable techniques be employed for these examinations in order to initially yield an adequate data base reflecting the true integrity of a component in its preservice condition and subsequently to provide viable capability of detecting, evaluating, and analyzing the impact of service-induced flaws resulting from component fatigue or other factors due to service life.

Southwest Research Institute (SwRI) has been performing ultrasonic examinations of stainless steel welds on nuclear power plant components for approximately 15 years. During this time, approximately 60 examination activities at 28 nuclear facilities involving over 12,000 stainless steel welds have been conducted. In order to address the very broad subject of nondestructive examination of stainless steel components, we have limited our paper to consideration of several specific areas, including intergranular stress corrosion cracking, austenitic piping, advances in ultrasonic transducer design, and the potential impact of new technology, such as adaptive learning techniques and acoustic imaging systems.

INTERGRANULAR STRESS CORROSION CRACKING

In late 1974, intergranular cracks were detected in the stainless steel recirculation bypass piping and core spray piping of a number of operating boiling water reactor (BWR) nuclear plants by either the drywell leakage monitoring system or by a combination of inservice nondestructive evaluation (NDE) techniques such as visual inspection, ultrasonic testing (UT), liquid penetrant testing (PT), or radiographic testing (RT). Metallurgical examination of cracked pipes removed from these plants indicated that the cracking appeared to be caused by a mechanism of intergranular stress corrosion. Such a mechanism is reported to result from a complex interaction among stress (including fabrication and duty-related stress), sensitization, and an oxygenated high-purity aqueous environment.[1]

In late 1974 and early 1975, in order to determine the nature and extent of the problem, the Nuclear Regulatory Commission (NRC) called for special inspections of all BWR's in operation. The primary volumetric nondestructive examination technique for these inspections was pulse-echo ultrasonics.

Although the special inspections revealed few additional cracks, the industry was still interested in upgrading present inservice inspection techniques for detection of the specific type of intergranular stress corrosion-initiated cracking encountered in BWR piping. This concern first arose in 1974, when cracks in piping were discovered by means other than ultrasonics after the pipe had been inspected according to the American Society of Mechanical Engineers (ASME) Code requirements applicable at that time.

Metallurgical examinations have shown that many of these intergranular stress corrosion cracks (IGSCC) are found near the heat-affected zone on the inside surface of the pipe. In some cases, the inside surface contains significant changes in geometry--pipe counterbore, grinding marks, and weld anomalies such as drop-through and mismatch--

that are near the location of the stress corrosion cracks. The shapes of these geometrical changes provide ideal reflectors for ultrasonic energy.

These geometric reflectors tend to mask the reflection of ultrasonic energy from the crack to the extent that, in some cases, careful analysis must be used to distinguish the crack signals from the geometric signals. Instead of presenting the essentially straight and highly reflective surface of classic fatigue cracking (on which many NDE techniques are based), IGSCC's present a very diffuse face which follows the grain boundaries in the material and tends to provide rather poor ultrasonic response relative to typical machined calibration reflectors. This reflective characteristic of IGSCC's compounds the inspection problem because it results in lower probability of detection at equivalent fatigue crack sensitivity levels.

> For austenitic stainless steel welds, the microstructure dominates the wave propagation response. The weld zone is characterized by large columnar gains (dendrites) which result from the slow cooling and directional solidification during welding. Wave velocities and propagation modes within these dendrites exhibit anistropy, which is describable by wellknown relations derived for face-centered-cubic crystal structure, i.e., austenite phase. These variations in wave velocity with crystallographic direction produce large acoustic impedance differences at grain boundaries. Impedance differences also exist near the edges of dendrites due to composition gradients. These gradients form during cooling when alloy content changes as lower melting point microconstituents occupy volume between dendrites.
>
> A substantial portion of the ultrasonic beam is reflected by the local impedance changes at grain boundaries. These reflections develop signal responses often described as "grain noise."[2]

Present inservice inspection practice, as required by the American Society of Mechanical Engineers (ASME) Code, Section XI, uses pulse-echo ultrasonics to detect and define the nature of flaws in primary system piping. This technique is the most appropriate volumetric inspection method presently available. However, time has not permitted a full evaluation of detection probability for the specific type of intergranular stress corrosion cracking observed in BWR piping.

To achieve the best possible inspection, it is imperative to know how efficiently inservice inspection can detect the presence and degree of through-wall penetration of IGSCC's.[3] The through wall penetration of these flaws during operation is not a critical safety issue because of the properties of the material of construction. The flaw size required for unstable flaw growth in response to piping loads is so large compared with the flaw depth which penetrates the pipe wall that detectable leaks should occur before loss of coolant accident flaw initiation criteria can be attained (leak-before-break). All of the flaw experience to date with flaws in both laboratory specimen and operating reactor austenitic stainless steel piping justify this assumption. Nevertheless, such penetration represents a considerable loss in availability when repair and cleanup operations require plant shutdown. Thus, the present ultrasonic inspection practice used for stainless steel piping should be evaluated extensively to define limitations and effect improvements.

In order to define the ability of present ultrasonic and radiographic methods to detect IGSCC's, the Electric Power Research Institute (EPRI) recently completed a project that included a round robin evaluation of cracked and uncracked pipe samples by five groups who are currently providing inservice inspection services to the commercial nuclear power industry.

SwRI participated in this project (EPRI, TPS-75-609). In addition to evaluation of the flaw detection and analysis capability of each inspection group, the variables studied included Code interpretation, procedures, techniques, standards, equipment, and training. The details and preliminary results of this program are presented in EPRI Report NP-436-SR.[4] The practical realism of the study was enhanced by using actual cracked bypass and core spray lines obtained from two BWR's. The flawed pipe sections were radioactive and, therefore, had to be examined under simulated field conditions in accordance with radiation safety work procedures. The overall results of the program indicate that ultrasonic examination is a viable inservice volumetric inspection method for the detection of IGSCC's.

The majority of the ultrasonic inspections found indications in essentially all of the flawed pipes (as determined by destructive examinations of specimens), but interpretation varied as to whether further confirmation was needed to establish the importance of the indications.

The response from a coherent reflector, such as a lack of fusion, may be 25 times greater than that obtained from an IGSCC. This is not surprising, due to the vastly different nature of the respective reflectors. The lack of fusion is localized to a single plane which serves to reflect the ultrasound in a coherent, nondistorted manner. The IGSCC, on the other hand, tends to scatter the reflected signal in a random manner because of its many-faceted morphology. Evidence suggests that we are not looking at a single crack surface, but rather trying to gain a reflection from a spongelike region which does not yield coherent reflections.

It is evident from the experiences derived during this overall specimen characterization that the reliable detection of an IGSCC in its early stages of development will have to be done using highly specialized techniques which stretch the current capabilities of NDE technology.

In a follow-on effort, SwRI participated in EPRI Project TSP 15-26,[3] "Analysis of the Fundamental Ultrasonic Response from Stainless Steel Stress Corrosion Crack Specimens." With the availability of actual crack specimens removed from operational environments, the basic character of the reflected ultrasonic energy from the stress corrosion crack could be studied. The data collected from these specimens consisted of oscilloscope recordings of the radio frequency (RF) waveform and a linear spectrum analysis of the signal reflected from the crack specimens. On a given crack specimen, data were collected for a variety of transducer types and frequencies and flaw detector instrumentation. Transducers and instrument types were provided that displayed various bandpass characteristics classified as wide, medium, and narrow bandwidth. At each data point for a crack specimen, a data point was recorded on the side-drilled hole in an IIW block, a typical calibration standard used in ultrasonic examination of piping. Data from the IIW block allow a direct comparison of a given system's performance on a typical calibration reference standard versus its performance on a stress corrosion crack. The crack orientations that were studied during the course of this program ranged from circumferential and longitudinal to an angular orientation of approximately 45 degrees with respect to the weld centerline.

The spectrum analysis and RF waveform analysis clearly showed that the material's metallurgical structure filters and distorts ultrasonic frequencies higher than 1.8 MHz (Figure 1). When frequencies greater than 1.8 MHz are used, the material greatly attenuates the higher frequency content of the beam and passes only the lower frequency components.[3]

The bandwidth of the search unit proved to be an important characteristic. Those search units that displayed narrow bandwidths, regardless of their frequency, performed better than the wide bandwidth search units.[3]

The data produced by this study program indicate there are certain instrument characteristics that will enhance the detection of stress corrosion cracking in stainless steel piping. Of the instrument types used in the study, those incorporating tuned, multicycle pulsers and medium bandpass receivers provided somewhat better signal-to-noise ratio and frequency fidelity. Medium bandpass receivers as used in this study had a Q of 1.0 and exhibited a 6.0 dB bandpass at 1.0 MHz, 2.25 MHz, and 5.0 MHz as follows:

Frequency	Bandpass
1.0 MHz	0.1 to 1.5 MHz
2.25 MHz	1.0 to 3.5 MHz
5.0 MHz	2.0 to 8.0 MHz

From the search unit's point of view, the data would indicate the material's microstructure and the crack interface form a rather effective "low bandpass filter." For the search unit design to accommodate this ultrasonic transmission characteristic, the operating frequency should be 1.5 MHz with a medium to narrow bandwidth, i.e., Q of 6 to 8.[3]

Figure 2 shows a typical RF wave form and linear frequency analysis of the search unit. This design was found to provide improved performance in beam penetration characteristics and in the detection of flaws.

SEARCH UNIT DESIGNS

In a follow-on effort to the EPRI TPS 15-26 program, SwRI designed and fabricated search unit configurations, incorporating the operating characteristics as indicated by the program results. Both single-element and dual-element, "pitch-catch" units were produced. The operating frequency

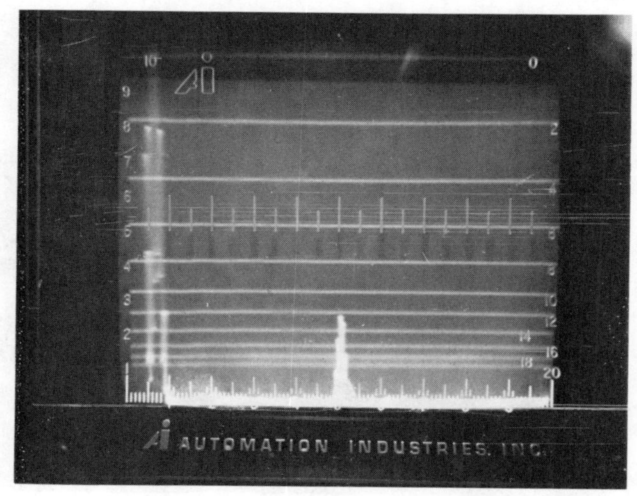

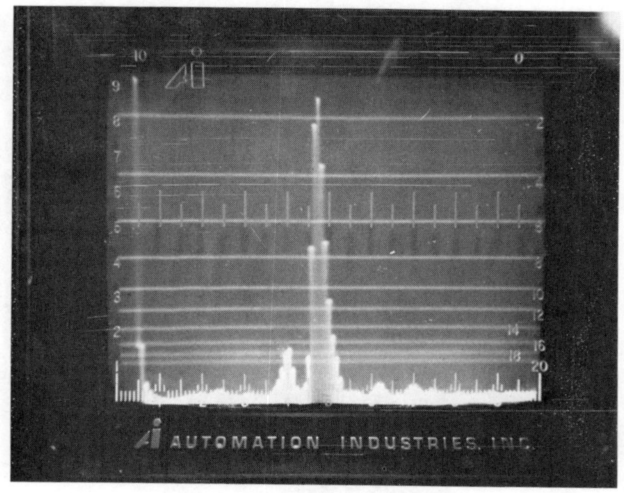

Figure 1. Comparison of Signal Amplitude Returned From a Given IGSCC Specimen

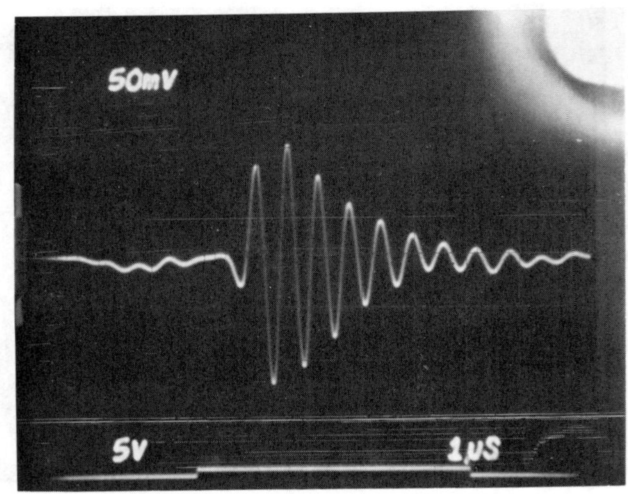

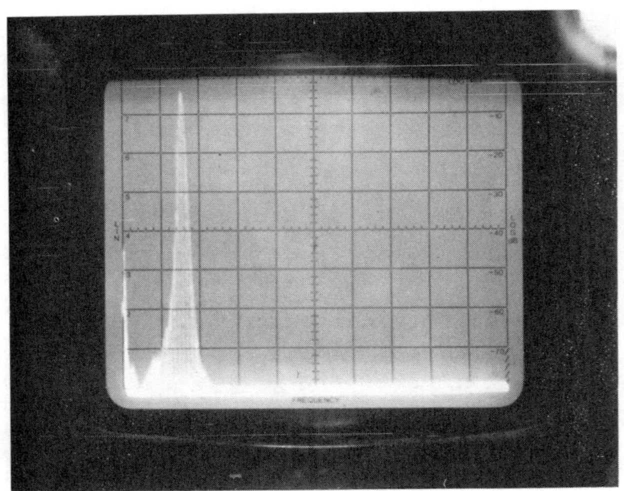

Figure 2. Typical RF Waveform and Linear Spectral Analysis of Search Unit Design That Provided Improved Performance on IGSCC Specimens

was 1.5 MHz with medium to narrow bandwidths and shear wave angles of 45 and 60 degrees. The single-element units were made at 3/8- and 1/2-inch diameters. The dual-element units contained 3/8 x 3/8-inch transmit and receive element sizes with a shoe convergence angle of 5 degrees. The shoe convergence angle provided a focusing effect that optimizes the beam's intensity at a distance of 0.78 inch in stainless steel.

The evaluation of these search unit designs was performed on the same IGSCC specimens that were used during the TPS 15-26 program.

During the evaluation, a calibration standard (SS, 10-inch diameter, Schedule 80, 3/32-inch, 1/2T hole) was used to establish a three-point distance amplitude correction (DAC) curve for each search unit prior to its evaluation on the crack specimens. The crack specimens used during the evaluation were all 304 Stainless Steel, 10-inch diameter, Schedule 80, and contained transverse and circumferential stress corrosion cracks (Figure 2). The results of this evaluation correlated well with the postulated bandwidth and center frequency requirements as indicated by the EPRI TPS 15-26 program.[3]

The performance of each prototype search unit was compared to search unit configurations commonly used in field examinations of piping of this size and schedule. The performance of each search unit was expressed in percent DAC for the signal amplitude received from each crack specimen.

The 3/8- and 1/2-inch diameter, single-element units provided a signal amplitude increase from the cracks of 150 to 300 percent DAC. The dual-element units displayed an increase of 450 percent DAC. Figure 2 shows a typical signal amplitude comparison from a given IGSCC specimen. The specially designed 3/8-inch dual element, 1.5 MHz search unit response is 11 dB greater than the 3/8-inch diameter, 2.25 MHz unit. Additional search units of this design are presently undergoing extensive field evaluation in preservice and inservice examinations.

AUSTENITIC PIPING AND MATERIALS

In early 1973, SwRI began a developmental program to evaluate the metallurgical conditions contributing to the difficulty in examining some austenitic materials in an effort to develop a more viable ultrasonic technique for performance of these examinations. Through the course of the research program, it was realized that most of the austenitic materials, in fact, are dendritic in metallurgical structure. This dendritic structure serves as a diffracting and diffusing medium to normal ultrasonic examination techniques utilizing shear or transverse wave energy. Several examinations utilizing shear energy at frequencies varying from 1 to 5 MHz were performed. Results of these examinations revealed that a shear mode of propagation is very limited in its capability to penetrate and resolve flaws located in austenitic materials. Subsequently, research activities were directed towards the use of a longitudinal or compressional mode of sound propagation for use in examination of austenitic and stainless materials. Through exhaustive studies, it was determined that use of a dual transducer technique utilizing a longitudinal mode of propagation at a frequency range between 1 and 2.25 MHz yielded the best results for both penetration and resolution of flaws. The longitudinal or compressional mode of sound propagation appears to be more successful in penetration since it is less receptive to dispersion and diffraction by the dendritic structure of typical austenitic materials. Additionally, use of a dual transducer technique, whereby one transducer serves solely as a transmitter and the second transducer serves solely as a receiver, allowed a multitude of variations to the inspection technique with regard to interpretation and analysis. Some of these variations include the capability to selectively shape the driving pulse waveform in the transmitter unit and concurrently to introduce preamplifiers and frequency bandpass filters in the receiver network to enhance and improve the signal-to-noise ratio. This technique has been refined over the past four years and has proven quite successful in examining austenitic materials ranging in thickness from 1/2 to 2.5 inches. This examination technique was used extensively in examination of pressurized water reactor main coolant loop systems constructed of forged and centrifugally cast stainless steel materials, such as the Ringhals Unit 2 Reactor in Sweden, the two Donald C. Cook plants in the United States, and others.

Research by other organizations, such as Westinghouse; Tekniska Rontgencentralen AB (TRC) of Stockholm, Sweden; Rontgen Technische Dienst (RTD), and others, has revealed similar results and the necessity to utilize the longitudinal or compressional mode of sound propagation to provide adequate penetration and resolution into stainless steel materials. Research by these organizations, conducted independently, has provided a significant experimental data base supporting the use of the compressional mode of sound propagation. This becomes increasingly important for the examination of stainless steel weldments in which inconel buttering is utilized.

One of the main problems in examination of austenitic stainless steels is the procurement of satisfactory material for use as calibration standards. The calibration standard is, of course, the basis for a reliable ultrasonic examination. Should the calibration standard not exhibit the same acoustic properties as the test material, the examination results may not be meaningful. One of the problems encountered in the ultrasonic examination of austenitic materials has been the assumption of similar acoustic properties for nearly all austenitic materials. Research has proven that even material of similar composition, such as 304 or 316, can have acoustic properties drastically altered by heat treatment or method of processing. Sensitization of austenitic materials is a common cause for drastic alteration of acoustic properties. This results from carbide precipitation at the grain boundaries during sensitization which can significantly alter the scattering potential of the metallurgical structure to the propagation of sound. High concentrations of carbon precipitates at the grain boundaries serve as localized diffusers to ultrasonic waves passing through them, thereby dispersing the sound beam and reducing the penetration potential. The sensitized area adjacent to a weld fusion line may cause a lack of ultrasonic penetration and, subsequently, a poor examination of the weld and adjacent material. As an example of the acoustic differences which can occur through different processing of the same material composition, wrought 316 stainless steel pipe may respond well to a typical shear wave examination using 2.25 MHz transducers. On the other hand, a reflector of similar size would probably not be seen in cast 316 stainless steel pipe of the same size when utilizing the same calibration standard. Efforts are under way at this time to upgrade ultrasonic techniques for application on austenitic materials. Some problems have been solved. Others remain to be defined prior to researching for a solution. The recent rash of cracking in piping and fitting materials of austenitic components in typical BWR plants has provided the industry with some actual worst-condition flaws to evaluate from an NDE standpoint.

An area that continues to play a large role in the definition of flaws in austenitic as well as other materials is weld geometry. The standard 14-degree or 4 to 1 chamfer, commonly found adjacent to the weld prep on the inside diameter of most piping, is the direct cause of the vast majority of reflectors found during ultrasonic examination. These conditions can very easily mask a true flaw indication, especially under one or more of the acoustic conditions previously discussed. It is highly desirable for obtaining meaningful results from an ultrasonic examination that weld preparation design undergo a change. With the difficulties we presently face in resolving problems in austenitic materials, it is absolutely essential that we eliminate all meaningless reflectors to the greatest degree possible.

To summarize, more development must go into codes and specifications to ensure adequate calibration standards are used for examinations. With the availability of actual crack samples removed from operational environments, development of ultrasonic techniques for austenitic materials should move forward rapidly. Finally, efforts must be put forth to alter design of pipe weld preparation geometry in order to eliminate non-flaw reflectors to the greatest degree possible. Research and development actively continue in these areas. A significant amount of work has been accomplished by the Swedish inspection agency, TRC, jointly with SwRI, in the use of longitudinal wave transducers, focused beam transducer elements, and evaluation of metallurgical structure and welding methodology as relates to the ultrasonic inspection of stainless steel weldments. Interim results from these programs are presented in quarterly and final phase reports provided to the Swedish State Nuclear Inspectorate.[4]

SPECIALIZED TECHNIQUES

Although a great deal of research and activity is being expended in the advancement of conventional ultrasonic techniques for the examination of stainless steel materials, several specialized techniques exist which show promise for both the examination of these materials and the interpretation of flaw indications found during those examinations. Among these new and/or specialized techniques are the recently developed adaptive learning networks and the many various systems available in acoustic imaging.

The adaptive learning networks presently developed provide a system by which the RF signals from flaws are recorded. Various parameters surrounding the flaw indication, such as its frequency, center frequency, bandwidth, pulse duration, waveform, and so forth, can be evaluated and assigned what is termed a "feature vector." The term "feature vector" is used to identify specific elements or parameters, such as those mentioned above, associated with the return echo from a flaw or other reflector. Since each discrete type of reflector will have some unique parameters distinguishing it from another type of reflector, by evaluating many of a family of parameters associated with each reflector and assigning a discrete identifying term to each parameter (feature vector), the accumulated data can then be compared. The term "feature vector" thus essentially assigns a retrievable designation to each of many elements of parameters which, taken as a whole, describe the ultrasonic echo from a reflector. This is further explained by Dr. Joseph L. Rose.[5] Comparison of a multitude of such indications from real flaws will establish a family of feature vectors that can be attributed to specific characteristics of an indication from a real flaw. This can be compared obviously to indications received from geometrical reflectors or other extraneous or spurious indications which will not exhibit the same feature vectors. Through this process of selectively choosing significant parameters or feature vectors from real flaw indications from a few samples, the adaptive learning network can then be applied to identify and rationalize the true nature of a multitude of unknown reflectors. This concept has been demonstrated by several investigators in a laboratory environment. Efforts continue at this time through research and development programs, sponsored primarily by EPRI, to advance the practical state-of-the-art with this methodology.[6] This process, called adaptive learning, has been utilized with a great deal of success over the past year. Continued research and development will broaden the application and increase the reliability with which flaws can be analyzed and detected.

Several systems and/or techniques are available in the acoustic imaging field which show great promise towards assisting the analysis and interpretation of ultrasonic data. These systems include acoustic holography, and some of the more recently developed, conventional ultrasonic systems providing three-dimensional imaging capabilities. While these systems may not necessarily advance the problems associated with penetration of the metallurgical structure, it is hoped that they can significantly increase the ease with which flaw signals can be interpreted. The three-dimensional imaging systems are the best example of such technology. It should be pointed out that, technically, the capability to view an image of a flaw can vastly improve the reliability and accuracy of interpretation of that defect. One of the major problems in examination of austenitic stainless steel materials continues to be one of interpretation to resolve material noise from flaw indications. Some of the more recently developed acoustic imaging systems have unique gates by which flaw signals can be interpreted more easily. As a simple example utilizing a conventional C-scan system, an image can be stored considering the time of flight, flaw location, and flaw characteristics by performing a conventional C-scan of the component. The parameters associated with indications which comprise the image stored can then be treated by computer logic to provide the capability of the three-dimensional viewing. In other words, a plan view representation of a flaw can be oriented to the viewer in a three-dimensional plane to more easily facilitate analysis and interpretation of the existence of the flaw and its location. Acoustic holography techniques, additionally, are being researched to provide for greater resolvability of austenitic stainless steel materials in order to provide a similar three-dimensional view of flaws. A great deal of work has been accomplished, primarily by Babcock and Wilcox, in the application of acoustic holography to thick section nuclear components. Application of the technology to stainless steel components has been limited due to problems of lack of penetration, high noise associated with stainless and austenitic metallurgical structures, and proper interpretation of holograms due to that noise. Further research continues in an attempt to resolve these practical problems. Continued research in these areas, especially continued research and development towards the integration of these specialized techniques with more conventional ultrasonic techniques discussed herein, hopefully will eventually lead to a hybrid inspection system capable of examining austenitic stainless steel materials and providing a better and more easily interpreted data display.

SUMMARY

In conclusion, we have discussed briefly many of the inherent practical problems associated with ultrasonic examination of stainless steel components in the nuclear industry. Research over the past few years has identified and isolated many of the variables contributing to the difficulty of inspection, such as metallurgical structure, welding practices, transducer designs, and weldment configuration. Additionally, research during this period has produced some significant improvements in ultrasonic examination capability, notably in transducer design improvements; use of selected frequencies; and application of new, specialized techniques. Much of this research, however, has clearly identified the fact that no single panacea has been put forth to solve the overall problem of austenitic stainless steel inspection. Continued research and experimentation in the areas discussed herein will hopefully advance the state-of-the-art closer to the ultimate solution. In short, more work is sorely needed.

REFERENCES

1. E. R. Reinhard, "A Study of Inservice Ultrasonic Inspection Practice for BWR Piping Welds," Electric Power Research Institute NP-436-SR, August 1977.

2. S. M. Mech and T. E. Michaels, "Development of Ultrasonic Examination Methods for Austenitic Stainless Steel Weld Inspection," Materials Evaluation, July 1977.

3. J. T. McElroy, "Detailed Analysis of the Fundamental Ultrasonic Response Data from Stainless Steel Stress Corrosion Crack Specimens," Southwest Research Institute Project 17-9183, Electric Power Research Institute TPS 15-26, June 1976.

4. Final Report, Phase 2, "Evaluation of Ultrasonic Testing in Austenitic and Inconel Weld Materials," Bo Sundstrum (TRC), Amos Greer (SwRI), May 24, 1974. (This program now being managed by Sveriges Tekniska Kontrollinstitut, Stockholm, Sweden.)

5. Joseph L. Rose, Phillip W. Mast, and Ludwig Niklas, "The Potential of 'Simulearning' in Flaw Characterization," July 1975.

6. EPRI/SwRI RP-892 Program, "Ultrasonic System Optimization Study."

ULTRASONIC INSPECTION OF AUSTENITIC WELDS

J.R. Tomlinson
A.R. Wagg
M.J. Whittle
N.D.T.A. Centre, C.E.G.B.

1. INTRODUCTION

Ultrasonic inspection of austenitic welds has long been considered difficult. High levels of scatter and attenuation often prevent observation of defects; beam splitting and skewing effects and spurious echoes from the weld hinder interpretation. While the problems themselves are well documented, hitherto there has been little progress in determining in detail which aspects of the weld structure are responsible. Until the fundamental aspects of ultrasonic propagation in austenitic structures are properly understood we can do no more than devise inspection techniques empirically. The danger in this approach is that a technique which is satisfactory on one weld may fail completely when applied to another, even though the second has been fabricated using apparently identical procedures.

This paper describes work carried out at the N.D.T. Applications Centre of the Central Electricity Generating Board to investigate the influence of austenitic weld structure on ultrasonic propagation. In Section 2 we describe the structure of austenitic welds and contrast this with the more amenable structures found in ferritic welds. Section 3 describes the results of experiments to explore the influence of the structure on ultrasonic attenuation, velocity and beam bending. The use to which these observations can be put in practical weld inspection is described in Section 4. Finally, Section 5 discusses the prospects for austenitic weld inspection and the restrictions and requirements imposed upon plant designers by the limitations of the inspection method.

2. THE STRUCTURE OF AUSTENITIC WELDS

The grain structure of austenitic welds has been discussed in detail by Baikie and Yapp, and the following section summarises their work (1).

The grain structure of austenitic welds differ markedly from that in ferritic welds. In both cases, the solidification process during welding initially produces a columnar grain structure in each weld bead. Grains grow along the maximum thermal gradients in the bead along a ⟨100⟩ crystallographic axis. Growth in this particular direction is faster than in other directions and this leads to the rapid disappearance of unfavourably oriented grains. Deposition of subsequent weld metal reheats the bead and, in the case of a ferritic weld, the columnar grain structure is destroyed by the austenite-ferrite phase transformation that occurs as the solid cools. No such transition occurs in the austenitic alloys considered here and consequently the columnar grain structure survives. Furthermore, each new weld bead remelts the surface of the preceding beads and the new grains grow epitaxially on the existing ones. Consequently grains several centimetres in length can be produced as shown in Figure 1. This shows a macro-section of a fillet weld made by the manual metal arc (M.M.A.) process with AISI 316 weld metal.

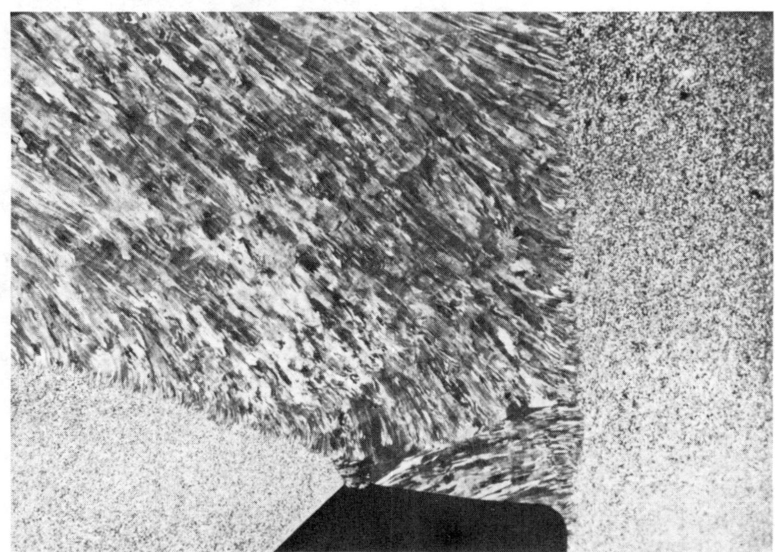

Fig. 1. Columnar Grains in a 316 Weld (1.2 x Full Size)

Austenitic welds, therefore, contain long columnar grains with the major axis along a <100> crystallographic direction. Such a structure is said to possess a fibre texture.

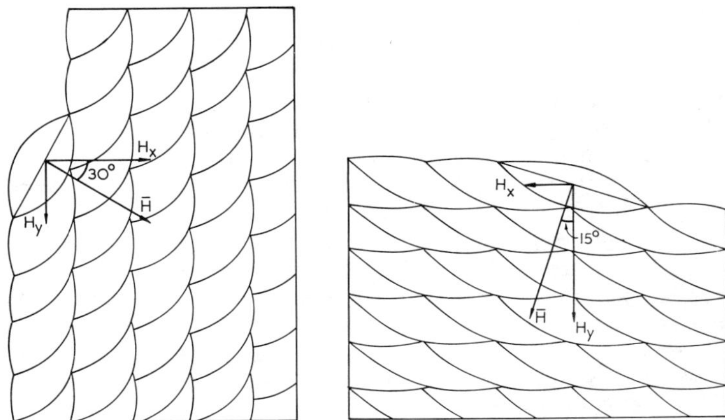

Fig. 2. Heat Flux in a) Horizontal - Vertical Weld and b) Downhand Weld.

The local direction of growth of the columnar grains is governed by the thermal gradient within the weld bead. This in turn is influenced by the welding procedure and, within each bead, the gradient is determined by the solid metal in contact with the bead as it cools. For example, consider weld metal deposited in horizontal runs onto a vertical plate - the horizontal-vertical welding position - as illustrated in Figure 2a. It can be seen that each weld bead is deposited into the corner formed by the preceding bead and the previous layer of beads. Heat therefore flows out of the bead in the two directions shown and the resultant thermal gradient, and therefore the direction of crystal growth, is at an angle to the horizontal. In practice the welder must deposit sufficient metal in each run to enable the following bead to be deposited on the 'ledge' formed by the preceding run and we find that the angle the grains make to the horizontal is typically about 30°. The angle of grain growth can be predicted by similar reasoning for other welding techniques. If the base plate is rotated away from the vertical, the height of the step into which new weld metal is deposited becomes progressively smaller as the plate approaches the horizontal until the downhand position shown in Figure 2b is reached. When the plate is horizontal, the direction of grain growth is found to be about 15° from the normal to the plate.

3. ULTRASONIC PROPERTIES OF AUSTENITIC WELD METAL

The previous section described how the metallurgical properties of the austenitic alloys give rise to a weld metal which is macroscopically anisotropic. We now examine the effect of this anisotropy on the propagation of ultrasonic waves through such material.

3.1 Attenuation

Our initial observations were made on machined blocks of 316 weld metal fabricated using a variety of welding techniques. With a 2.5MHz short pulse (single cycle) compression wave probe it was found that the attenuation was variable in the range 0.1-0.4dB/mm. Examination of the cause of the variation suggested that the orientation of the ultrasonic beam relative to the fibre axis was a major factor. The elastic properties of a material with a perfect fibre texture have circular symmetry about the fibre axis. Consequently the variation of properties with direction can be examined by observations in any plane containing the fibre axis. Measurements were accordingly made on cylinders cut from weld pads with the grain axes along a diameter.

The cylinders (50mm long and 38mm in diameter) were mounted on a table in a tank of water with their axes vertical. Ultrasonic transmitter and receiver probes (2.5MHz, 15mm diameter) were arranged so that the sound beam passed horizontally through the cylinder. Both the cylinder itself and the receiver probe could be rotated about the axis of the cylinder. With this arrangement the amplitude of the signal transmitted through the cylinder and the time of flight could be measured. By comparing results from weld metal cylinders with those obtained from equiaxed, fine grain ferritic or austenitic cylinders, it was possible to obtain values for the apparent attenuation and the velocity in the weld metal as a function of beam orientation.

Figure 3a shows the variation in transmitted signal as a cylinder of weld metal was rotated, with the receiver probe in line with the transmitter. The apparent attenuation varies from a value of 0.4dB/mm along, or at right angles to, the grain axes, down to 0.05dB/mm at an angle of approximately 45° to the grain axes. This curve relates to a cylinder cut from a downhand weld pad of 316. Very similar curves (see Figures 3b-e) have been obtained from horizontal-vertical and vertical-up 316 welds, a downhand Inconel 182 weld and a casting of alpha brass. Negligible variations with orientation were found with the fine grained ferritic and austenitic cylinders or with a columnar grained aluminium casting (aluminium has a very low elastic anisotropy).

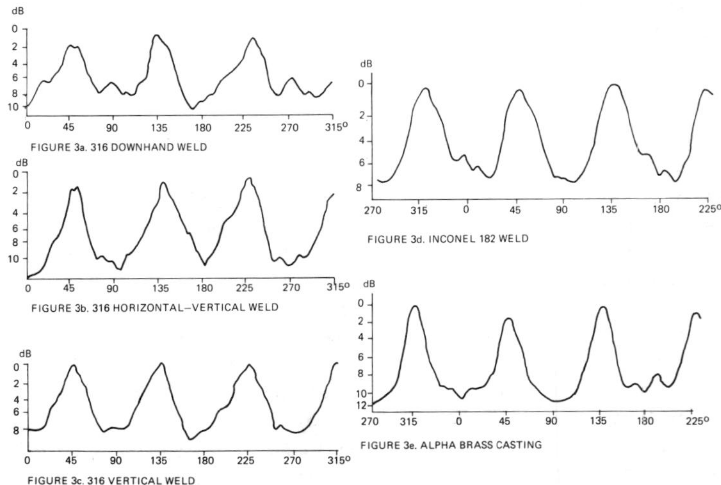

Fig. 3. Variation of Transmitted Signal with Direction Relative to Columnar Grain Axes.

It may therefore be concluded that the dependence of the transmitted signal on beam orientation arises as a result of the macroscopic anisotropy of the assembly of grains and is not related to the composition of the weld metal or the welding position.

The origin of the variation in through transmitted signal was further clarified by examining the variation in signal as the receiving probe was traversed around the sample, for different orientations of the sample. Results are shown in Figure 4 for the incident beam at 0°, 45° and 22.5° to the grain axes. The quantitative interpretation of these curves is complicated by the diverging lens action of the steel cylinder immersed in water. However, the effect of this divergence is only to amplify any variations in the beam width and direction that may take place within the cylinder. Thus we may deduce from the results shown in Figure 4 that the beam width is much smaller when the beam axis is near 45° to the grain axis than when the beam axis coincides with that of the grains. Also, at an intermediate orientation the beam is skewed off and does not pass along a diameter. Both of these effects contribute to the variations of signal shown in Figure 3.

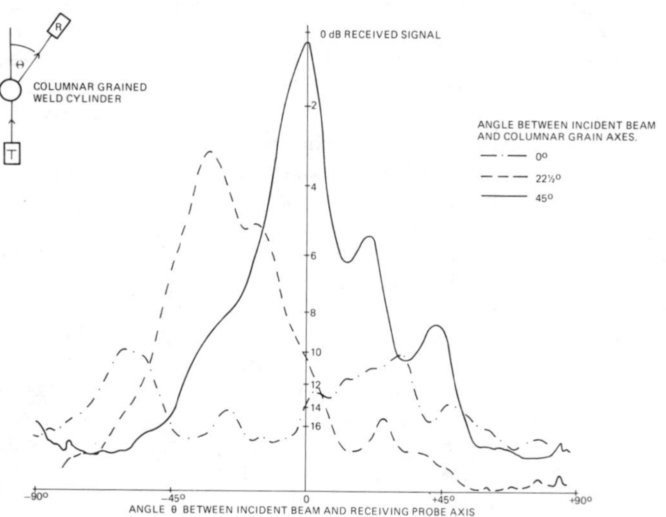

Fig. 4. Distribution of Sound Emerging from Cylindrical sample for 3 Incident Directions (316 Weld Metal).

In addition there could be a dependence of the scattering on the orientation of the beam. Indeed our earlier interpretation of the results shown in Figure 3 was entirely based upon such an effect. However, measurements of the amplitude of the back-scattered waves showed that this varied only slightly as the sample was rotated in the immersion tank. At 2MHz no significant variation with sample orientation could be detected. With a 5MHz probe the average grass level was at a minimum with the beam along the grains and rose by only about 4dB when the beam traversed the grains at right angles. Thus the major variation in signal to grass ratio in this weld metal arises from variations in signal, not in the grass.

3.2 Velocity

Measurements of compression wave velocity are shown in Figure 5. Reynolds has shown how the elastic properties of a fibre textured cubic material are related to the elastic constants of the cubic crystal, and Musgrave has given an expression for the velocity as a function of direction in a medium of the relevant symmetry (2,3). The curves shown in Figure 5 were calculated using this theory. In the absence of data specific to 316, the elastic constants obtained by Bradfield in 18/12 austenitic steel and Salmutter and Stangler in 12/12 austenitic were used (4,5). There is good qualitative agreement between the experimental curves and the theory with a 10% variation between the minimum velocity along the grains and the maximum which occurs at approximately 45° to the grain axis. The clear similarity between the form of this curve and that for the transmitted amplitude invites the hypothesis that the two are related. This was tested by examining the transmission through a single crystal of Nimonic 80A which was machined with a ⟨100⟩ direction along the cylinder axis. The observed variation in transmitted signal is shown in Figure 6. This is clearly similar to those shown in Figure 3, providing conclusive evidence that a variation in scattering from the grain boundaries was not the major cause of the variation in signal transmitted through weld metal.

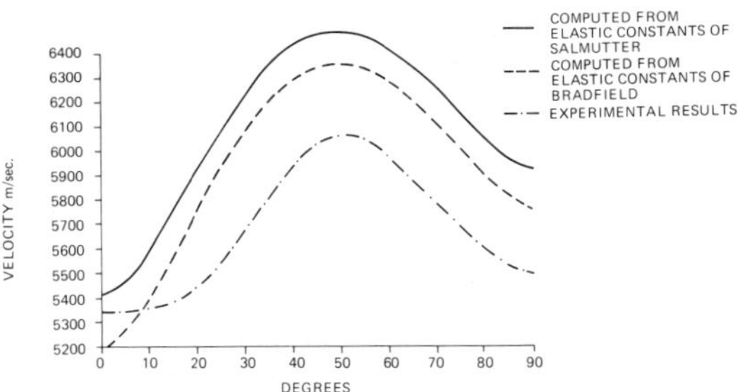

Fig. 5. Variation of Compression Wave Velocity with Direction Relative to the Fibre Axis in 316 Weld Metal.

3.3. Propagation of Waves in Anisotropic Media

It is believed that the observations described above may be explained in terms of the propagation of waves in a homogeneous, anisotropic material. While it is undoubtedly

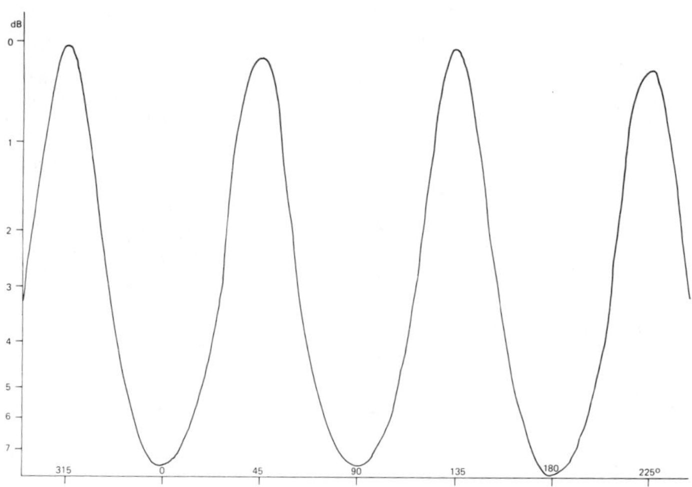

Fig. 6. Through-Transmitted Signal in a Single Crystal of Nimonic 80A vs Angle between Wave Normal and (100) Axis.

the case that there is significant scattering in these coarse-grained materials, it seems probable that the orientation-dependence of the apparent attenuation is not primarily governed by this phenomenon. An important feature of the propagation of waves in an anisotropic medium is that, in general, the ray direction - the direction of energy travel - is not perpendicular to the wave front. This phenomenon gives rise to a skewing of the beam off the axis of the transducer. Using the theory of Musgrave and the elastic constants of Salmutter and Stangler, the magnitude of this skewing has been calculated as a function of beam orientation.

Experimental confirmation of these calculations could not readily be obtained from the water immersion experiments referred to earlier because the large velocity difference between water and steel leads to strong refraction effects. These were minimised by inserting the test cylinder into a matching cylindrical hole in a mild steel block. The small gap between the cylinder and the block was filled with oil to permit ultrasonic transmission. From the displacement of the receiver probe necessary to maximise the signal the angle of skew in the weld metal was calculated. The experimental observations and the computed curve are shown together in Figure 7. The largest angle of skew observed

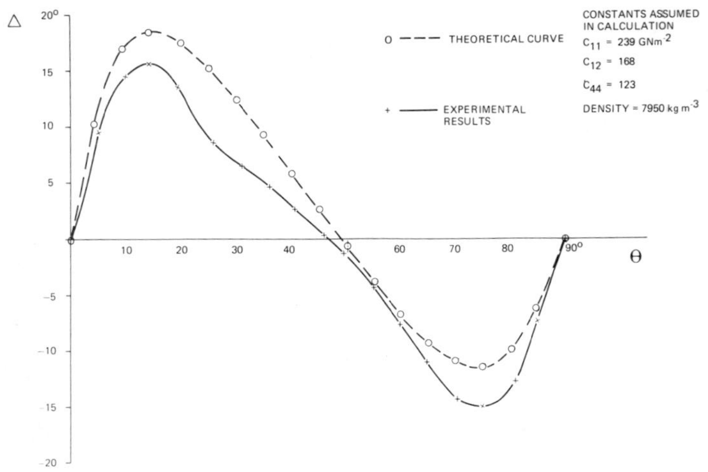

Fig. 7. Angle of Skew (Δ) vs Angle (θ) between the Wave Normal and the Columnar Grain Axes (Compression Waves).

for compression waves is about 15°, but for shear waves the calculations show that beams may deviate by up to 50° from the expected direction of propagation. It is clear that with deviations of this order of magnitude a shear wave inspection would be extremely difficult to carry out, and even a compression wave inspection will need extra care in interpretation of the position of indications.

The variation of the skewing angle with direction leads to variations in beam width. Thus, with the beam directed along the grain axis, all non-central rays are skewed outwards (see Figure 7) and a widened, lower intensity beam results. The converse happens when the beam is directed along the direction of maximum velocity. This problem has been examined by Papadakis who showed how with approximations the resultant beam shape could be calculated (6). Unfortunately, except along directions of high symmetry the calculations are very complex. Computation of the beam shape for the present material is currently in progress. Nevertheless it is clear that significant variations in beam spread occur for waves travelling in different directions.

To summarise, our present understanding of the characteristics of ultrasonic propagation in austenitic weld metal is based on an appreciation of the importance of the anisotropy of this material. This gives rise, we believe, both to variations in the beam width and to deviations of the beam axis which are more significant than grain boundary scattering in determining the amplitude of transmitted signals. As the effects of these beam variations are in many respects indistinguishable from those of an orientation-dependent attenuation caused by grain boundary scattering (which may also be present), we will throughout the rest of this paper refer to the consequences of all these effects in terms of attenuation.

4. WELD INSPECTION

The results presented in the previous section show that the apparent attenuation in an austenitic weld depends on the angle between the beam and the columnar grains. A wide variety of weld materials and samples welded under different conditions have been examined and it is clear that these effects are a general property of austenitic welds which arise from the intrinsic fibrous structure of the weld metal. We may not as yet fully understand the detailed causes of this behaviour but it is sufficiently well established empirically for it to be used to develop improved inspection procedures for specific weld geometries. We now illustrate the way in which this can be done by two examples. The first concerns a fillet weld in the boilers of an AGR (Advanced Gas-Cooled Reactor) and the second covers the general category of butt welds.

4.1. Inspection of an Austenitic Fillet Weld

In certain designs of AGR, an austenitic fillet weld is used as a closure between two coaxial tubes in wrought 316 austenitic steel. The original weld design is shown in Figure 8a. Early work showed that the weld metal could be far more readily inspected using compression waves than shear waves. This result was not unexpected since theories of scattering in equiaxed structures show that when either shear waves or compression waves are propagating in a multi-grained material, the bulk of the energy scattered by the grains is in the shear mode (7). Consequently, in pulse-echo ultrasonic inspections the signal to back-scatter (grass) ratio from defects is higher for compression wave transducers since these are unable to respond to the larger part of the scattered energy which is in the shear mode. A further advantage of using compression waves is that beam skewing effects are smaller than for shear waves, as discussed earlier.

The weld was therefore redesigned as shown in Figure 8b to enable the fusion faces to be inspected at right angles using normal compression wave probes placed on the two external surfaces.

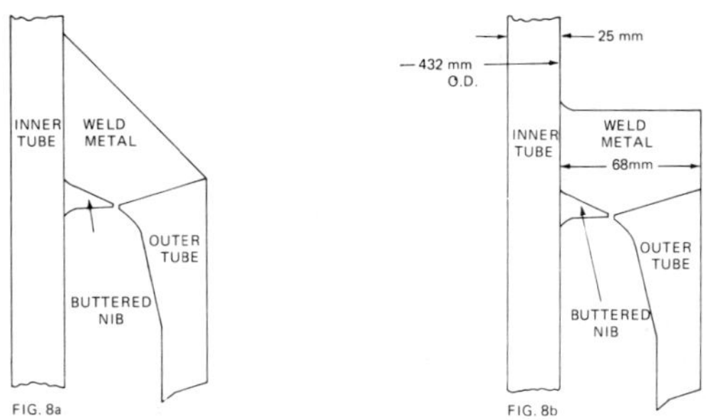

Fig. 8. Original (a) and Modified (b) Weld Profiles.

A further improvement in the signal to scatter ratio was obtained by using short pulse probes. This is because the scatter at any point on the flaw detector screen arises from energy scattered from the insonified volume of metal at that range. The insonified volume has a cross-section transverse to the beam set by the beam spread at the appropriate range and a depth along the beam determined by the pulse length. Reducing either the beam spread or the pulse length therefore reduces the number of scatterers contributing to the grass and improves the signal to grass ratio. In practice, pulse length is more readily reduced than beam spread. Other authors (8,9) have discussed the improvements brought about by using focussed probes either singly or in pairs to reduce beam spread, but in our case the additional practical difficulties were not felt to justify their use.

The results described in the preceding Section demonstrate that minimum attenuation is obtained when the angle between the ultrasonic beam and the columnar grains in the weld metal is 45°. In the present closure weld therefore the weld attenuation for the beams described above can be minimised if the weld grains grow at 45° to the tube axis. The main obstacle to achieving this aim around the full weld circumference is that the welds must be made with the tubes horizontal and consequently the welding position varies around the circumference. If, for example, the weld beads are deposited as shown in Figure 9a, at the

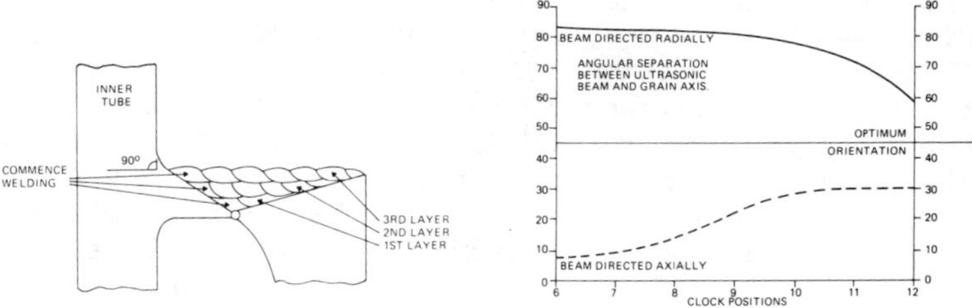

Fig. 9. (a) Deposition of Weld Layers at 90° to Tube Axis
(b) Resultant Angles between Beam Axes and Grain Axes around Weld.

top of the weld (12 o'clock position), the welding is of the horizontal-vertical type, whereas at the 3 o'clock position it is vertical-up and so on. This causes the grain orientation relative to the tube axis to vary systematically around the weld. Figure 9b shows the predicted variation of the angle between the grain axes and the ultrasonic beam if the weld metal is deposited in layers normal to the tube axis as shown in Figure 9a. The angle between the grain axis and beam is markedly different from the optimum value around the entire circumference and such a weld is uninspectable. However, when the weld beads are deposited onto a surface at 55° to the tube axis as shown in Figure 10a, the average departure from the optimum of the angle between the grains and the beam is minimised. (see Figure 10b). Such a welding technique is therefore best for these particular welds and has been shown to be both practicable and capable of producing inspectable welds.

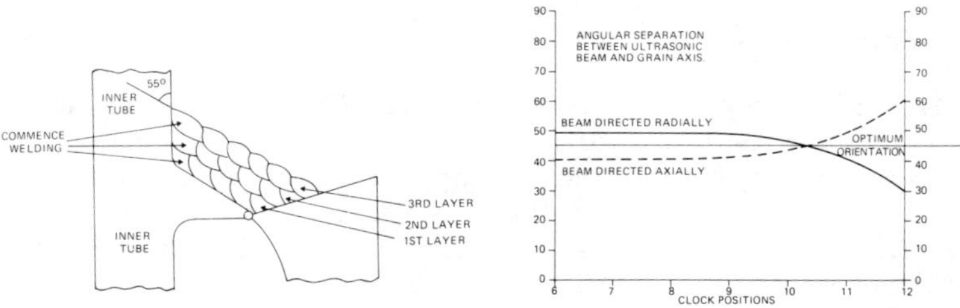

Fig. 10. (a) Deposition of Weld Layers at 55° to Tube Axis
(b) Resultant Angles between Beam Axis and Grain Axes around Weld.

Figure 10b shows that, even with the optimum welding technique described above, the grains are still distributed about the best orientation. The attenuation of the weld metal varies in a corresponding way around the weld circumference and as a result the sensitivity of the test to defects is also variable. It has therefore been necessary to compare the results of calculations for maximum tolerable defect sizes with those for minimum detectable defects. Fortunately the significant defects can all be found with a high probability of success even with the highest attenuation likely to occur.

A further problem is that of measuring defect sizes. The conventional defect sizing techniques involve scanning the probe to determine the position of the edges of the defect. The probe is moved until the signal from the defect has fallen by a prescribed amount from its maximum. In the present case this method is unusable due to fluctuations in attenuation as the probe is scanned. Experience with other weld configurations indicates that this is likely to be a widespread problem. A range of techniques has been developed to enable the size of defects in different parts of these particular welds to be measured. For example, when inspecting the fusion face to the inner tube, we can use obscuration of the echo from the inner wall of the tube. Also, the relative heights of the defect and back wall echoes provide a guide to defect size for small defects that do not completely obscure the back wall echo. This technique is independent of weld metal attenuation since the beam has traversed the same path through the weld for both echoes. The only difference is that the beam has passed through the tube material for the back wall echo. However the tube material, although austenitic, has been forged to a fine equiaxed grain structure and therefore has negligible attenuation. A similar technique can be used for defects growing from the weld root.

For defects on the fusion face to the outer tube no back wall echo is available and so we must use a beam scanning technique in spite of the deficiencies pointed out above. Fortunately defects on the fusion face are also accessible to an angled beam of shear waves directed through the forged material of the outer tube (see Figure 8b). The fine equiaxed structure of the forging permits shear waves to penetrate through to the fusion face. However, attenuation within the weld is so high that the technique is limited to defects in the immediate vicinity of the fusion face.

It will be evident from the discussion above that not only is the detection of defects in austenitic welds troublesome, but also that the difficulty of measuring defect size is increased by the variable attenuation found in these welds. Nevertheless, by controlling both the weld shape and welding procedure, a useful inspection technique has been developed.

4.2. Inspection of Austenitic Butt Welds

A programme of work has been commenced to establish the possibilities and limitations of ultrasonic inspection of butt welds in austenitic material. These are in widespread use and feature particularly in current fast reactor designs. Although this development is aimed primarily at welds between identical materials, it is equally applicable to transition welds and, indeed, the example to be discussed here is such a weld between 316 stainless steel and 2¼%Cr 1%Mo ferritic steel. In the ultrasonic inspection of a transition weld there may be complications, arising from the interface between the dissimilar metals, that are absent in a non-transition weld. Our experience suggests that these effects are of less significance than the properties of the weld metal itself.

The object of the exercise was to decide which of two alternative weld designs would be preferable for non-destructive examination. Both employ a weld metal that retains a columnar grain structure down to room temperature - in one case 316 stainless steel, in the other Inconel 182. Thus the problems of inspecting these welds will be identical in most respects to those to be expected for any austenitic butt weld.

Trial welds were made between flat plates 50mm thick, the surfaces of which were subsequently machined to remove the weld root and cap. Cylindrical holes 3mm in diameter with their axes along the direction of welding were then drilled in the weld metal and at the fusion faces, to serve as target reflectors for the ultrasonic beam. The inspection was carried out using short pulses of compression waves with beams normal to the plate surfaces and at 45° to them. The two weld designs and their effect on the propagation of the ultrasonic beams are now considered in turn.

The first design examined used a single-sided weld with

a J-preparation. Inconel 182 was deposited with the M.M.A. (manual metal arc) process. The weld cross-section and target reflectors are shown in Figure 11b; Figure 11a shows the etched macro-section of the weld which reveals the grain structure characteristic of an austenitic butt weld made with the M.M.A. process.

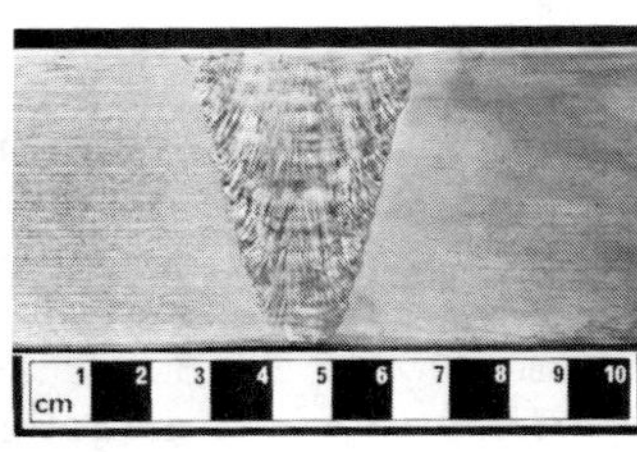

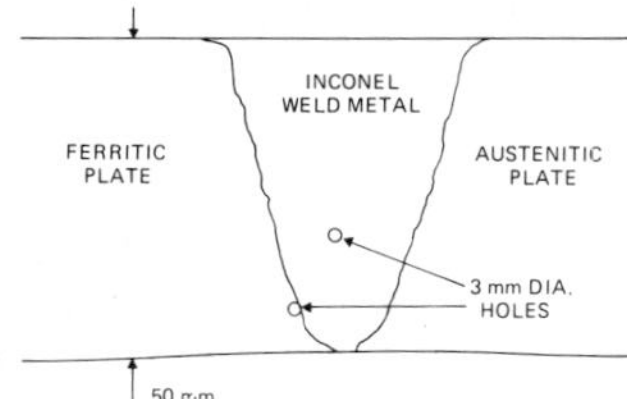

Fig. 11. Inconel Filler Transition Weld (a) Etched Macro-section (b) Location of Target Holes.

In the fillet weld discussed above the volume of weld metal was large compared with the area of weld at the fusion faces and so localised cooling effects at the fusion boundaries did not unduly affect the orientation of the columnar grains. The grain orientation was, therefore, substantially uniform throughout the weld. In butt welds such uniformity is prevented by significant cooling effects at the fusion faces. The maximum thermal gradient is normal to the weld surface in the centre of the weld but normal to the fusion face at the edges. Moreover, rapid cooling at the fusion faces results in smaller grains in this region than elsewhere.

With a normal beam directed down the centre of the weld a remarkably low attenuation (0.16dB/mm) was observed with a short pulse 5MHz probe. The drilled holes could be detected at all positions within the weld metal with at least 12dB signal to grass ratio, both with normal and 45° compression probes, though with the latter higher electronic gain was needed to compensate for the lower sensitivity of the probe. However, although the holes could be readily detected, the positioning of the reflectors was in some cases in error due to the skewing of the beam discussed above. This effect was further examined by positioning transmitter and receiver probes on opposite faces of the plate and scanning the receiver along a line perpendicular to the weld for a series of transmitter positions along a similarly oriented line. With the transmitter placed on the parent material or on the weld centre-line the maximum through-transmitted signal was obtained with the receiver directly above the transmitter, but for all other transmitter positions under the weld the maximum signal was received with the second probe not directly above the transmitter. In certain positions the transmitted beam was bifurcated and there were two maxima in the received signal. It is believed that this unusual phenomenon arises from the inhomogeneous nature of the weld metal in a butt weld discussed above. The non-uniformity of the grain alignment enables sound to find more than one path between the two surfaces of the welded plate. A similar 'double image' of one of the cylindrical holes has also been observed.

The results for this weld design may be summarised by saying that the target holes could be detected, but that positioning and sizing of real defects would be difficult because of the skewing effects.

The second weld was made with the English Electric Mark III design. An asymmetric weld preparation was used (see Figure 12a) and, after buttering the ferritic face with 2¼%Cr 1%Mo weld metal followed by a layer of 316 weld metal, the joint was made with 316 filler. The M.I.G. (metal inert gas) process was used to make this weld and it is believed that this was responsible for the grain structure observed (Figure 12b). Instead of the long grains growing epitaxially across many weld beads with a well-defined orientation usually seen in the M.M.A. welds, this weld had 'fans' of grains within each bead which only occasionally extend into the next layer of beads. It is believed that this is caused by the higher and more localised heat input resulting from the M.I.G.

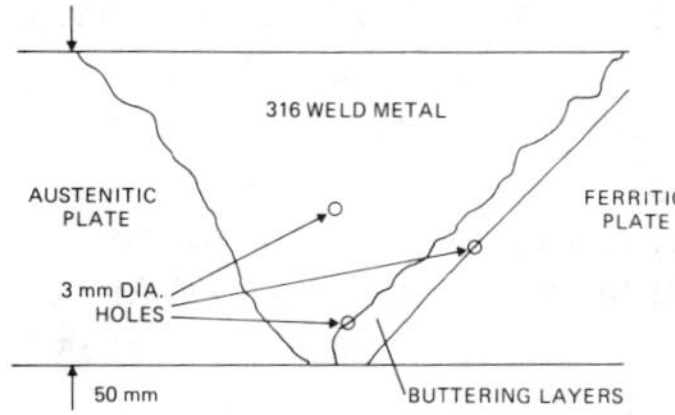

Fig. 12. English Electric Mark III Transition Weld
(a) Target Reflectors (b) Etched Macro-section.

process which causes a deeper remelting of the underlying beads. This in turn means that the liquid-solid boundary is more curved than in a bead deposited by the M.M.A. process and consequently the grain growth direction is not so well defined.

The effect of the less ordered grain structure is clearly seen in the ultrasonic behaviour of this weld design. The attenuation measured with a beam normal to the plate surface was high - 0.55dB/mm. Drilled holes in the lower half of the weld could not be detected above the grass. On the other hand there was no evidence of beam skewing. This pattern of response may be understood if this weld is regarded as having a grain structure intermediate between the highly ordered columnar grains of the fillet weld and the equiaxed grain structure found in a forging. In a forging there is no texture and consequently no macroscopic anisotropy; in welds made by the M.M.A. process there is a strong fibre texture and consequently a strong elastic anisotropy. In the former case the disorder leads to high scattering and hence high attenuation. In the latter structure the increased long-range order reduces the scattering but leads to significant modifications of the ultrasonic beam. A M.I.G. weld apparently occupies an intermediate position. Our understanding of the characteristics of M.I.G. - welded material is as yet incomplete and work on this is continuing but on the basis of the above tests the first of the two designs is to be preferred for non-destructive testing.

5. DISCUSSION

The preceding sections have discussed the difficulties in ultrasonically inspecting austenitic welds and highlighted some of the fundamental causes of such problems. These advances in our understanding of ultrasonic propagation in austenitic welds have enabled improved inspection techniques to be developed. It is clear that further significant improvements in austenitic weld inspection will arise from recognition of the dominant role played by metallurgical structure. In some cases inspection may be improved by choice of technique to suit the existing structure. In others it may be possible to optimise the structure and geometry of the weld itself as for the fillet weld discussed in Section 4.

A further benefit of this approach is that the important metallurgical influences are recognised and controlled. A great danger in the purely empirical approach to austenitic weld inspection is that a technique developed on a test weld may be useless when applied to an apparently similar weld due to changes in the structure arising from an uncontrolled welding parameter. This is also the objection to an approach based solely on improved inspection methods. Techniques such as focussed or short pulse probes or signal processing may be valuable to improve signal to scatter ratios, but can only be applied with confidence once the important metallurgical variables have been recognised and controlled.

At the time of writing it seems likely that some kind of inspection technique can be devised for most austenitic welds. However the sensitivity to defects will usually be inferior to that achieved in ferritic welds. Also, there will be additional difficulties of defect size measurement. These limitations have two important implications for plant designers. Firstly, since inspection cannot be relied upon to detect very small defects, we cannot use ultrasonics as a quality control tool to reject small, innocuous defects when they occur in such quantity that they represent bad welding quality. Consequently the customary post-fabrication acceptance standards in which small defects are unacceptable are likely to be impractical in many cases. Acceptance standards based on rejecting only those defects which have a significant effect on the integrity of the plant will have to be used. Such 'fitness-for-purpose' standards will usually be based on fracture mechanics analyses.

A further point is that, with a limited inspection sensitivity, no general guarantee can be given that even the larger defects, which fracture mechanics show to be unacceptable, can be detected. With ferritic welds it is possible to assume in most cases that, provided the weld geometry, position and thickness are suitable, the significant defects can be detected with a high probability of success. This assumption cannot be made for austenitic welds. The plant designer must identify those welds which must be inspected because the consequences of failure are unacceptable. The maximum tolerable defect sizes must then be calculated and compared to the sizes which can be detected ultrasonically. Calculation of the detectable sizes will involve consideration of the welding metallurgy, weld geometry and position and will often require construction of weld test blocks. It will be clear that this is not a trivial matter and is likely to occupy considerable time. This means that the designer of plant containing austenitic welds which must be inspected should initiate the calculations and experiments discussed above at the earliest possible stage if serious delays are to be avoided. Indeed, the results of these calculations may influence the design itself if the intolerable defects cannot be detected.

6. CONCLUSIONS

1. Strong $\langle 100 \rangle$ fibre textures are produced in austenitic welds with the $\langle 100 \rangle$ fibre axis parallel to columnar grain boundaries.

2. The orientation of the fibre axis is a function of the welding procedure and can be varied over a wide range by simple modifications to the weld procedure to improve ultrasonic inspection.

3. Ultrasonic propagation depends strongly on the orientation of the ultrasonic beam relative to the fibre axis. The beam width varies, being smallest when the sound propagates along a direction of maximum velocity. The signal to grass ratio is also a maximum when the beam is along such a direction (approximately 45° to the fibre axis). The beam axis, in general deviates from the direction it would have in equiaxed austenitic steel.

4. Consideration and control of the weld structure must precede attempts to improve inspection sensitivity by the use of special transducers or signal processing methods.

5. Sensitivity to small defects is, in general, unlikely to be as high for austenitic welds as for ferritic even when the weld structure has been optimised. Austenitic welds should, therefore, be examined to acceptance standards based on fracture considerations.

6. Plant designers should take account of the limitations of ultrasonic inspection when applied to austenitic structures at the earliest possible stage.

ACKNOWLEDGEMENTS

The authors wish to acknowledge many helpful discussions with Dr. B. L. Baikie of this laboratory and Mr. D. Yapp of the C.E.G.B. Marchwood Engineering Laboratory and for permission to make use of their work prior to publication. We also wish to thank our colleagues of the N.D.T. Applications Centre for assistance and in particular Mr. A. C. Johnson who carried out much of the experimental work reported here. Thanks are due to the Director General of the North Western Region of the Central Electricity Generating Board for permission to publish this paper.

REFERENCES

(1) Baikie, B. L. and Yapp, D., 1977, "Oriented structures and properties in type 316 stainless steel weld metal. Proc. Sheffield Int. Conf. on Solidification and Casting, Sheff. Univ. 18-21 July 1977, Metals Society (to be published).

(2) Reynolds, W.N., 1968, "Uniaxial textures in cubic materials." Phil. Mag. 18, 1155-9.

(3) Musgrave, M.J.P., 1954, "On the propagation of elastic waves in aelotropic media. II Media of hexagonal symmetry." Proc. Roy. Soc. 226A, 356-366.

(4) Bradfield, G., 1964, "Comparison of the elastic anistropy of two austenitic steels." J. Iron Steel Inst. 198, 616.

(5) Salmutter, K. and Stangler, F., 1960, "Elasticity and plasticity of an austenitic Cr Ni steel." Zeits. F. Metallkunde 51, No. 9, 544-548, C.E.G.B. Translation No. 2066.

(6) Papadakis, E.P., 1966, "Ultrasonic diffraction loss and phase change in anisotropic materials. J. Acoust. Soc. Am. 40, 863-876.

(7) Papadakis, E.P., 1965, "Revised grain - scattering formulas and tables." J. Acoust. Soc. Am. 37, 703-717.

(8) Neumann, E., Wustenberg, H., Nabel, E., Leisner, W., 1975, "Ultrasonic N.D.T. of welds in austenitic steels," Int. Conf. on Quality Control and N.D.T. in Welding, London, Nov. 1974, Welding Institute.

(9) Saglio, R., et al, 1973, "Progress in automatic ultrasonic control of austenitic and stainless steels with the aid of focussed transducers." Proc. 7th Int. Conf. on N.D.T., Warsaw 4-8 June 1973. Eng. Trans. AEC-TR-7584.

DEVELOPMENT AND IMPROVEMENT OF ULTRASONIC TESTING TECHNIQUES FOR AUSTENITIC NUCLEAR COMPONENTS

E. NEUMANN, M. ROEMER, T. JUST, K. MATTHIES,
E. NABEL, E. MUNDRY
BUNDESANSTALT FUER MATERIALPRUEFUNG (BAM)
BERLIN, GERMANY

1. INTRODUCTION

During recent years an ultrasonic testing technique for austenitic welds has been developed in the Federal Republic of Germany. This has come about because the licensing procedure for the sodium cooled fast breeder reactor (SNR 300) requires ultrasonic testing during production. Furthermore, acceptance testing and in-service inspections are conducted using ultrasound during shut down periods.

2. ON GRAIN STRUCTURE AND ULTRASONIC ATTENUATION IN AUSTENITIC WELD METAL

The difficulties in testing austenitic welds with ultrasound arise from the special grain structure in austenitic weld metal.

The limited reliability of the ultrasonic inspection of austenitic welds is caused by the high attenuation of ultrasound due to grain scattering. Fig. 1 shows the frequency dependence of the ultrasonic attenuation α [dB/mm] of austenitic steels and of grey cast iron for pressure waves. Both axes are in a logarithmic scale.

Whereas the ultrasonic attenuation of fine grain austenitic base material is found to be as low as in common ferritic steels, the attenuation in weld material of the X6 CrNi 1811 steel (comparable with AISI 304) and of austenitic cast material CF8 is a magnitude higher.

For comparison the attenuation behaviour of grey cast iron is shown in Fig. 2. For testing frequencies greater than 2 MHz, the US attenuation of this material is the same as in austenitic cast material. But for testing frequencies around 1 MHz the US attenuation is very much reduced and of the same order as that of common rolled steel materials. Thus, one can improve the testability of this type of grey cast iron by choosing a lower testing frequency, although one has to put up with a decreased resolving power due to the greater wavelength of ultrasound.

In the case of austenitic welds having a fibrous structure with long dendrites, we found that the attenuation remains very high at frequencies around and below 1 MHz the frequency dependence of the ultrasonic scattering apparently being of another type. Because of the decreasing resolving power one cannot apply transducers with a much lower testing frequency.

For the inspection of these austenitic steel welds, special techniques had to be developed which are effective in materials with a rather high attenuation coefficient (of the order of $\alpha \sim 0.3$ dB/mm) which results in a low level of flaw echo signals and a high noise level due to grain scattering.

The ultrasonic testing technique was first adapted to weld joints of the unstabilized austenitic steel (X6 CrNi 1811) used in the sodium-cooled fast breeder reactor, SNR 300. The ultrasonic inspection results depend on (1) actual weld structure (2) type of the austenitic steel (3) geometrical conditions and (4) length of the necessary sound paths. But the results which are reported here for the steel X6 CrNi 1811 can in our opinion be generalized to other types of coarse grain materials having the same or a lower ultrasonic attenuation. Tab. 1 gives an overview of components built of austenitic steel types, whose weld joints have been subjected to ultrasonic testing.

3. THE TRANSMITTER-RECEIVER TECHNIQUE FOR TESTING COARSE GRAIN MATERIALS

Different approaches to develop testing techniques for scattering materials are possible as for instance:

- Limitation of the sensitivity range of the ultrasonic probes. This is obtained by the transmitter-receiver technique with separated ultrasonic transducers and by focusing probes.
- Use of short, broadband ultrasonic pulses which have a low coherence length. This reduces interference effects between different grain echoes.
- Use of longitudinal ultrasonic waves. A simple mathematical estimation using the scatter coefficients of ultrasound shows that the signal to noise ratio obtainable with longitudinal waves should be higher than with transverse waves. This has been described in (1) and conforms to testing experiences of the last years.

3.1 OPERATING PRINCIPLE OF THE TRANSMITTER-RECEIVER TECHNIQUE

The main part of our investigations was concerned with further development of the transmitter-receiver technique. The operating principle is shown in Fig. 2. The main lobes of the two transducers (transmitter and receiver) of the probe are inclined toward one another. Their crossing point is the sensitive area of the probe. The advantage of this method over the normal "pulse-echo" technique is that only ultrasound coming from this area is registered with a significant sound pressure. Such an acoustical localization leads to a reduction of structural scatter echoes. If the sensitive range is reduced, the signal to noise ratio increases.

The sensitive range of the probes (defined by the crossing beams) is placed in the far field of the transducers. The required transducer dimensions are calculated using a fixed zone width of 15 mm (lateral and azimuthal directions). The signal strength should be down approximately 6 dB at the zone edges.

The ultrasonic testing technique employing a transmitting transducer and a separated receiving transducer is very sensitive to the shape of curved coupling surfaces. The curvature of the surface of a specimen has the effect of a lens on the ultrasonic lobes. This has to be taken into account when designing the ultrasonic probes. This means that components with complex shaped surfaces must have their own especially tailored ultrasonic probes.

Calculation of the transducer dimensions is through application of the Kirchhoff diffraction theory. This calculation is complicated and several severe approximations are necessary. We have developed a computer program which calculates these data for a given sensitive area of the probe and a given coupling surface geometry of the specimen.

3.2 PROPERTIES OF THE TRANSMITTER-RECEIVER PROBES

3.2.1 DEFINITION OF TESTING ZONES

Because of the limitation in the sensitive range of the probes for the testing of thick walled parts, several angular transmitter-receiver probes have to be employed whose sensitive ranges cover the whole weld under inspection. Fig. 3 shows an example of dividing the weld seam in testing zones. The weld in a 60 mm thick plate has been divided into four testing zones in depth. The ultrasonic testing with transmitter-receiver probes is performed within half the skip distance. The angle of incidence has to be adapted to the zone depth and, if possible, to the preferred orientation of the expected flaws. For the surface testing zone, the angle of incidence is chosen to be 70°. Experiments with test-flaws near the surface (for instance with slots of 1 mm height and width), have shown that a reflector near the surface is detectable with a sufficient signal to noise ratio. The angles of incidence for the second and the third testing zone are smaller to reduce the lengths of the sound paths. For the testing zone at the opposite surface an angle of 45° has been chosen, in this case exploiting the reflection of an angle mirror for flaw detection.

3.2.2 SIZE AND POSITION OF THE SENSITIVE RANGE

Fig. 4 shows an example of the size and the position of the sensitive range in a specimen for a transmitter-receiver probe measured relative to the outer dimensions of the probe housing. The crossing point of the acoustical axises represents the maximum of the sensitivity. The other lines are the levels -3 dB, than -6 dB and the outer line is -12 dB. The test reflector for these measurements was a 3 mm flat bottom hole. These graphs may serve as a basis for determining the scanning pattern and the depth range of the testing zones in the specimen for automatic testing purposes.

3.2.3 DISTANCE-AMPLITUDE-CORRECTION CURVES

Fig. 5 shows an example of a distance-amplitude-correction (DAC) curve of a probe with plane coupling surface obtained at a 3 mm flat bottom hole as a function of reflector depth and projection distance. Especially in cases of complex shaped surfaces of the specimen, the DAC-curves have to be measured with a suitable reference specimen of equal scale containing test reflectors. Fig. 6 shows an example of such a reference specimen used for the ultrasonic testing of an austenitic valve housing.

We measured the same dependence of the echo height when varying the diameter of the FBH serving as a test reflector as that found with the usual ultrasonic probes working with one single transducer i.e., doubling the diameter of the flat bottom hole increases the echo height by 12 dB.

3.3 ON THE ADAPTATION OF THE PROBE'S COUPLING SURFACE TO THE SPECIMEN'S SURFACE AND ON THE SPECIMEN'S SURFACE FINISH

Special probes with adapted coupling surfaces are necessary as for specimens with curved surfaces. The test results gradually deteriorate the more a specimen's radius R and probe radius r deviate from each other. For the case of tubes, Fig. 7 gives the allowable variations of tube radius R for a given probe radius r. We assume gaps of Δs = 0,1 mm and 0,2 mm. According to our existing provisional experience the gap may not be much greater than Δs = 0,2 mm. Taking into account that the probes should be supported at the edges one has to choose probe radius r< specimen radius R when testing from the outer surface and the opposite relation: probe radius r> specimen radius R when testing from the inner surface. To determine the allowable gap size further experimental investigations are necessary.

The surface finish of the specimen has considerable influence on the testing results. The surface should be smooth-cut and there should be no long-wave ripples with a depth of more than 0,2 mm (produced for instance by manual grinding). Fig. 8 shows an example of such a testing surface with a weld joint. The surface is smooth-cut satisfactorily but is rippled too much by grinding manually.

3.4 BROADBAND ULTRASONIC PULSES

Further improvement in suppression of reflections from coarse grain, combined with a desirable high resolution, is obtained by employing short, broadband ultrasonic pulses.

When the difference of sound paths of different grain echoes is of the order of, or greater than, the coherence length, interference effects between different grain echoes are no longer appreciable. These interferences effects between grain echoes result in the enhanced "grass" on the screen of the ultrasonic flaw detector and beyond that may produce ghost echoes looking like "Christmas trees" on the screen. The ghost echoes cannot be discriminated so easily from echoes of real flaws. By using broadband pulses you avoid these difficulties significantly.

To further increase the effective bandwidth of the ultrasonic testing system we used methods of signal processing, namely inverse filtering (or deconvolution), of the ultrasonic signals. By inverse filtering the transfer characteristics of the ultrasonic testing system are broadened. The effect is a further reduction of the coherence length of the grain scatter echoes resulting in a lowered grass level on the screen (2).

3.5 TYPES OF TRANSMITTER-RECEIVER PROBES FOR TESTING OF LONGITUDINAL AND TRANSVERSAL FLAWS

The transmitter-receiver probes for the surface near two testing zones are suitable for the inspection of longitudinal as well as of transverse defects in welds. For the lower testing zones, we propose a transducer where the transmitting and the receiving part of the probe are far more separated so that they are arranged on both sides of the weld seam (Fig. 9). By this measure long sound paths in the scattering weld metal are avoided.

The data of the probes are given in tab. 2. Part II of tab. 2 yields data of probes for testing of transverse defects, they are denoted by a capital B. The transmitter-receiver probes for curved testing surfaces are smaller (35X35X30 /mm/). The data of these probes are given in tab. 3.

Fig. 10 shows a comparison between the transmitter-receiver-technique with other ultrasonic testing techniques.

The probes compared are a single transducer angle beam probe operating with 1 MHz-shearwaves, a focused beam probe containing a lens and a transmitter-receiver probe the latter two probes operating with 2 MHz-longitudinal waves. The three probes are comparable regarding the wavelength and the beam angle $\alpha \approx 60^{\circ}$. The focused beam angle probe and the transmitter-receiver probe were especially adapted to the testing of austenitic welds. Both have a defined range of high sensitivity, operating with broadband ultrasonic longitudinal wave pulses.

Fig. 10 shows as an example echo indications from a 3 mm cylindrical hole in an austenitic weld detected by the different probes. The echo indication obtained by a conventional single transducer probe operating with shear waves is apparently inferior to the other indications, having a higher noise level due to grain scattering and greater pulse length and therefore a poorer resolution.

5. APPLICATION OF THE TRANSMITTER-RECEIVER PROBES IN THE WORKSHOP

In tab. 1 several examples of application of transmitter-receiver probes are listed. In all cases test specimens with equal geometry and from same material served to demonstrate the testability by ultrasound with especially tailored transmitter-receiver probes. Kind, size and position of the reference reflector for sensitivity calibration were given by testing instructions.

The probes were tested first on a weld in a test specimen of the steel X6 CrNi 1811 with intentionally produced natural flaws (Fig. 11). As this test specimen until now could not be examined by a destructive inspection, we are only able to compare the results of the ultrasonic inspection with the results of the radiographic inspection. The agreement of these test results is good and it is shown that especially little flaws as single pores are found by ultrasound with a signal to noise ratio of at least 10 dB.

At the moment, we are designing a test system of ultrasonic probes for the ultrasonic inspection of the vessel of the sodium cooled fast breeder reactor, SNR 300 including the nozzle welds. Fig. 12 and Fig. 13 show the sort of weld joints we have to do with. For the nozzle welds one has to use probes with adapted curved coupling surfaces.

Fig. 14 shows the weld joint of the inlet nozzle of an austenitic pump barrel. For the ultrasonic testing of this nozzle weld we had to use probes with cylindrical coupling surface. Fig. 15 shows the weld joint in the spherical part of the pump barrel. In spite of the large diameter of 1561 mm it was necessary to use an ultrasonic probe with an adapted spherical coupling surface. This is in agreement with the suggestion that the probes should be better supported at the edges than at the middle part otherwise the wrangling of the probes is leading to fluctuating and false indications. The sensitivity calibration was done with a 3 mm FBH. Agreement with the results of the radiographic inspection was obtained and beyond that several hitherto unknown reflectors were found, which were not obtained by radiographic inspection. The amplitudes of these indications were as much as 6 dB lower than the amplitudes of the reference reflector 3 mm FBH.

Fig. 16 shows the flange of a steam generator tube plate with varying radius from 99 to 150 mm. For ultrasonic testing of subcladding cracks in transversal and longitudinal orientation we had to use four transmitter-receiver probes with curved coupling surfaces. Sensitivity calibration was done with a 2 mm flat bottom hole, 10 mm below the cladding surface of a test specimen. The ultrasonic results were correlated with the results of a destructive examination (grinding down of the cladding). Defects with 8-14 mm length and 4-5 mm depth could be correlated with ultrasonic indications with amplitudes between -10 dB and +8 dB above echo height of the reference reflector indication and with signal to noise ratios between 10 and 18 dB.

Fig. 17 shows a circumferential weld joint between ferritic and austenitic material the ferritic part of the compound tube being clad. The ultrasonic testing is performed from both the inner and the outer surface. Severe difficulties with ultrasonic testing arise from the root of the weld because there are joining ground material, austenitic weld material and cladding material. With transmitter-receiver probes adapted to the surface it was possible to detect a 2 mm FBH with a at least 10 dB higher amplitude than the amplitude of the root indication. Fig. 18 shows the probe adapted to the inner coupling surface of a test specimen. The cladding on the ferritic part is discriminable as well as 2 and 3 mm flat bottom holes as reference reflectors.

The following figs. 19 and 20 are further examples of weld joints in austenitic components. Special transmitter-receiver probes had to be constructed for ultrasonic testing of these weld joints too. In the case of a test weld of the austenitic steel X10 NiCrAl Ti 3220 there has been a correlation of ultrasonic indications (equivalent reflector size 2-3 mm, S/N = 12 dB)with hot cracks approximately 2x2 mm found by metallographic destructive inspection. But these results have to be scrutinized further more.

Tab. 1: Examples of Applications of Transmitter-Receiver-Angle-Probes (SEL-Probes) on Austenitic Components

Component	Materials	Weld, Weld method	Geometry of the components	Type of inspection	Calibration of sensitivity, S/N-ratio
SNR 300-Vessel and Test-welds	1.4948 X6 CrNi 1811	Double-V welds, V-welds, Manual electrode welded, Submerged arc welded	plain $40 \le d \le 60$ mm	Weld, Longitudinal flaws, transverse flaws	3 mm dia. FBH in weld metal, S/N = 20 dB
Pump housing	1.4948 X6 CrNi 1811	V-welds	cylindrical: $600 \le \varnothing \le 1400$ mm spherical: $\varnothing$ 1600 mm, $20 \le d \le 35$ mm	Weld, Longitudinal flaws	3 mm dia. FBH in base metal, S/N $\ge$ 24 dB
Steamgenerator, Flange of tube plate	Ferritic base material, CrNi-cladding (Inconel)	Hand welded cladding	cylindrical: $186 \le \varnothing \le 288$mm $6 \le d \le 9$ mm	Subcladding cracks	2mm dia. FBH below cladding, transv. flaws: S/N = 16 dB, longit. flaws: S/N=12-14dB
Control rod stud (Compound tube)	Ferrite (St52.4) cladded, Austenit 1.4550	V-weld, Manual electrode welded	cylindrical: $\varnothing$ = 112 mm d = 24 mm	Weld, Longitudinal-flaws	2 mm dia. FBH insonified through weld seam, S/N =16-18 dB
Heat exchanger	1.5662 X8 Ni 6	V-weld, Submerged arc and inert gas welded	cylindrical (plain) $\varnothing$ = 2500 mm d = 28 mm	Weld, Longitudinal-flaws	Testing according to ASME-code: 4 mm dia. cylindrical hole, S/N $\ge$ 30 dB
Valve housing	X20 CrMoV 121 and GS-12Cr Mo 910 or 10CrMo910	V-welds, Manual electrode-welded	cylindrical $\varnothing$ = 250 mm d = 38 mm	Weld, Longitudinal-flaws	3 mm dia. cylindrical hole through weld metal, S/N $\ge$ 14 dB
Preheater for chemical industry	X10 Ni Cr Al Ti 3220	U- and V-welds, Submerged arc and TIG-welded	cylindrical and spherical $380 \le \varnothing \le 1200$ mm $28 \le d \le 40$ mm	Weld, Longitudinal flaws, Transverse flaws	3 mm dia. flat bottom hole through weld metal, S/N $\ge$ 12 dB

Tab. 2, part I

SEL-probes with plane coupling surface (AE)

(A: austenitic; E: plane)

Frequency: f = 2 MHz

Distance between sound exit point and probe's housing edge: x = 20 ± 5 mm

Type of probe	SEL 70 AE1.7	SEL 65 AE2.20	SEL 60 AE3.30	SEL 45 AE3.30	SEL 45 AE4.40
dimension BxLxH (mm)			40x40x30		
testing zone	1	2	3	3	4
angle of incidence α^o	70	65	60	45	45
depth of maximal sensitivity b_o (mm)	7±2	20±5	30±8	30±8	40±10
projection distance at maximal sensitivity a_o	20±5	40±5	50±10	30±8	40±10
applicable in the depth range	0-12	10-30	20-50	20-50	30-70

Tab. 2, part II

SEL-probes with plane coupling surface (AE)

(A: austenitic; E: plane)

Frequency: f = 2 MHz

Distance between sound exit point and probe's housing edge: x = 20 ± 5 mm

Type of probe	SEL 60 AE3.30B	SEL 45 AE4.40B	SEL 45 AE 2.20*)
dimension BxLxH (mm)	60x40x30		40x40x30
testing zone	3	4	2
angle of incidence α^o	60	45	45
depth of maximal sensitivity b_o (mm)	30±8	40±10	20±5
projection distance at maximal sensitivity a_o	50±10	40±10	20±5
applicable in the depth range	20-50	30-70	10-30

*) For testing of the second zone with another angle of incidence

Tab. 3

SEL-probes with curved coupling surface

Type: ARI, ARA; AKI, AKA
(A: austenite; R: tube; K: sphere (double curved),
I: testing from the inner side
A: testing from the outer side)

dimension BxLxH:	35x35x30 mm
frequency:	f = 2 MHz
distance x:	x = 17,5 ± 5 mm

1. No curved coupling surface in the direction of sound beam
Type: ARI, ARA

testing zone	1	2	3	3(4)
angle of incidence α °	70	65	60	45
depth of maximal sensitivity b_o	7±2	20±5	30±8	30±8
projection distance a_o at maximal sensitivity	20±5	40±5	50±10	30±8
applicable in the depth range	0-12	10-30	20-50	30-70

2. With curved coupling surface in the direction of sound beam
Type: ARI, ARA; AKI, AKA

testing zone	1	2	3	3(4)
angle of incidence α° *)	60-70	60-65	45-60	≤ 45 **)
depth of maximal sensitivity b_o	7±2	20±5	30±8	40±10
projection distance a_o at maximal sensitivity **)	20-30	30-55	30-50	30-45
applicable in the depth range	0-12	10-30	20-50	30-70

*) dependent on radius of curvature

**) The angle of sound impact at the opposite surface of specimen should be between 55° and 45°. The angle of incidence α is choosen adequately.

REFERENCES

(1)-Bundesanstalt für Materialprüfung, partial report "Ultraschallprüftechnik" in the final report RS 143 - "Null- und Wiederholungsprüfungen von austenitischen Reaktorbehältern und -komponenten mit Ultraschallverfahren"; INTAT 77.19/23.77019.3 (Interatom, Bergisch Gladbach 1, Germany), March 1977

-H. Wüstenberg, T. Just, W. Möhrle, J. Kutzner, Zur Bedeutung fokussierender Prüfköpfe für die Ultraschallprüfung von Schweißnähten mit austenitischem Gefüge; Materialprüfung 19 (1977), p. 246

(2) E. Nabel, T. Just, "Diskussionsbeitrag zur Verbesserung des Signal-Rausch-Verhältnisses bei der Ultraschallprüfung grobkörniger Werkstoffe" in "Neuere Verfahren zur Analyse von Ultraschall-Befunden", Deutsche Gesellschaft für Zerstörungsfreie Prüfung e.V., DGZfP 1977-500, p. 119, Berlin, 24.2.1977

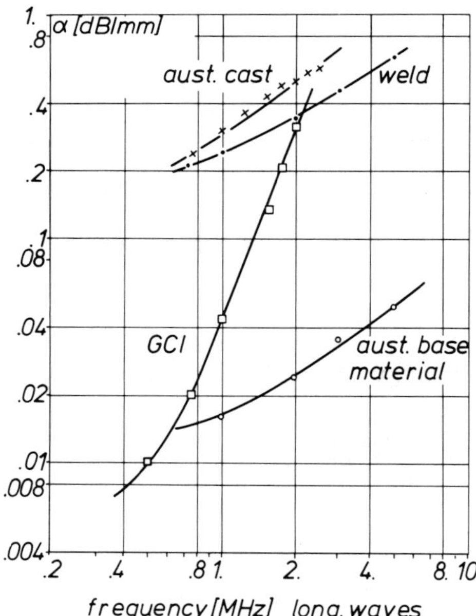

Fig. 1

Ultrasonic Attenuation of austenitic steels and grey cast iron as a function of frequency for longitudinal Waves

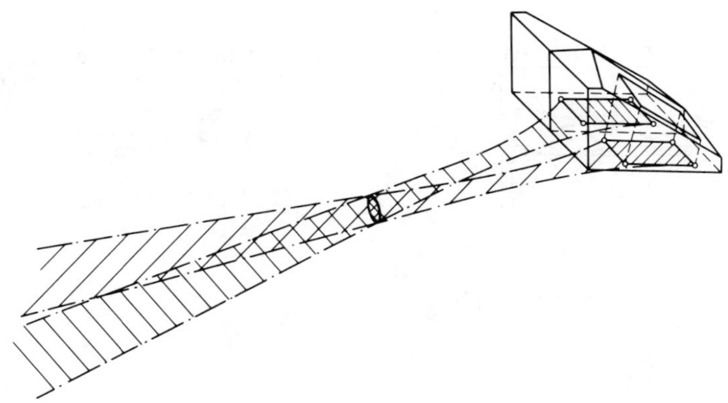

Fig. 2: Lobes and sensitivity range of a transmitter-receiver probe

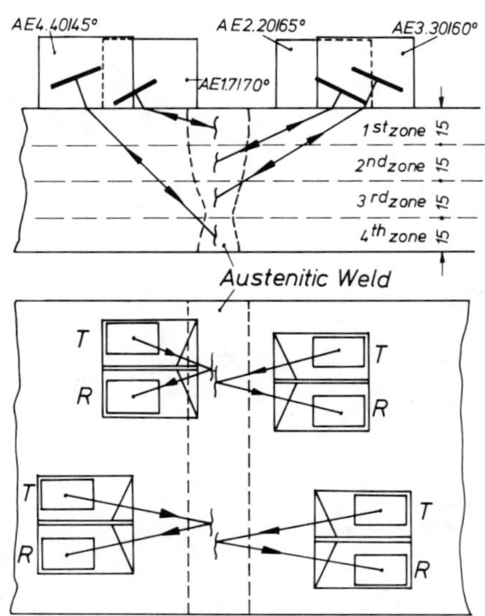

Fig. 3: Division of a weld seam into testing zones

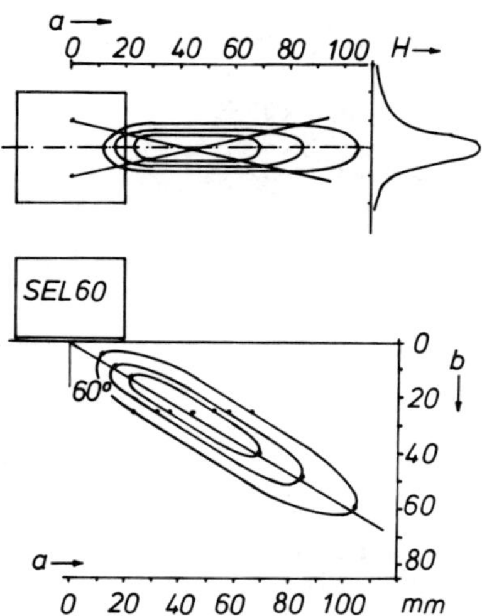

Fig. 4

Size and position of the sensitivity range of the probe SEL 60 AE 3.30

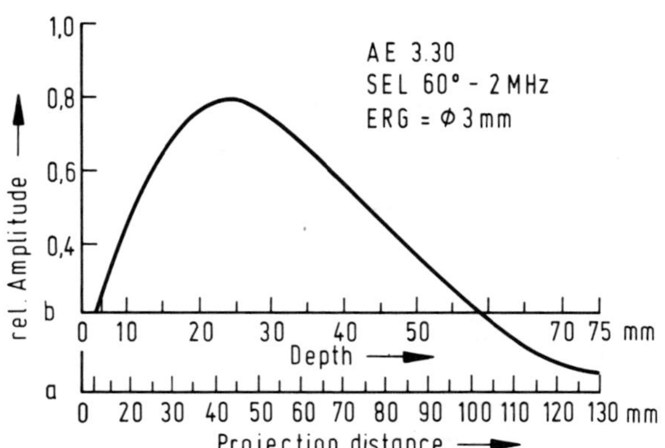

Fig. 5: DAC-curve of a transmitter-receiver probe with plane coupling surface SEL 60 AE 3.30

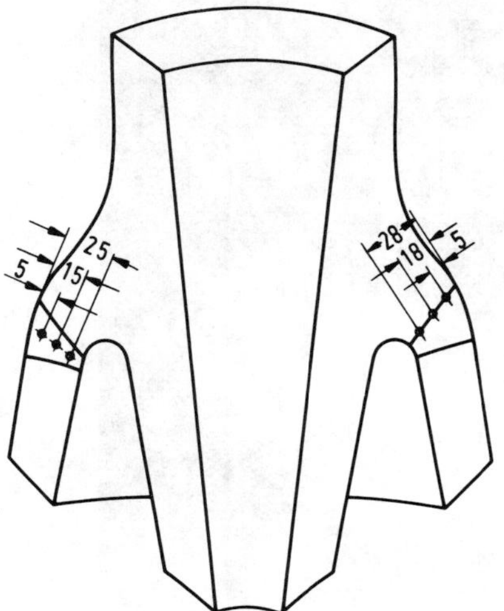

Fig. 6
Reference specimen with test reflectors

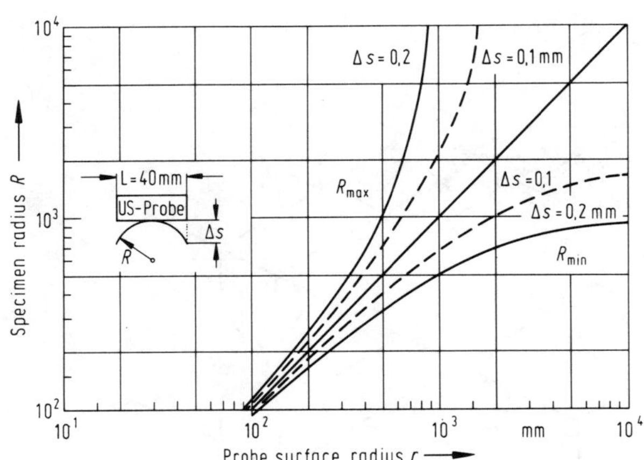

Fig. 7: Relation between tube radius and probe radius calculated from geometrical considerations

Fig. 8: Example of a testing surface with a weld joint

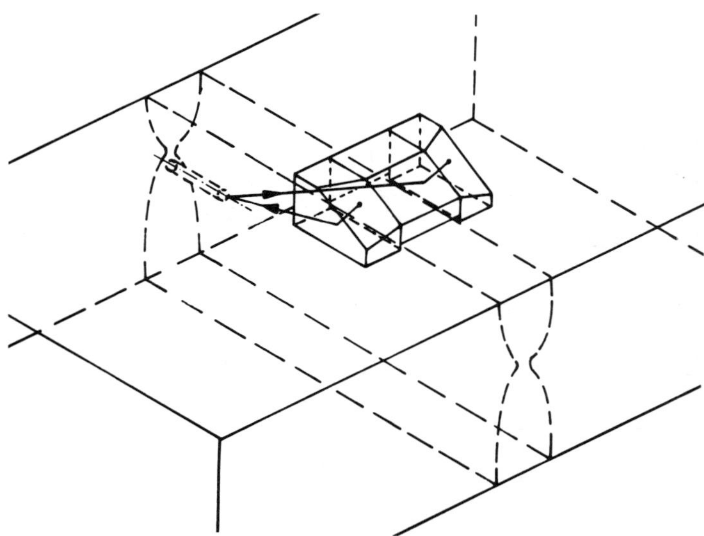

Fig. 9: Draft of a transmitter-receiver probe for transversal flaw detection

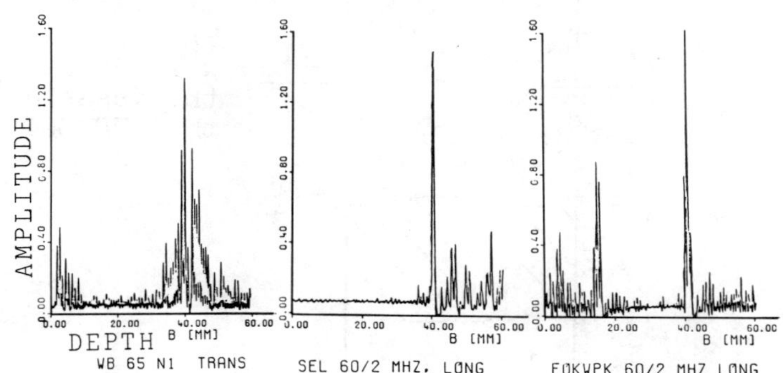

Fig. 10: Echo indications of a 3 mm Ø cylindrical hole in an austenitic steel weld with different angle probes

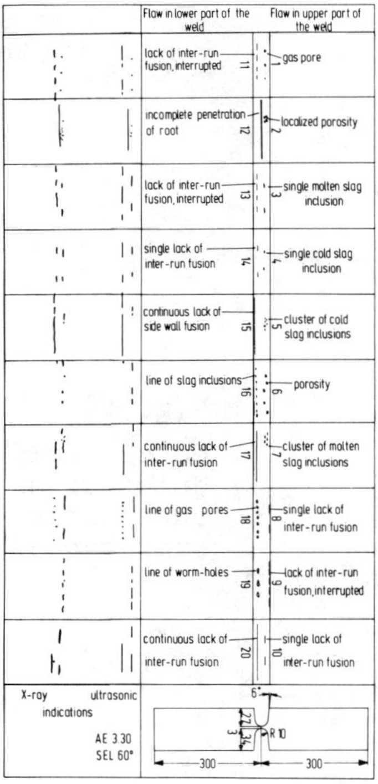

	Flaw in lower part of the weld	Flaw in upper part of the weld
	11 lack of inter-run fusion, interrupted	1 gas pore
	12 incomplete penetration of root	2 localized porosity
	13 lack of inter-run fusion, interrupted	3 single molten slag inclusion
	14 single lack of inter-run fusion	4 single cold slag inclusion
	15 continuous lack of side wall fusion	5 cluster of cold slag inclusions
	16 line of slag inclusions	6 porosity
	17 continuous lack of inter-run fusion	7 cluster of molten slag inclusions
	18 line of gas pores	8 single lack of inter-run fusion
	19 line of worm-holes	9 lack of inter-run fusion, interrupted
	20 continuous lack of inter-run fusion	10 single lack of inter-run fusion
X-ray ultrasonic indications AE 3.30 SEL 60°	6° 22 3 34 R 10 300	300

Fig. 11

Test specimen of the steel X6 CrNi 1811 with intentionally produced natural flaws

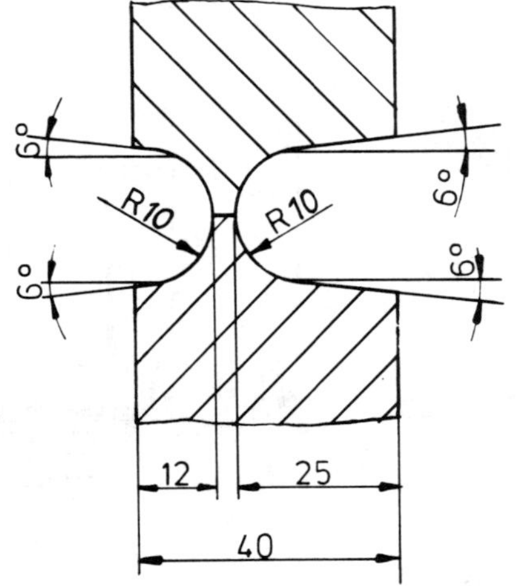

Fig. 12

Example of a weld joint of the vessel of the SNR 300

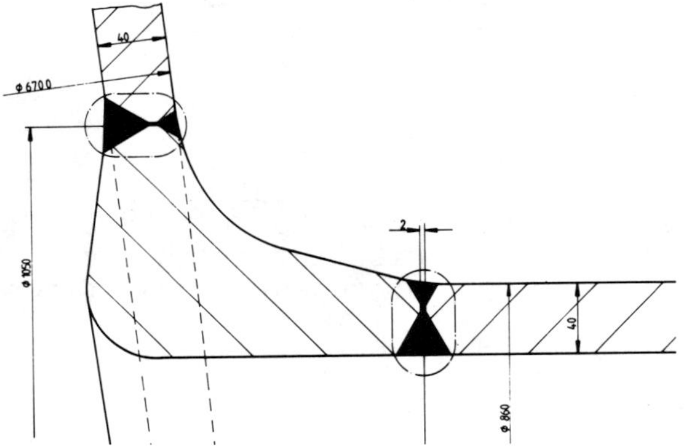

Fig. 13

Example of a nozzle weld joint of the vessel of the SNR 300

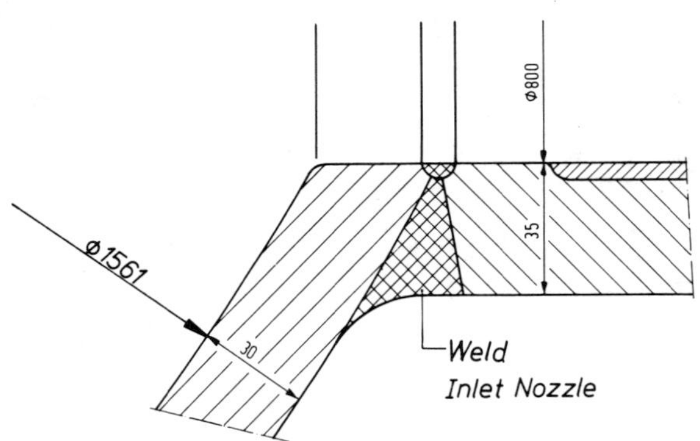

Fig. 14

Nozzle weld joint of an austenitic pump barrel

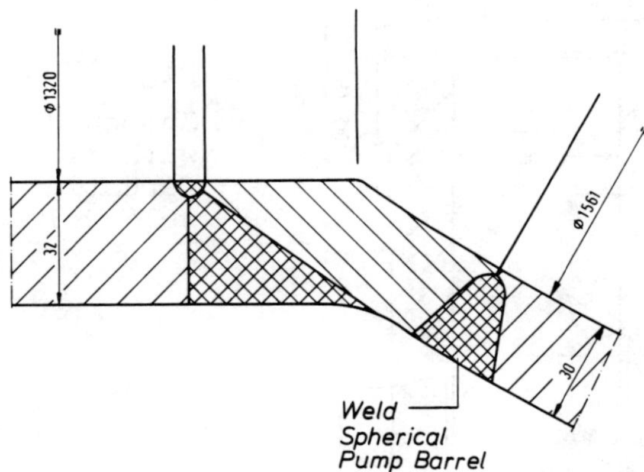

Fig. 15: Weld joint in the spherical part of an austenitic pump barrel

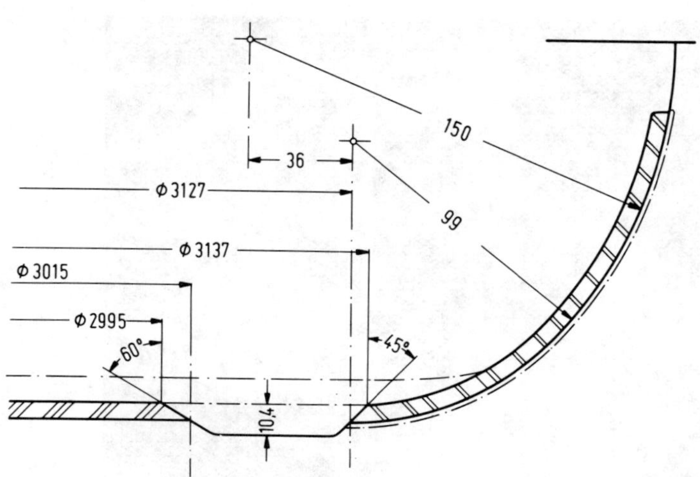

Fig. 16: Flange of a steamgenerator tube plate

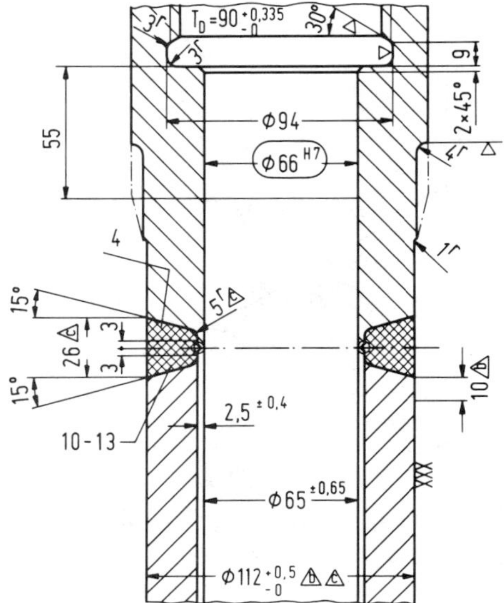

Fig. 17

Compound tube with circumferential weld joint

Fig. 18: Transmitter-receiver probe adapted to the inner coupling surface of a test specimen from the compound tube

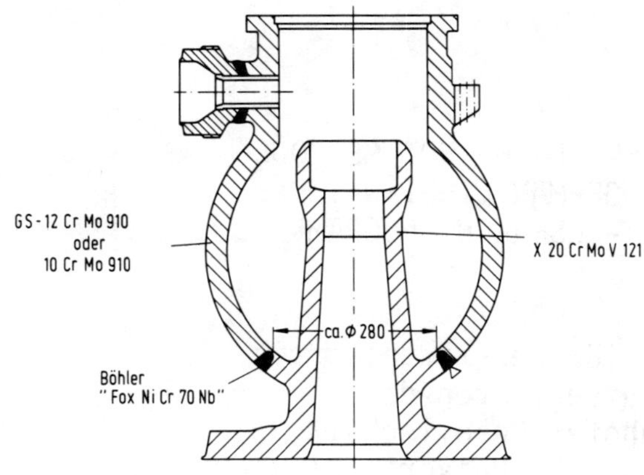

Fig. 19: Valve housing

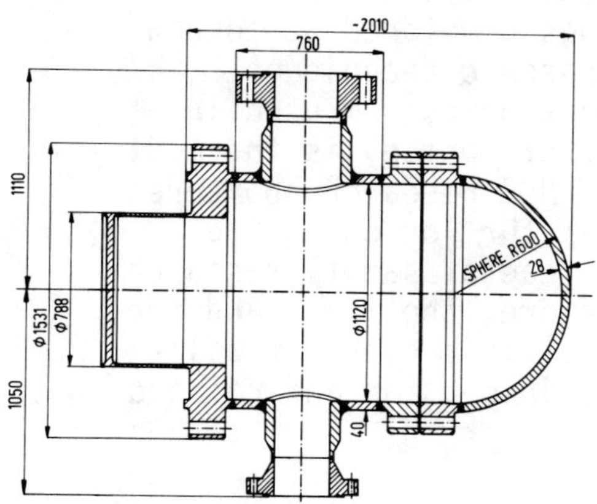

Fig. 20: Preheater for chemical industry

AN OVERVIEW OF THE NDT METHODS IN THE GERMAN LIGHT-WATER SAFETY-RESEARCH PROGRAM

P. Höller

Institut für Zerstörungsfreie
Prüfverfahren der
Fraunhofer-Gesellschaft
Saarbrücken, Universität

The most important contribution of NDT to the safety technology of nuclear power plants is the in-service inspection of the primary circuit especially the reactor pressure vessel. All approaches developed up to now extensively use ultrasonic techniques.
The in-service inspection system used in the Federal Republic of Germany is the most important product of NDE research sponsered by the Bundesminister für Forschung und Technologie for the reactor safety research programme /1/. Therefore, the first and the most extensive part of this overview will deal with in-service inspection and related research.

In the present in-service inspection systems for ultrasonic testing classical techniques are in use; pulse-echo, tandem and through-transmission. The design of the probe array shown in Fig. 1 is derived from experience gained in weld testing plus physical considerations of scattering, diffraction and reflection caused by weld defects. Scattering at small defects and diffraction at the corners of larger defects are covered by the pulse-echo technique. Several tandem

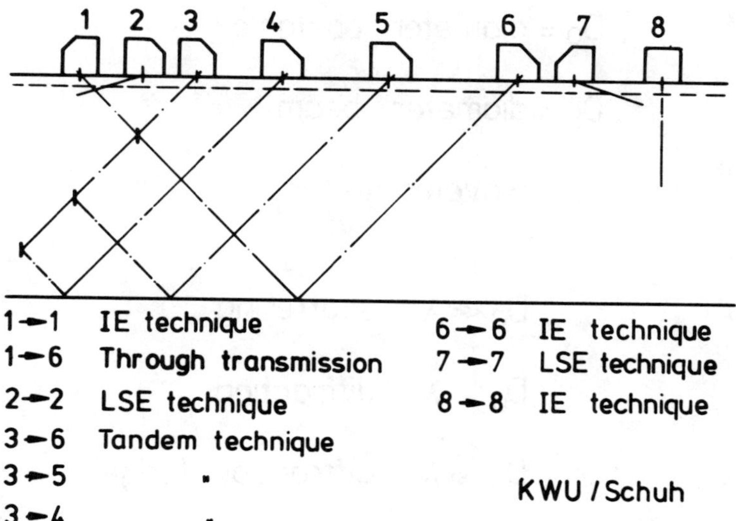

Fig. 1
Probe array for testing pressure vessels

probe-couples are devoted to geometrical reflections of larger defects parallel to the weld-base metal-interface. Up to 12 probes are necessary to cover the wall of a PWR pressure vessel.

It is surprising experience to find that the same problem has lead experts of different countries to quite different solutions regarding the ultrasonic testing of weld in reactor pressure vessel. In view of this fact it will be expanded on the thougths which lead to the probe array in Fig. 1.

The behaviour of an obstacle in the oscillating stress field of an ultrasonic wave depends mainly on three parameters of the ultrasonic field and the obstacle:

- the diameter of the obstacle,
- the wavelength (not so much the frequency) of the ultrasonic wave,
- the diameter of the ultrasonic beam.

D_o = diameter obstacle

D_b = diameter beam

λ = wavelength

$D_o \ll \lambda$ scattering

$D_o \approx \lambda$ diffraction

$D_o \gg \lambda$, $D_b > D_o$ } diffraction (edge) + reflection

$D_o \gg \lambda$, $D_b < D_o$ } reflection

Fig. 2
Categories of the ratio $\frac{D}{\lambda}$

According to these three parameters the reaction of the obstacle to the stress field of the ultrasonic wave as in Fig. 2, can be divided into four categories:

- category No 1:
  The diameter of the obstacle is small compared to the wavelength and the beam diameter.
  In this case the obstacle oscillates in the stress field of the wave and emits a scattered radiation in all directions.

- category No 2:
  The diameter of the obstacle is of similar size to the wavelength but smaller than

the beam diameter. The reaction of the obstacle to the incident wave is much more complicated than in case No 1. The ultrasonic wave is diffracted at the obstacle. Resonance-like phenomena are to be observed. Small changes of the ratio "diameter of obstacle to wavelength" may significantly change the effect. The directivity of the obstacle is not great. The diffracted and scattered radiation is emitted over a wide angle.

- category No 3:
  The diameter of the obstacle is greater than the wavelength, but smaller than the beam diameter. In this case a diffracted wave from the corner of the obstacle is observed as well as geometrical reflection from flat area of the reflectors. The directivity of the reflected radiation increases with the size of the reflector.

- category No 4:
  The diameter of the reflector is greater than the beam diameter. In this case, at least towards the center of the beam, the reflector behaves like an infinitely large one. Fully geometrical reflection occurs.

Under practical circumstances the defect in a RPV-wall is considered serious when its size is of the order of cm^2. Most of the ultrasonic testing is done in the far-field. Therefore, such a defect belongs to category 3. Pulse-echo and/or tandem are adaptable; "and" makes the system redundant; it becomes even more redundant with the additional use of the through-transmission technique.

The more the situation approaches category 4, the more the test system should rely on geometrica reflection.

A defect of category 3 for the far-field of a piston type probe may belong to category 4 if focussed probes are used, as is the practice in France /2/.

All this can easily be demonstrated in experiments and has been investigated thoroughly Practical experience only can be achieved when extensive crack growth occurs during operation. Fortunately, this has not yet happened. Therefore, experience is restricted to only a few cases in the lower area of category 3.

A weak point of ultrasonic testing is the dead zone or a zone of minor sensitivity near the incident surface. This zone is smaller for direct coupling than for free incidence in water (immersion technique). Since the cracking potential at surfaces is much larger than in the interior of the wall, one has to contemplate complementary methods to cover the surface, or to up-grade ultrasonic systems to solve the problems in this zone. Research and development to both alternatives has been done and continues.

A special transmitter-receiver probe has been developed by Wüstenberg /3/ and is in use to cover the interface "cladding - base material" and also a part of the cladding itself. The probe is shown in Fig. 3 and 4. It emits longitidunal waves at nearly grazing incidence. Focussing in the azimuthal direction is achieved by an inclination of transmitter and receiver in the plane parallel to the surface.

Research is going on towards more powerful solutions. Longitudinal creep waves emitted from an angle probe adjusted just above the critical angle for the longitidunal waves propagate along the surface and therefore cover the surface area /4/. The same holds for shear horizontal (SH) waves as shown in

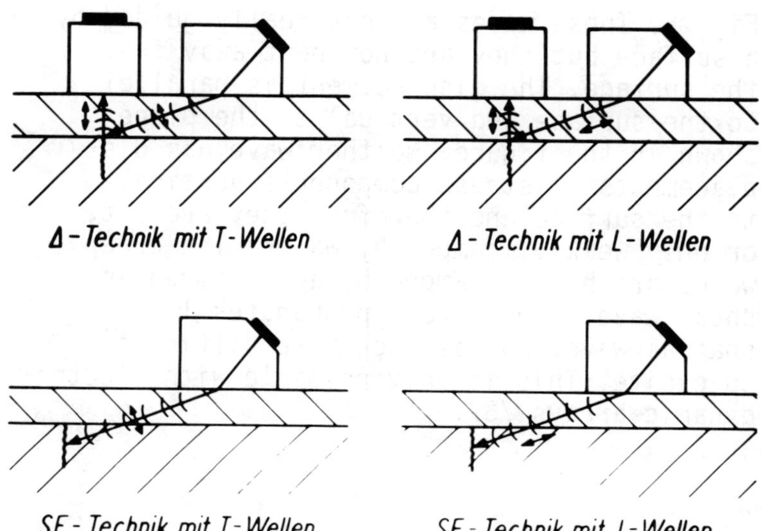

Fig. 3
Transmitter-receiver probe for testing the interface "cladding - base material"

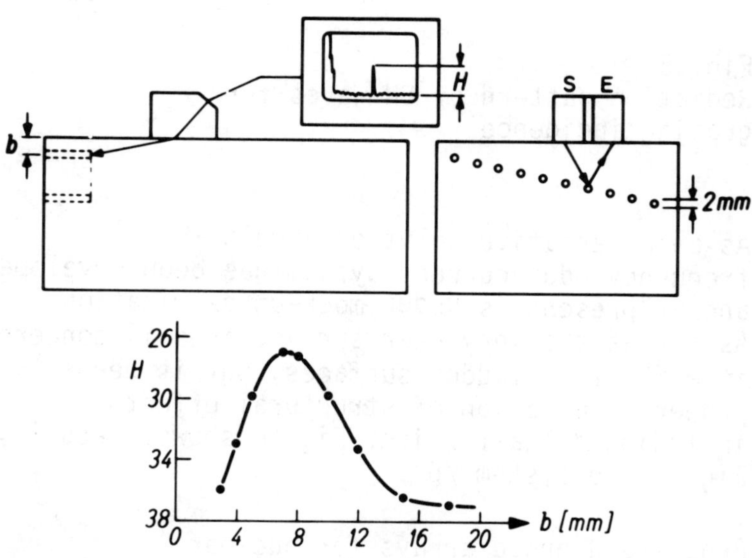

Fig. 4
Adjusting of the transmitter-receiver probe

Fig. 5. These waves are not really guided by a surface but they are not bent away from the surface. The displacement is parallel to the surface and vertical to the plane shown in the figure. Neither wave has displacements or stress components vertical to the surface and therefore they are not, or only weakly, damped by water. The creep waves are highly damped by an accompanying shear wave. The SH-wave propagates like a spatial wave, but is much more difficult to excite. This is only possible with electrodynamic arrays /5/.

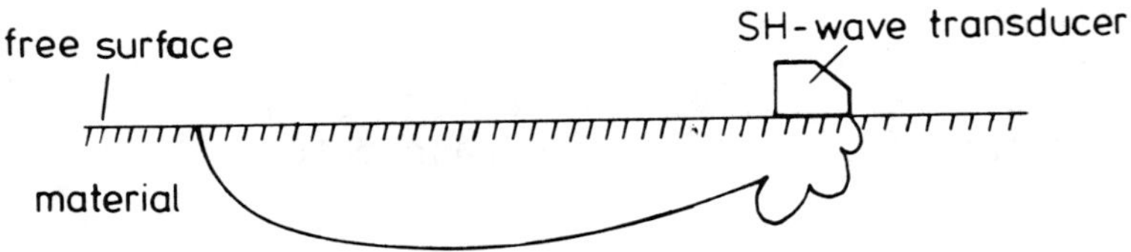

Fig. 5
Rediation pattern of SH-waves for grazing incidence

As an alternative solution a multi-frequency eddy current system has been developed and at present is under mock-up examination. As far as the very near surface area is concerned it works for cladded surfaces, but it needs proper elimination of structural effects including delta-ferrite. Fig. 6 shows a result, Fig. 7 the system /6/.

Probes and probe arrays for nuclear application need very careful design. For this several computer codes based

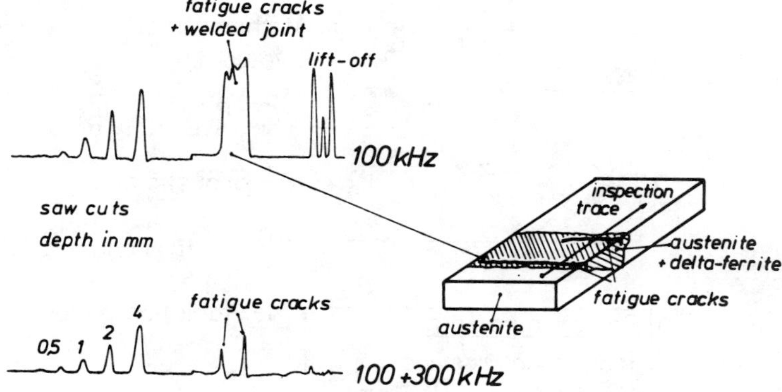

Fig.6
Result of multi-frequency eddy-current test of a weld

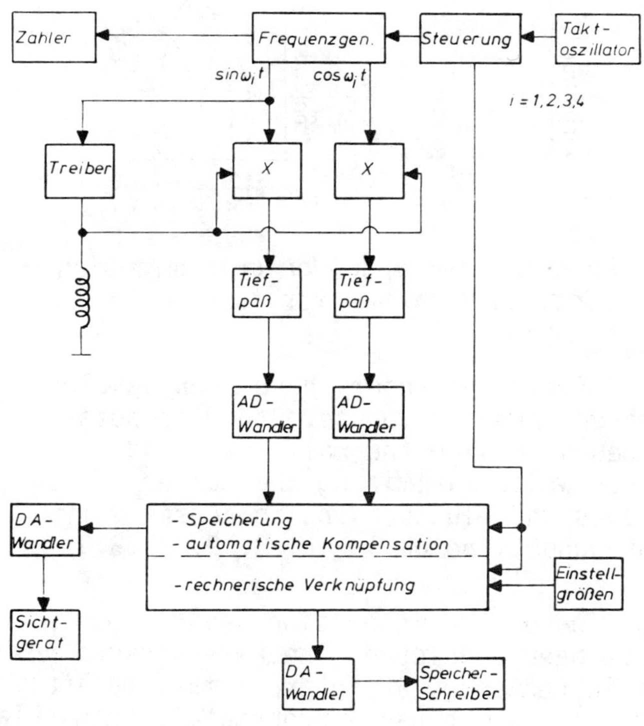

Fig. 7
Schematic diagram of the multi-frequency eddy-current system

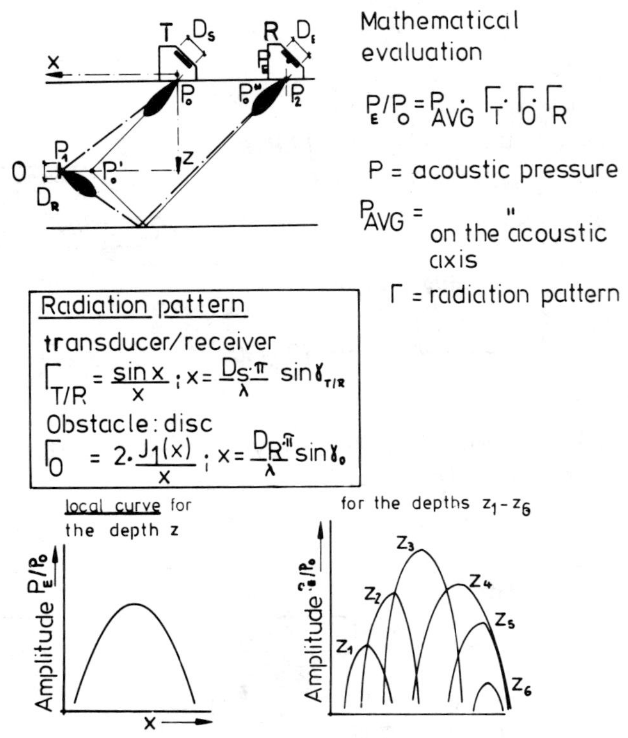

Fig. 8
Mathematical model on determination sensitivity zones for tandem

on diffraction theory have been developed, proved, and put to use. Other computer codes subdivide the wall into tandem zones while minimizing the number of probes and fluctuations in sensitivity throughout the wall Fig. 8, 9. /7,8,9/

For in-service inspection several manipulators have been developed. The PWR-pressure vessel is inspected by an internal mast manipulator, Fig. 10, which has appropriate submanipulators and system mountings to cover the tricky areas like nozzle corners, bottom-cap, etc. There are also special manipulators for top-cap and tubing.

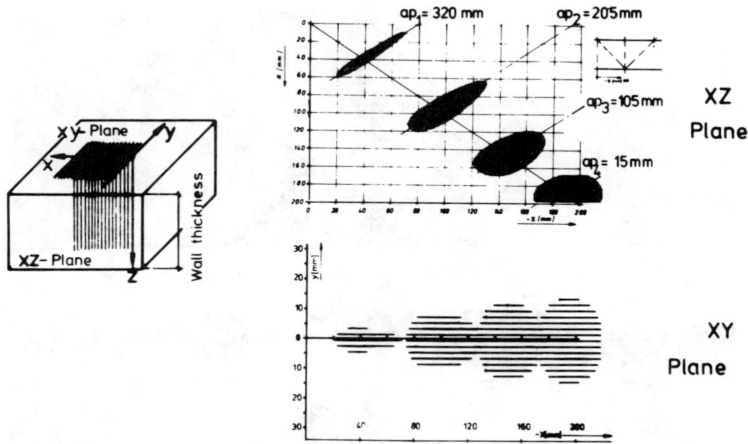

Fig. 9 Tandem zones

Fig.10
Central mast manipulator

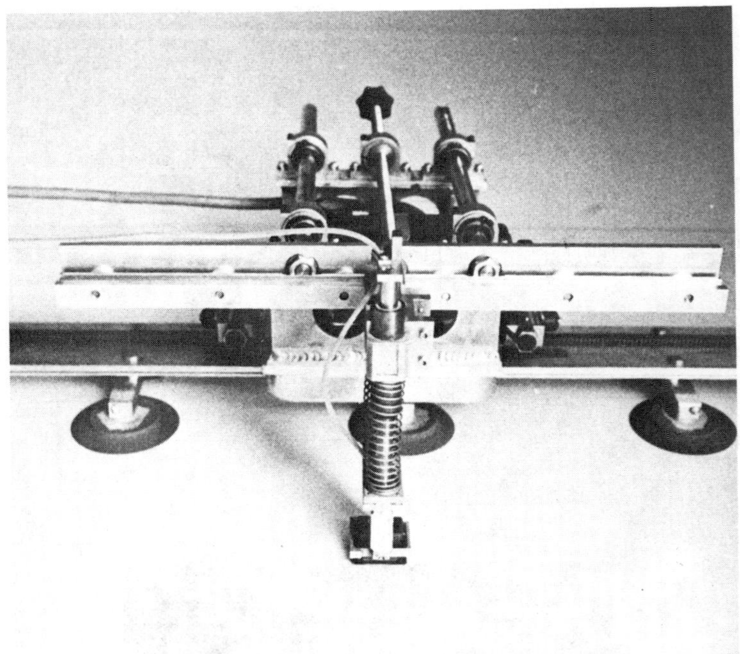

Fig. 11
Rails system

The BWR pressure vessel is inspected from the outside. The testing system is guided by rails Fig. 11.

In order to be able to pick up and process the great amount of data being produced by automatic testing /11/, multi-channel ultrasonic instrumentation has been developed. This controls the different testing modes, quantitizes the measured amplitudes, and transfers the position of the testing system and any measured data to a computer system. The computer collects basic data and processes the measured data for each testing mode to final values for volume elements of the order of some cm^3. One of the important data processing steps is an averaging algorithm to improve signal-to-noise ratio /12/.

The sentitivity of the over-all system for pulse-echo corresponds to a flat bottom hole of 3 mm diameter vertical to the beam; in the tandem case to a 10 mm circular disc vertical to the surface of the wall ± 10o inclination /13/.

It is a basic point that reflectors defining the detection limit or pick-up limit are planar (like cracks); they are category 2 for pulse-echo and 3 for tandem. We are not expecting much correlation between real crack- or defect size and the corresponding FBH-diameter.

Present research on automatic or semi-automatic testing is concerned with

- detection
- reconstruction
- redundance
- mock up experiments

To improve detection, especially in coarse grained materials, several averaging approaches are under investigation. The main noise to be eliminated is caused by grain scattering. Scattered radiation is not stochastic but coherent. Therefore, the scattered radiation has to be made stochastic either by wobbling of frequency or by moving the probe or the beam. Fig. 12 shows that the signal from a reflector remains unchanged but the grain scattering becomes incoherent when the probe oscillates during the measurement. An improvement of signal-to-noise ratio of the order of 20 dB can be obtained; Fig. 13 /14/.

With regard to reconstruction the principle thoughts and ideas followed by the present research are:

- any reconstruction should approach a true image of a defect as far as possible

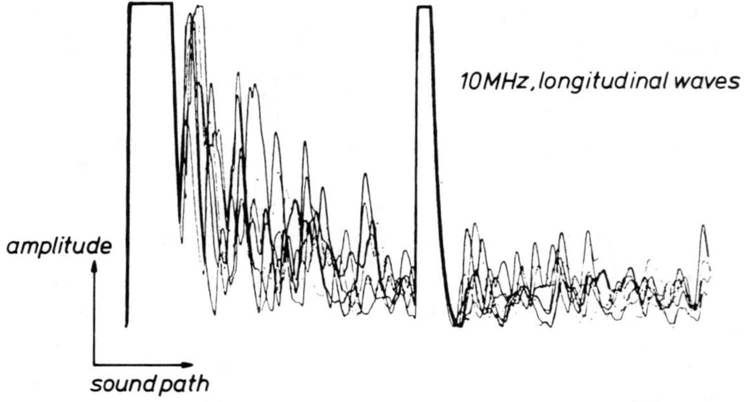

Fig. 12
Sacttering signals and backwall echo for eight A-scans

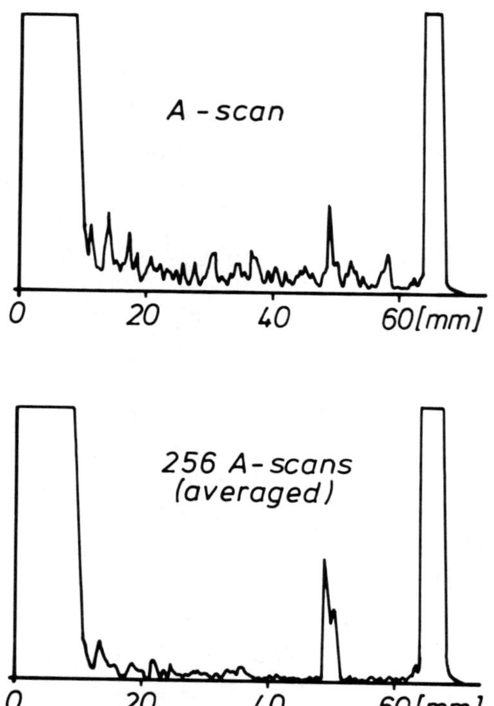

Fig. 13
Spatial averaging

- for basic reasons of physics it is impossible to perform such a reconstruction by processing amplitude data only. In addition to amplitude we need time delay or phase
- the users would not like to have large apertures, that means,sampling procedures have to be used
- any reconstruction improves redundance. This is a byproduct, because by definition imaging means that the radiation from one point of the object is concentrated to an appropriate point in the image or the reconstruction.

In an overview it is not possible to explain all the details but only touch upon some points.

Measuring time-delay and amplitude during scanning, the results are locus curves for time-delay and amplitude; Fig. 14. For one angle but two incidence directions four locus curves for time delay and amplitude are obtained. There is twofold access to the reflector; direct or by back-reflection. During automatic scanning, signals are obtained from the main lobe and from the side lobes. Considering two perpendicular plane of incidence, any defect is seen from 8 directions. There can be even more, if we apply more than one angle or more than two incident planes.

The objective is to collect all these data during dynamic testing and to try to reconstruct the reflector from them. There are two main steps to that aim:

- Compare the locus curves with those obtained by diffraction theory from typical regularly shaped reflectors; for instance globe, disc, cylinder, strip /15/. Decide by pattern recognition methods what reflector

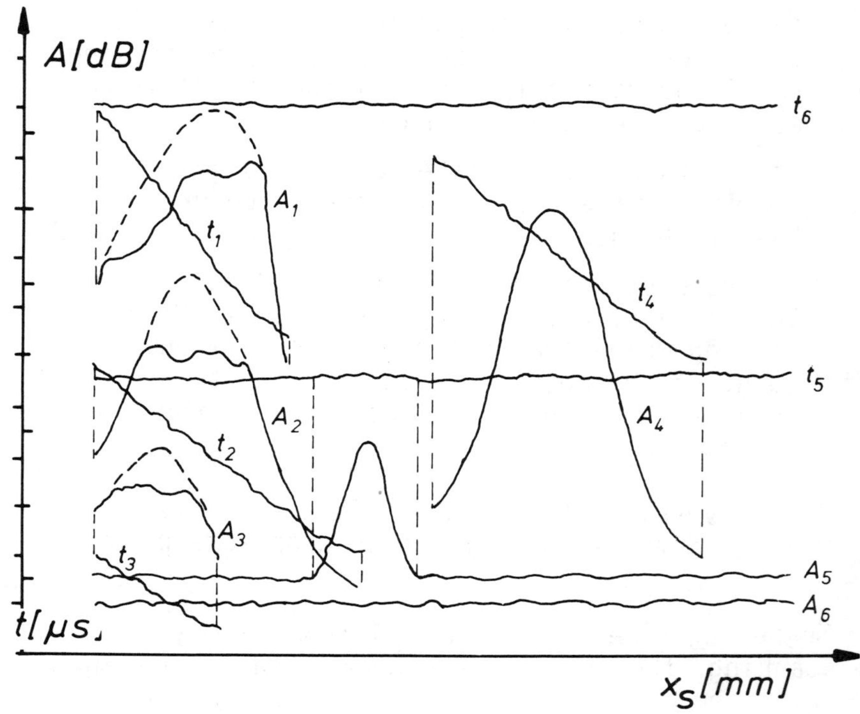

Fig. 14
Time-delay and amplitude locus curves

fits best /16/. Determine the parameters of the appropriate model, Fig. 15. For the planar models not only the width or diameter but also the inclination is obtained. The agreement is very reasonable for the models. We have not much experience with defects as yet.

- By means of the time-delay curves the edges of the reflector can be reconstrutcted; Fig. 16 /17/

One particular point should be mentioned. If there are several reflectors at the same time in the beam, they can be separated by the time-delay-locus

MODEL	MEASUREMENT		RECONSTRUCTION		DGS-METHOD	
	D_R [mm]	ϱ [°]	D_R [mm]	ϱ [°]	D_R [mm]	ϱ [°]
SPHERE	3 ⌀	-	3,25 ⌀	-	1,2 ⌀	-
CYLINDER	4 ⌀	-	3,9 ⌀	-	4,5 ⌀	-
STRIP	6	35	6,3	39	3,8	-
DISC	6 ⌀	40	5,1 ⌀	43	1,8	-

Fig. 15
Results of reconstruction

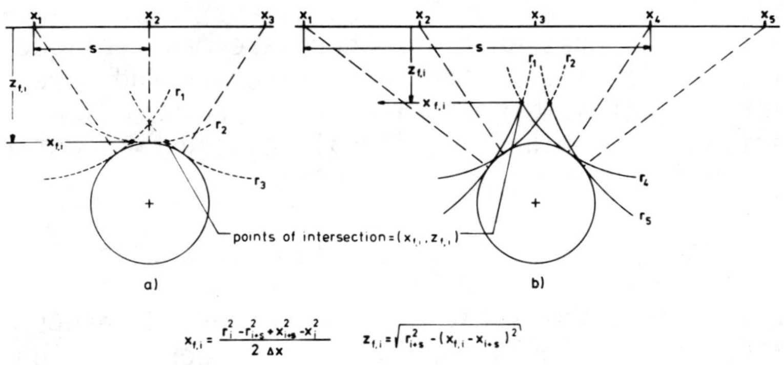

Fig. 16
Schemata of flawborder reconstruction

curves, because the time-delay for each locus of incidence is different for each reflector. Also any noise can be separated from reflections because sources of noise are randomly distributed. By selecting only amplitudes which belong to time-delay curves of a particular reflector better resolution as well as elimination of background is achieved.

Testkörper 2 (50-52) Längsfehlerprüfung , Ebene 110 bis 220

Fehlerbestätigung durch a,b,c,d,e,f,g,h,i,j,k : 211und 221

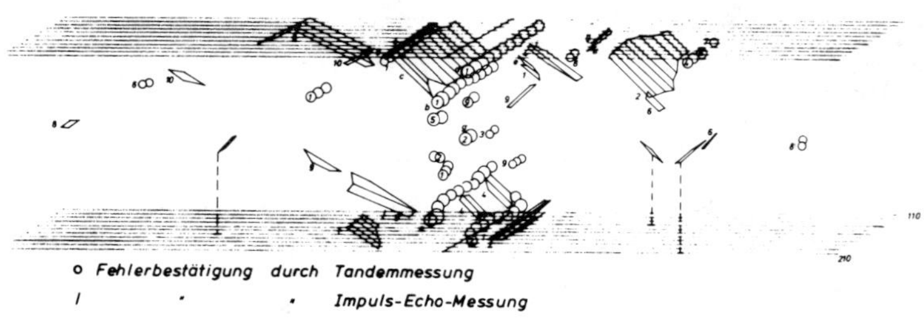

Fig. 17
HSST-Specimen, reconstructed with time-delay locus curves

Fig. 17 is one of the HSST specimens in the time-delay locus presentation. There are many oblique lines indicating particular reflectors and moreover a great number of randomly distributed or horizontally orientated signals caused by scattering in the interior or at the surface

With regard to the reconstruction by amplitude and phase the IzfP in Saarbrücken applies the equipment of Holosonics and, moreover, develops mathematical models for numerical reconstruction /18/. Their approach seems useful for an analytical application; Fig. 18, 19. The numerical reconstruction is - no doubt - possible for the linear as well as the two-dimensional scan. One of the advantages of the numerical reconstruction is that it is easier to adapt to curved surfaces, Fig. 20. Synthetic aperture and numerical holography are considered as competitive procedures /19, 20/.

There are two important and basic facts which are equally valid for synthetic aperture, holography and amplitude-time-delay reconstruction:

Fig. 18
Equipment of Holosonics

- from wide beam probes high spatial resolution is obtained;
- redundance with respect to deflection is increased.

Two new types of ultrasonic probes are being developed:

- electrodynamic transmitters and receivers for several wave modes /21/; some of them have not been used in ultrasonic testing up till now;
- (phased) arrays /22/.

The electrodynamic probes are available on a laboratory level for longitudinal and torsional tube waves and for shear- and shear-horizontal waves. Their

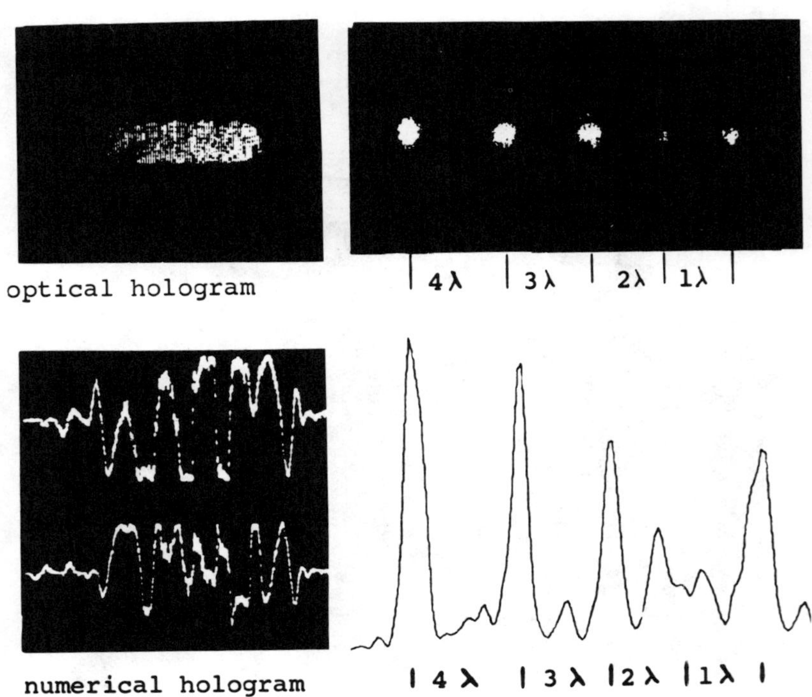

Fig. 19
Shear wave holography and resulotion measurement
Pulse-Echo - 45^0, flaw depth 45 mm, flaw size 7 mm ø
frequ. 2,5 MHz

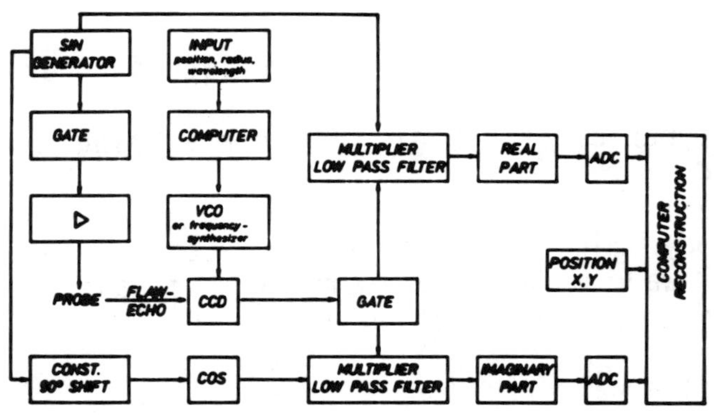

Fig. 20
Schematic diagram for ultrasonic holography on curved specimen

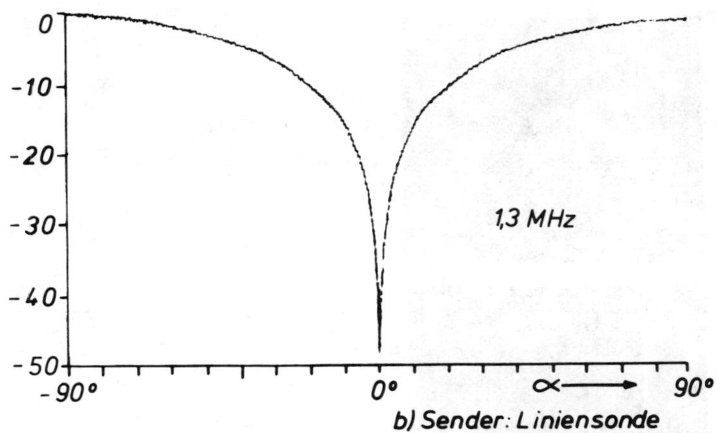

Fig. 21
Polarization of shear waves

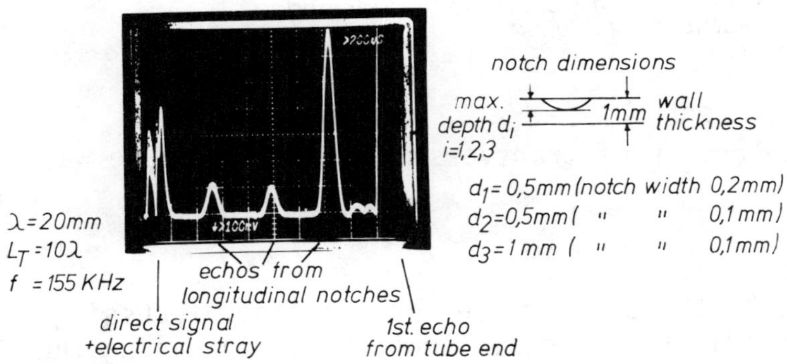

Fig. 22
Detection of three longitudinal notches with tube mode T(0,0) (austenitic tube, length 5 m, diameter 22 mm, wall-thickness 1 mm)

feasibility has been demonstrated. As shown in Fig. 21 a strong linear polarization of the shear waves is obtained. Some results on tube waves are shown in Fig. 22 and 23.

Work on piezoelectric and electrodynamic phased arrays has been started recently. According to

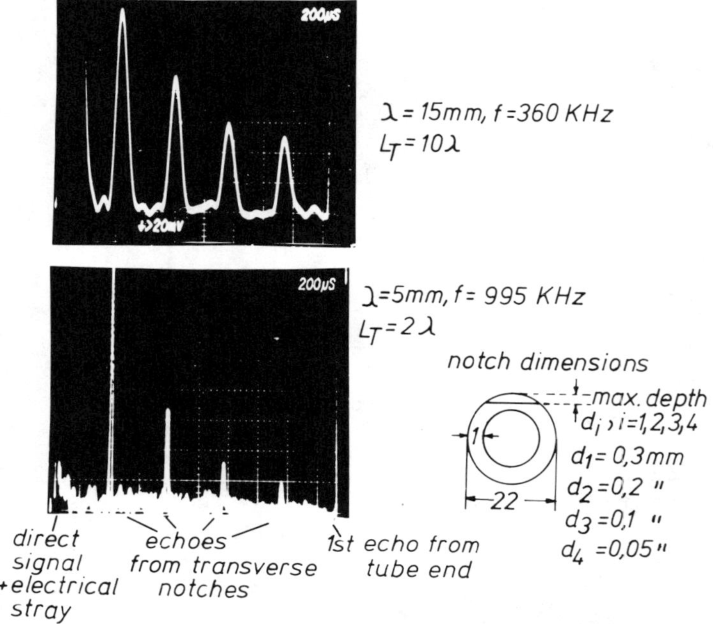

Fig. 23
Detection of transverse notches with tube waves

Fig. 24 the aims are sector scanning, focussing with adjustable focal length, and minimizing the side lobes by amplitude patterns.

Fixed focus ultrasonic probes with flat contact surfaces have been developed and are available /23,24/. The flat surface is very important for any non-immersing application. The following main applications are considered:

- testing the cladded surface layer of pressure vessels from the outside,
- high sensitivity testing for very small defects or inclusions,
- testing coarse grained material.

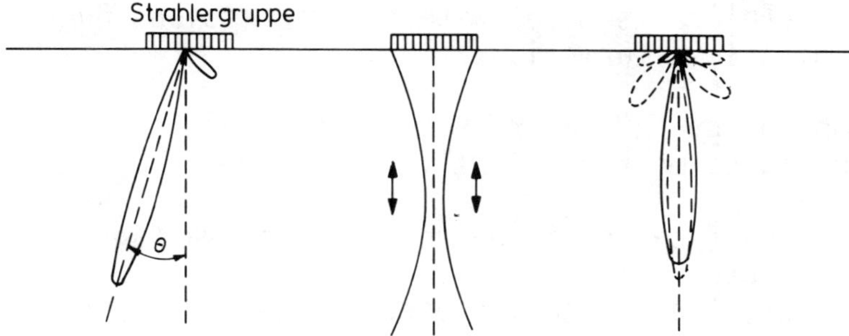

Fig. 24
Sector scan, focussing and minimizing side lobes

Acoustic emission

The terms characterizing the work going on during the last two years are:

- interpretation
- detection of crack growth during hydro-test
- correlation between crack growth and acoustic emission signals
- wave propagation
- signal analyses.

The first part of a project following the above mentioned terms is just finished. About 100 CT specimens of different types of steel, different heat treatments and structures were investigated. The results obtained in general are promising /25/. During the experiments the following acoustic emisson phenomena are observed consecutively:

- emissions due to friction between the fatigue crack faces,

- emissions resulting from plastic deformation in the plastic zone ahead of the crack,

- high energy acoustic emissions at the initiation

of micro-cracks mixed with burst emissions due to plastic deformation,

- high energy and low rate emissions when the crack propagates macroscopically.

This classification is on the basis of wattage (energy rate).
The first, the second and part of the third phenomena are not detected by presently available equipment, if the distance between source and probe is large. That means, they are to a large extent missed by present techniques using array lengths of some meters.

For the second phenomenon there is a very strong correlation with stress intensity factors. This correlation can be explained by fracture mechanics theory.

The crack propagation emission is reasonably well correlated with crack growth rates but the structure of the material is an important parameter.

This does not seem very new, but it had to be done, because for the acoustic emission technique consolidation is necessary. All of these experiments were done with full analytical instrumentation. The evaluation of the data obtained in three laboratories is in the final stage /25,26,27/.

The next step will be to do similar experiments with more realistic initial cracks, plus work on vessels.

It has been shown very clearly that acoustic emission is a good method for leakage detection /28/.

Electrical and magnetical methods

Eddy current testing is considered very important for thin tubing in steam generators as well as for surface testing of heavy components. Three different approaches are followed:

- pattern recognition of impedance locus curves and evaluating phase changes in the pattern;

- multi-frequency eddy current,

- wide-band testing.

The first approach has been used for years in the field. The development concerns data processing and automation /29,30/.

For multi-frequency testing a new system (hard- and software) has been developed. Its main features are a single coil, multiplexing frequency, a very fast electronic impedance-meter, a new algorithm for the evaluation, and very simple hardware to perform it /31,32/. Examples are shown in Fig. 25, 26, 27, a prototype equipment in Fig. 28.

Wide-band work has recently been started /33/.

For the most important cases there are computer codes to optimize a set of frequencies and to design coils.

The evaluation of multi-frequency signals to quantitize defects is a very particular point, because the contribution of the different frequencies to the final signal is accidental. There is in view a solution for this difficult problem.

Magnetic leakage testing is the most widely used NDT method. The quantitative description of defects by measured leakage flux is a physical problem no more difficult than many others already solved. The magnetic tape procedure seems promising. Reconstruction of a crack from a leakage tape is much easier than reconstruction of defects in ultrasonic testing. The IzfP in Saarbrücken is active in this field /34/.

Nearly the same comments could be made about the potential drop or current flux method. This very

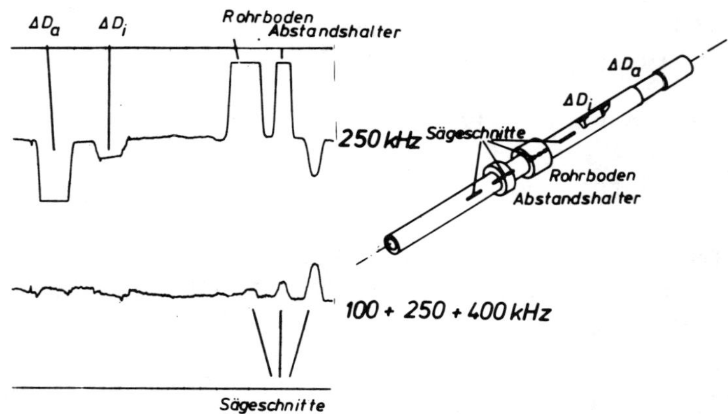

Fig. 25
Multi-frequency test of tubes

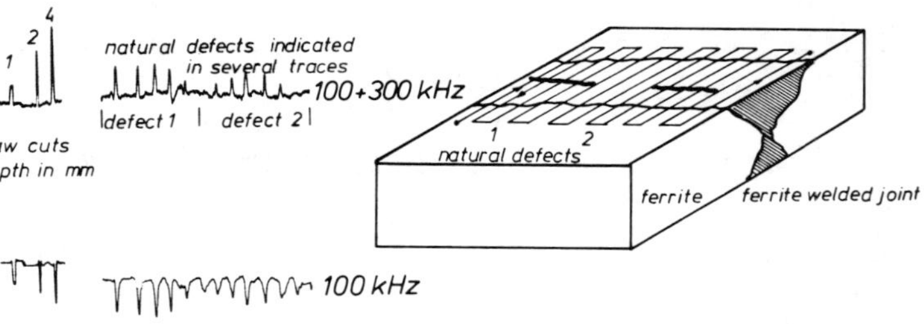

Fig. 26
Multi-frequency test of a weld

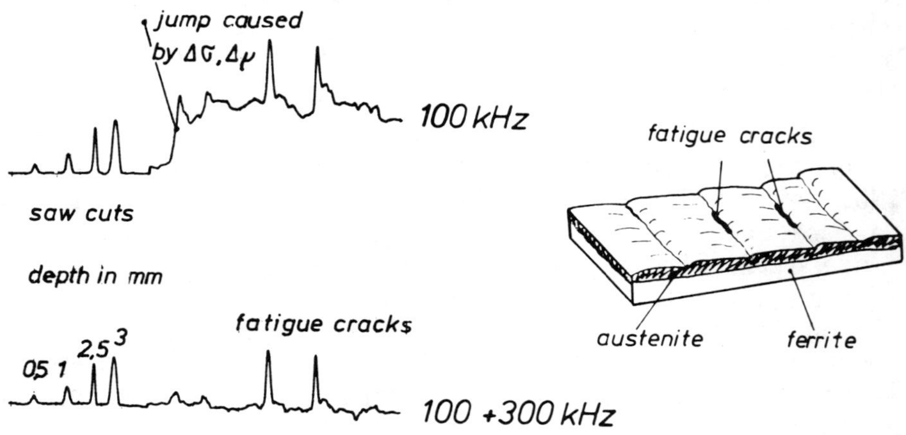

Fig. 27
Multi-frequency test of a cladding

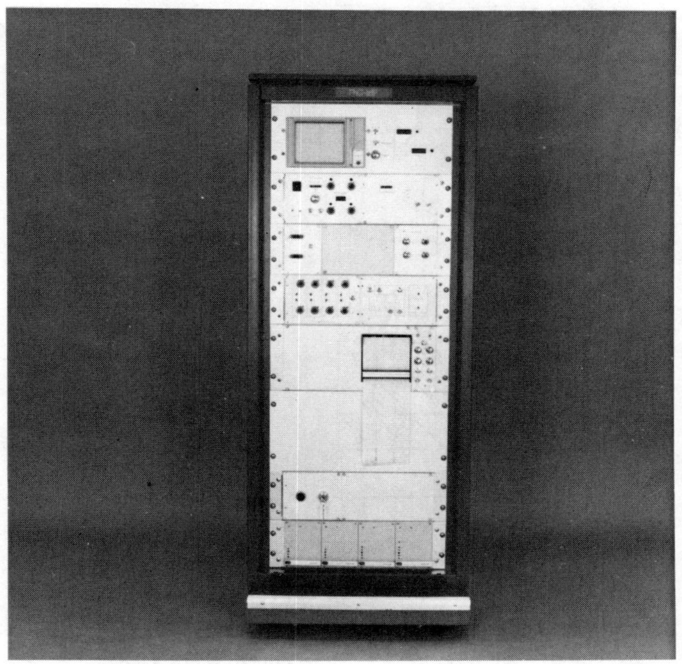

Fig. 28
Prototype equipment for multi-frequency testing

old and very effective technique of quantitizing large cracks has been treated mathematically and experimentally. Current flow can be calculated by the well-known finite element procedure. For DC it is a Poisson equation and the boundary conditions are the same as in stress analysis. What is coming out is shown in the figures 29 and 30 /35,36/.

At the end of this overview, there is a final point which seems very important; that is the extension of NDT research and development to mock-up experiments and prototype work.

There are two objectives:

- to learn how real defects response to NDE systems (hardware and software);

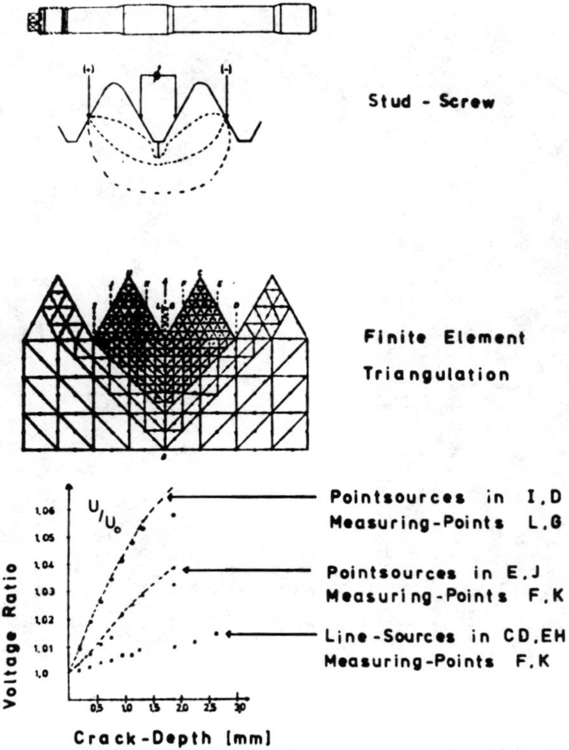

Fig. 29
Reactor pressure vessel stud-screw service problem

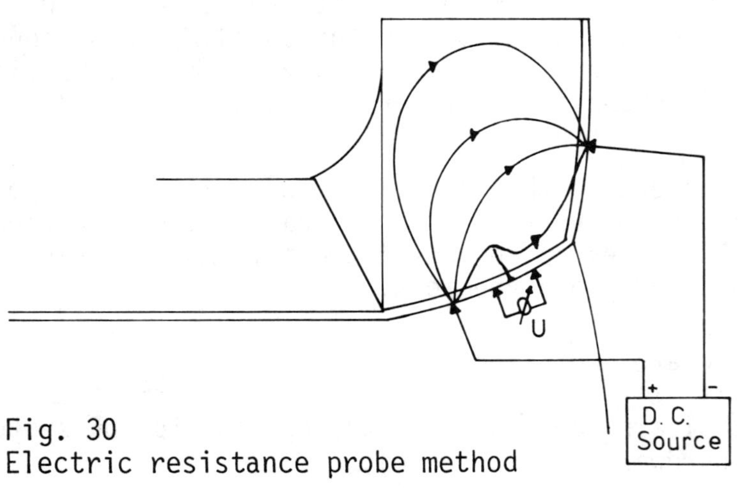

Fig. 30
Electric resistance probe method

- to test and prove any NDE system under environmental conditions typical of the practical case.

To cover these two objectives, in several laboratories mock-up experiments are made. Beside small blocks with holes and drillings, more and more actual large components are used as objects for experimental work /37/.

There are several large material research programmes mainly concerned with materials problems, metallurgy, ductility, etc., but involving also sub-programmes for NDE.

Two programmes are:

- the HDR /38/; the main object to be investigated is the pressure vessel of 100 MW BWR. In the NDE sub-programme testing systems including manipulators are applied and examined. The final aim is system analysis of any relevant testing procedure. This project is open for international cooperation.

- the component programme; Objects to be investigated are big components of different grades. At the beginning of this programme the NDE methods available at this time thoroughly will be proved to quantisize their ability to detect defects and to reconstruct them.

In other words, the HDR mainly is concerned to in-service inspection and the component programme more to pre-service inspection and analytical work in the in-service inspection area.

References

/1/ H.G. Seipel, D. Lummerzheim, D. Rittig, "German Light-Water Reactor Safety Research Program". Nuclear Safety 18 (1977) 6, p. 727-759

/2/ R. Saglio, M. Rule, "Focused Ultrasonic Transducers". Metallurgy and Nuclear Fuel Design Division, Technology Department, Fuel Elements and Structure Sections, Advanced Techniques Section, DMECN-DT-ECS-STA, August 1971

/3/ Bundesanstalt für Materialprüfung (BAM), "Dependence of the Scattering Indication Technique on the Surface Condition and on the Frequency". FRGMRT-RS 27/2, Technical Report, BAM, Berlin, May 1974

/4/ A. Erhard, H. Wüstenberg, E. Mundry, "90° Longitudinal Waves Probe for Detection of Near Surface Cracks". European Conference on Non-Destructive Testing, April 24-26, 1978, Mainz (FRG), Paper 10

/5/ C.F. Vasile, R.B. Thompson, "Periodic Magnet Non-Contact Electromagnetic Acoustic Wave Transducer. Theory and Application". IEEE Ultrasonics Symposium Proceedings 1977 (77 CH 1264-1SU), p. 84-88

/6/ R. Becker, P. Höller, "Results of Application of Eddy-Current Testing with a Multifrequency Method". 8th World Conference on Nondestructive Testing, September 6-11, 1976, Cannes (France)

/7/ G. Engl, W. Rathgeb, H. Wüstenberg, F. Walte, "Auslegung von Prüfköpfen und Prüfkopfsystemen für die automatische Ultraschallprüfung unter Benutzung beugungstheoretischer Modelle". European Conference on Nondestructive Testing, April 24-26, 1978, Mainz (FRG)

/8/ F. Walte, R. Werneyer, B. Horst, "Determination of Test Zones for the Ultrasonic Test by the Tandem Method". Materialprüfung <u>19</u> (1977) 5, p. 174-177

/9/ J. Kutzner et al., "Use Zoning, Sensitivity Calibration, and Probe Mounting for the Ultrasonic Test with the Tandem Method". Materialprüfung <u>17</u> (1975) 7, p. 246-250

/10/ H.J. Meyer, W. Rath, "Ultrasonic Apparatus for Repetitive Testing of Reactor Pressure Vessel". Nuclex Conference No. 9, 1972, Basel (Switzerland).

/11/ Krautkrämer GmbH, "Development and Construction of an Ultrasonic Test and Evaluation Electronic including Pre-Amplifier Box". RS-169, Technical Report, 1978, Köln (FRG)

/12/ A. Waas, "Automatische Ultraschallwiederholungsprüfung von Reaktordruckbehältern. Modell zur Verknüpfung der Teilvolumenintegrationsmethode mit der Ersatzfehlergröße". Annual Conference Non-Destructive Testing 1977, Bremen (FRG)

/13/ Instructions for Water Pressure Reactors. Chapter 4.1, Reactor Coolant Pressure Components (in German). RSK14-02/1, 30 Apr, 1974. Köln (FRG)

/14/ S. Kraus, K. Goebbels, "Signalmittelungsverfahren zur Unterdrückung des kohärenten Untergrundes bei der US-Fehlerprüfung grobkörniger Werkstoffe". 4. Zwischenbericht zu RS 102-16, IzfP-Bericht Nr. 760717-TW, Saarbrücken 1976

/15/ R. Werneyer, F. Walte, M. Klein, "Mathematical Model for the Reconstruction of Defects when Using Ultrasonic Pulse-Echo and Tandem Methods, and Results of Model Experiments" (in German). Materialprüfung 20 (1978) 2, p. 68-72

/16/ J.L. Rose, "A 23 Flaw Sorting Study in Ultrasonics and Pattern Recognition". Mat. Eval. 35 (1977) 7, p. 87-92

/17/ O.A. Barbian, U. Klocke, W. Lotze, "Border Reconstruction of Defects by Means of Time-Delay-Locus Curves - Results of Automatic Test of Specimens with Natural and Artificial Defects." European Conference on Nondestructive Testing, April 24-26, 1978, Mainz (FRG)

/18/ V. Schmitz, M. Wosnitza, "Experiences in Using Ultrasonic Holography in Laboratory and in the Field with Optical and Numerical Reconstruction". 8th Int. Symp. Acoustical Imaging, May 29-June 2, 1978, Miami, Florida

/19/ V. Schmitz, "Akustische Holographie. Labor- und anwendungstechnische Versuche." Abschlußbericht RS 102-20/2 (1977), IzfP Saarbrücken

/20/ J.R. Frederick, R.C. Fairchild, B.H. Anderson, "Improved Ultrasonic Nondestructive Testing of Pressure Vessels". Annual Report NUREG-0007 -2 (1977), University of Michigan

/21/ W. Mohr, W. Repplinger, "Contactless EMA-Excitation of Ultrasonic Bulk Waves. Part I". Materialprüfung 20 (1978) 4, p. 147-154;
Part II: 20 (1978) 6, p. 221-225

/22/ W. Gebhardt, F. Bonitz, "The Phased Array as a New Electronically Steered Probe in NDT". European Conference on Nondestructive Testing, April 24-26, 1978, Mainz (FRG), Paper 58

/23/ H. Wüstenberg, J. Kutzner, "Sensitivity Calibration for the Use of Focussing Probes in Ultrasonic Inspection at Planar and Curved Surfaces" (in German).
Materialprüfung 19 (1977) 10, p. 441-444

/24/ U. Schlengermann, "Focussed Sound Fields and their Applicability to the Nondestructive Testing of Materials". 8th World Conference on Nondestructive Testing, September 6-11, 1976, Cannes (France)

/25/ J. Lottermoser, E. Waschkies, P. Zenner, B. Voss, J. Götz, "Laboruntersuchungen zur Erarbeitung von Interpretationsmodellen für die Bewertung von Fehlstellen bei der Schallemissionsprüfung am Kernreaktor". IzfP-Bericht Nr. 780236-TW, 1978, Saarbrücken

/26/ T. Fischer, "Rißausbreitung und Schallemission in Schweißnähten des Stahls 22 NiMoCr 3 7". IfaM Technical Report (1977), IfaM, Bremen (FRG)

/27/ J. Eisenblätter, H. Jöst, "Schallemissionsmessungen bei bruchmechanischen Versuchen an Werkstoffen leichtwassergekühlter Reaktoren". Battelle-Bericht BF-R-62945-2, 1978, Frankfurt (FRG)

/28/ G.J. Dau, "A Review of On-Line Leak Detection Methods for Reactor Systems". 3rd Conference on Periodic Inspection of Pressurized Components, 1976, London. The Institution of Mechanical Engineers, p. 67-73

/29/ R.C. McMaster, Nondestructive Testing Handbook. Ronald Press Company, New York, 1963, Vol. II.

/30/ W. Stumm, "Verbesserung von Fehlererkennbarkeit und Bewertung mit Hilfe der Mustererkennung bei elektromagnetischen Prüfverfahren". European Conference on Nondestructive Testing, April 24-26, 1978, Mainz (FRG)

/31/ R. Becker, L. Regneri, "Anpassung elektrischer, elektromagnetischer und magnetischer Prüfverfahren für den Einsatz an Reaktoren". Technical Report No. 780832-TW, 1978, IzfP,Saarbrücken

/32/ K. Betzold, "Theoretische und numerische Untersuchungen zur Wirbelstromprüfung geschichteter Materialien mit koaxialen Innenspulen und Außenspulen". Technical Report No. 780725-TW, 1978, IzfP, Saarbrücken

/33/ G. Wittig, Grigulewitsch, "Untersuchungen zur

Anwendung des Impuls-Wirbelstromverfahrens an Prüfobjekten mit größeren Dicken". European Conference on Nondestructive Testing, April 24-26, 1978, Main (FRG)

/34/ G. Dobmann, "Streuflußverfahren". Technical Report No. 780333-TW, 1978, IzfP, Saarbrücken

/35/ R. Becker, G. Dobmann, P. Höller, "Contribution of the Electrical and Magnetic Methods in Nondestructive Determination of the Size of Surface Cracks in Welded Materials". IIW-Kolloquium, 1977, Copenhaguen, Denmark

/36/ G. Dobmann, "Finite-Elemente-Näherung zur Bestimmung von Potentialverteilungen für ein elektrisches Durchströmungsverfahren in der zerstörungsfreien Werkstoffprüfung". Kleinheubacher Berichte Nr. 21, 1978, FTZ-Darmstadt (FRG)

/37/ R. Becker, "Prüfung der Stutzenkanten des HDR-Behälters auf Oberflächenfehler mit dem zerstörungsfreien Wirbelstromverfahren". Technical Report No. 780209-TW, 1978, IzfP, Saarbrücken

/38/ W. Schmülling, G. Deuster, "Der Einsatz zerstörungsfreier Prüfverfahren am HDR". KFK (Kernforschungszentrum Karlsruhe) Nachrichten 9 (1977) 2, Karlsruhe (FRG).

EDDY CURRENT INSPECTION SYSTEMS FOR STEAM GENERATOR TUBING IN NUCLEAR POWER PLANTS

Clyde J. Denton
Zetec, Inc.

ABSTRACT

Eddy current inspection of steam generator tubing in commercial nuclear power plants has evolved from a simple manual effort to test two (2) tubes during 1970 to semi-automatic systems inspecting thousands of tubes today.

Although improvements have been made in the recording and interpretation of data as well as in mechanical fixturing, the basic eddy current inspection is still performed in the same way.

This paper describes the inspection system employed by all utilities in the U.S.A. and most utilities overseas.

Discussions are presented on the following topics:

1. General eddy current theory as it applies to the inspection of non-ferrous tubing.

2. Typical data collection instrumentation and its operation.

3. Basic data interpretation.

4. Mechanical fixturing systems.

DISCUSSION

A system employed to inspect steam generators uses eddy currents as the probing media to measure variations in the conductivity of the tube wall being tested.

An alternating voltage is impressed across two test coils. The magnetic field developed by current flow in the test coils causes eddy currents to flow in the tube wall. The magnitude and phase relationship of eddy current flow in specific concentric ring segments of the tube wall is dependent on the test frequency utilized and the electrical conductivity of the tube. The corresponding magnetic field caused by eddy current flow in the tube wall is out of phase with the field developed by the current in the test coils. Since these fields tend to cancel one another, the coil voltage is decreased and phase shifted in proportion to the magnitude of eddy currents in the test piece, thus the coil voltage, is dependent on the electrical properties of the tube being tested. The electrical properties of the tube which affect the flow of eddy currents are permeability and conductivity. In non-magnetic materials, such as Inconel and 300 series Stainless Steel, conductivity is usually the only significant variable. When the conductivity decreases due to a discontinuity in the tube wall, the coil voltage increases and phase shifts in direct relationship to the characteristics of the flaw. Thus, the amount of increase in the coil voltage and the phase change is related to the discontinuity.

The coil voltage is sinusoidal, thus it can be described with a single vector having magnitude and phase. The eddy current test system used in steam generator inspection provides a method for reading out the two quadrature components of the test coil voltage vector.

The two test coils are electrically connected in opposite legs of the balancing network in the eddy current instrument. Thus, the tube is being inspected by the differential technique. The differential technique decreases the affects of probe motion, temperature variations and geometry differences. However, gradual changes in wall thickness are not detected with the differential annular coil. In some cases, coils with increased separation, or single coils, are used to detect gradual wall thickness changes.

The electronic portion of the eddy current system contains five (5) separate instruments (see Figures 1 and 2). The eddy current instrument has a continuously variable frequency from 1 kHz to 2.5 mHz with a digital readout to indicate the operating frequency. The readout is accomplished on an X-Y memory oscilloscope which is an integral part of the instrument. The instrument has X-Y outputs of plus or minus 8 volts and a frequency response of DC to 100 Hz.

The output of the eddy current instrument is connected to a Two-Channel FM Magnetic Tape Recorder. The tape recorder also has input and output capabilities of plus or minus 8 volts and DC to 100 Hz frequency response. In addition to recording the X-Y channels, the tape recorder has a microphone to allow tape recording tube identification and other pertinent data. The circuits in the recorder are designed to allow voice insertion and retraction without interaction with the test data.

The output of the tape recorder is connected to the input of a Two-Channel Strip Chart Recorder. The strip chart recorder has a frequency response range from DC to 100 Hz and it is capable of displaying a voltage input of plus or minus 8 volts. The strip chart recorder provides three functions. First, it provides a permanent record which can be scanned rapidly for initial inspection results. Second, it provides the information used to determine the axial position of indications. Third, since it monitors the output of the magnetic recording, it assures that the recording equipment is functioning properly.

The fourth instrument is a Communications Amplifier which allows voice contact between four (4) stations with variable inputs and outputs for all stations. The amplifier contains high and low filters to decrease normal plant noise.

The fifth instrument is used to assist in data analysis and will be discussed at length later in this presentation.

The eddy current test system is normally used in conjunction with a mechanical system which is used to position the probe over a tube and then a pusher/puller is used to insert and withdraw the probe. The insertion rate is approximately two feet per second and the withdrawal rate is one foot per second. The inspection is performed during the retraction of the probe. Some inspection agencies perform a wall thickness measurement during insertion and a flaw inspection during retraction.

When the probe is inserted the proper distance, the tube number is written on the strip chart and the voice entry is made on the magnetic tape, then the probe is retracted while the recording systems are operating.

When the magnetic tape is completed, the tape and its associated strip chart records are taken to a remote location where they are analyzed by an ASNT-TC-1A Level IIA qualified interpreter.

The equipment used to analyze data consists of a tape recorder identical to the one used to record the data, and a vector analyzer which more realistically should be called an electronic protractor. The "analyzer" provides a rapid means of measuring the phase angle and amplitude of signals.

The basis for phase analysis eddy current testing can be simplified and explained as follows: Given four concentric tightly fitting tubes as shown in Figure 3, a fixed test frequency - low enough to penetrate their combined wall thickness, a single coil (Absolute) to allow measurement of nominal wall thickness, and starting with the probe in air, first the air vector is obtained. When the probe is inserted in the smallest diameter tube, eddy currents flow in the tube wall with a resulting magnetic field. The resultant coil voltage vector is decreased in amplitude and phase shifted. As the second tube is slipped over the probe area, the vector amplitude is further decreased and phase shifted. The current flowing in the second tube is a function of the magnetic field from the coil and the magnetic field associated with the current flow in the first tube. This process continues for each tube with the current flow in each tube dependent on the current flow in the adjacent tubes. The eddy currents are not affected (in a non-defective tube) by the laminar type tube to tube interfaces. Thus, this example can be expanded to include eddy current flow in a solid tube wall. The current flowing in any circumferential tube segment has its own distinctive phase and magnitude. The exact phase and magnitude at any point in the tube wall is dependent on the test frequency and the conductivity of the tube being tested. The eddy current test system's function is to detect variations in the magnitude and pattern of eddy current flow in the tube wall and display/record signals which relate to these variations.

When a differential probe is passed through a tube with a defect, the signal is formed as in Figure 4. Point 1 of Figure 4 is the signal from a good tube, Point 2 shows the first coil approaching the defect, Point 3 shows the coil directly centered in the defect, Point 4 shows the first coil leaving the defect and the second coil entering the defect, Point 5 shows the completion of the signal.

Figure 5 shows three defects tested at three different frequencies. The probe was a differential bobbin type and

and the two defects not penetrating through the wall are on the outside surface of the tube.

Figure 6 is essentially the same as Figure 5 except a different sample tube is shown. The sample is 7/8" x .050" wall thickness Inconel 600 and the inspection frequency is 400 kHz. The 400 kHz test frequency provides the maximum phase angle spread without rotating the 20 percent flaw into the horizontal probe motion signal. This choice of frequency provides the greatest phase angle measurement accuracy but the sensitivity to defects occurring on the outer tube wall surface is less than that obtainable at lower frequencies. This consideration is important when inspecting for small volume flaws in free standing tubing. When the tubing being inspected is adjacent to a carbon steel support member, the increased sensitivity gained by decreasing the frequency is lost in the increased interference from the signal caused by the supporting member.

Taking the data from Figure 5 and plotting a calibration curve of percent flaw penetration of the tube wall versus signal phase angle results in the data presented in Figure 7.

The eddy current test system has been shown to exhibit a long term two sigma measurement error of plus or minus 5° under actual field conditions. (Data collected from calibration standard signals during a 6-week inspection using multiple probes, calibrations and operators.) Plotting this information versus the calibration curves in Figure 7 results in the measurement error curves shown in Figure 8.

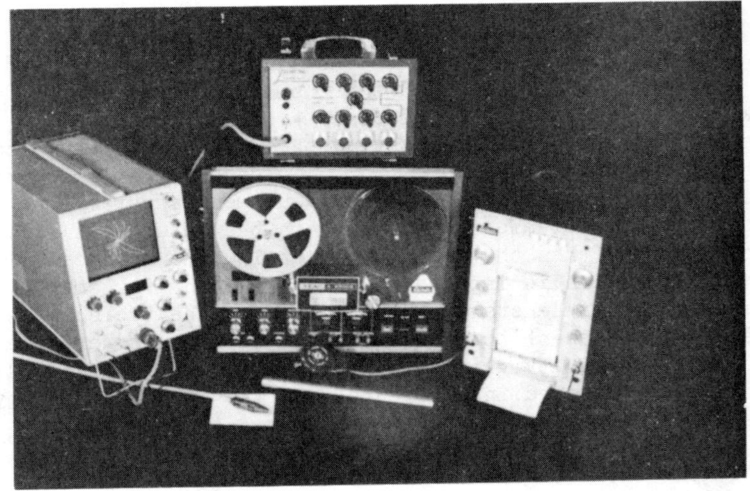

Fig. 1. Eddy Current Test System for Data Collection

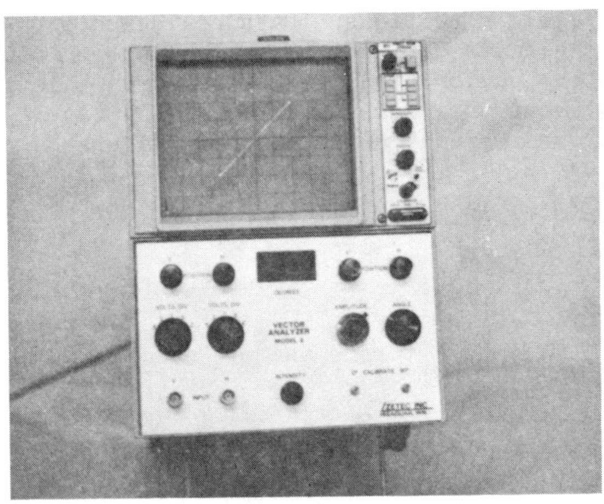

Fig. 2. "Vector Analyzer" Data Interpreting Aid

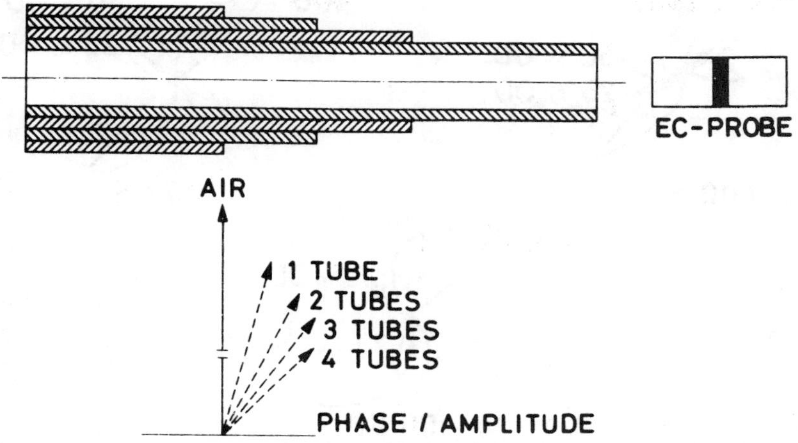

Fig. 3. Phase Relationships

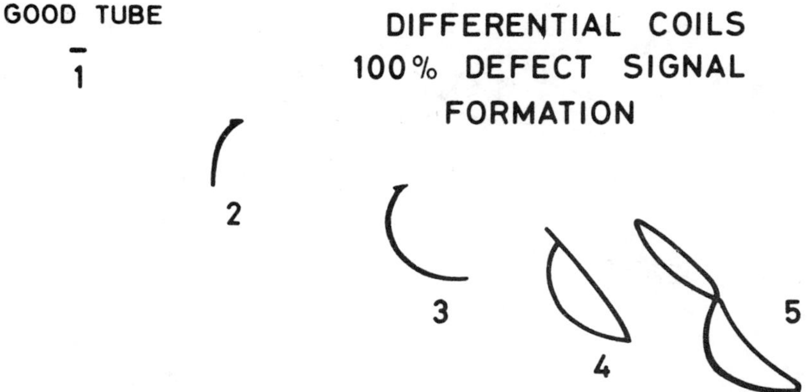

Fig. 4. Signal Formation

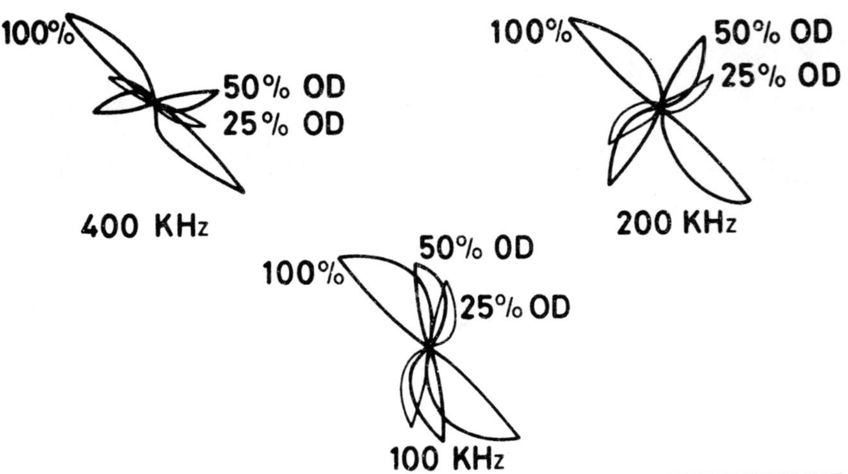

Fig. 5. Signal Phase Angle Comparisons At Three Frequencies

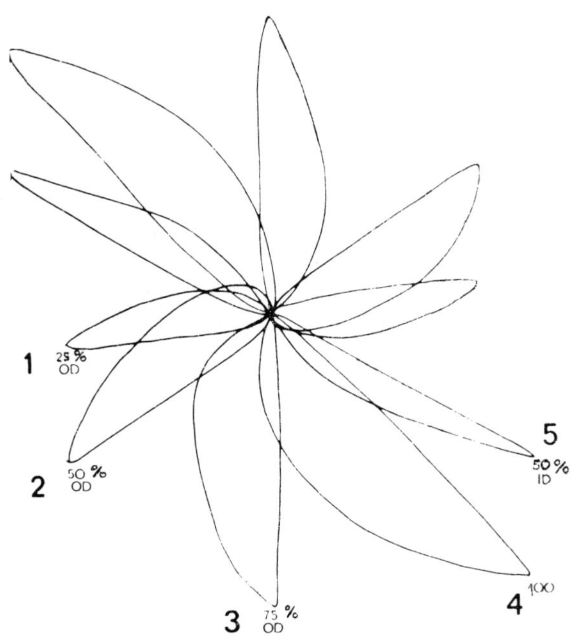

Fig. 6. Amplitude and Phase Angle Display of Sample Signals

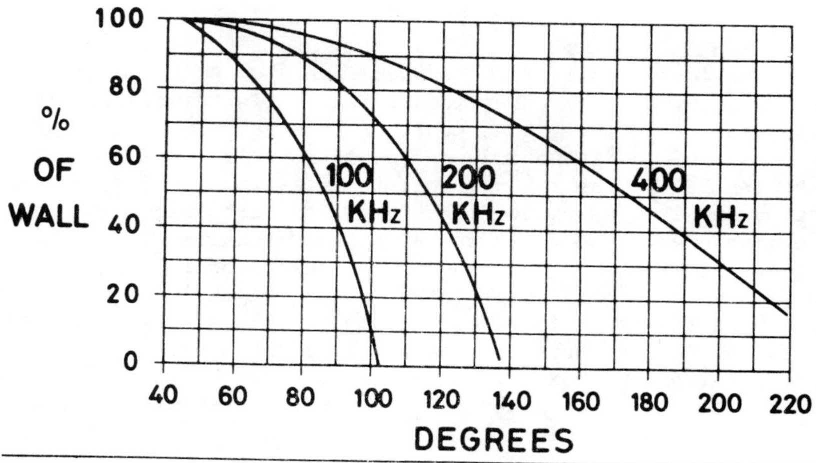

Fig. 7. Calibration Curves For Three Frequencies
400 kHz .050" Wall 304L Stainless Steel

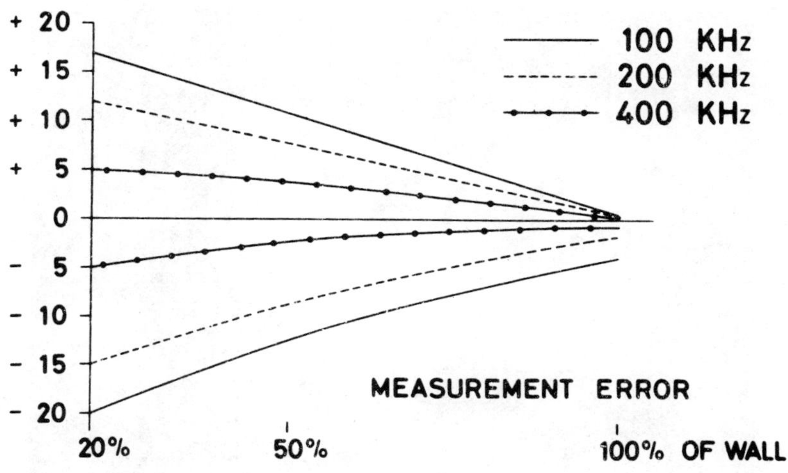

Fig. 8. Measurement Error Comparisons At Three Frequencies

The mechanical portion of the eddy current test system varies to accommodate the conditions imposed by the designs of the various steam generator vendors.

Basically, all of the systems function as follows: A template with tube number identification is temporarily installed in the steam generator. A rotatable circular fixture with a minimum of two independent motions is installed over the template. The fixture operator positions the probe guide tube, and its associated light and TV camera over the tube to be inspected. The probe pusher/puller mechanism is used to insert and retract the probe. Test speeds of 100 tubes per hour are achievable when the tube test length is in the order of 10 feet. Thus, it is obvious that fixture positioning time between adjacent tubes is relatively short. The complete data station and fixture control center can be operated up to 150' from the steam generator although shorter distances are recommended.

A brief pictorial presentation of mechanical systems now in use is given in Figures 9 through 11.

Fig. 9. Typical Probe Pusher/Puller With Take Up Drum

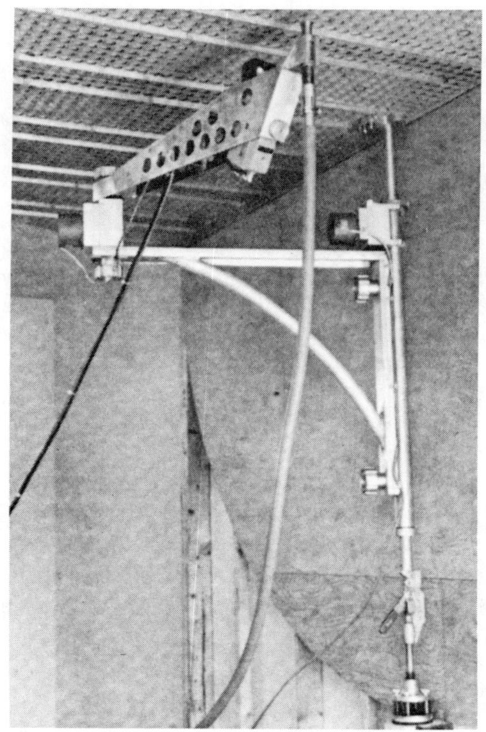

Fig. 10. Fixture For U-Tube Type Steam Generators

Fig. 11. Fixture For Straight Tube Steam Generators

MULTIFREQUENCY EDDY CURRENT EXAMINATION OF STEAM GENERATOR TUBES

R. Saglio, Intercontrole, Rungis, France
M. Pigeon, Intercontrole, Rungis, France
W. J. Kitson, Jr., Holosonics/Intercontrole, Richland, Washington U.S.A.

I. INTRODUCTION

The launching of the French program of PWR Nuclear Power Plants included, at the origin, a large amount of R and D work in the field of in-service inspection.

The actions which were undertaken may be split up into different levels:

- the regulatory level,
- the level of the R&D work, methods and devices which were studied to improve the reliability of NDT techniques,
- the design level, actions undertaken to facilitate or improve the testing during in-service inspection,
- the flaw evaluation level.

As the steam generator is the largest heat transfer area between the primary and secondary circuits, its integrity is of prime importance for the safety of the installation, either during normal operation or during potential accidents (LOCA or steam line rupture).

The typical flaws which can be found in such tubes are, for example:

- thinning and crud deposits (eventually magnetic crud),
- fatique cracking,

- cracks in the vicinity of the tube support plates or of the antivibrating bars,

- denting, alone, or accompanied with cracking.

Eddy current testing is the most frequently applied method used to test steam generator tubes, but a monofrequency apparatus does not have the capability to extract all the information content of the signal. As a matter of fact, the tube sheet, the tube support plates, the antivibration bars and the bend, give their contribution to the eddy current signal as well as other spurious parameters, such as, crud and deposits.

The solution lies in the use of a multifrequency eddy current device associated with combination devices. We will describe in detail the apparatus which has been developed and the results we are able to obtain in the field applications.

II. PRINCIPLE OF THE MULTIPLE FREQUENCY APPARATUS

Figure 1 is the representation of the eddy current signals appearing on the screen of the oscilloscope from:

- an internal defect Di
- an external defect De
- the support plate Sp

at three different frequencies. At the low frequency, f1, internal and external defects have the same phase and comparable amplitude. The amplitude of the signal of the support plate is great in comparison. At the medium frequency, f2, it appears as a phase difference between internal and external defects and a decrease in the ratio of amplitude between support plate and defect.

At the higher frequency, f3, the phase rotation between the external and internal defects is 90°. The ratio of amplitudes between the internal and the external defects is approximately 0.4 and the amplitude of the support plate signal is lowered in comparison with f2.

The use of a higher frequency than f3 will suppress completely the signal from tube support plates, but at the same time, it will also express the external defects with low depth and increased signals from internal defects.

Figure 2 shows how complete the signal can be when a defect is under a support plate.

The combination system of the IC3FA takes care of what we saw on Figure 1. For two frequencies, phase and amplitude obtained on different defects are different. If we are able to modify the phase, the shape and the amplitude of the support plate eddy current signals to have them equivalent at two frequencies, and after that present the resulting signal obtained by subtraction, the results will be nothing for the support plate and something for defects.

Figure 2b presents what is the result of such a combination made with the IC3FA.

Figure 3 presents the strip chart record before and after combination of a real steam genrator tube, (Fessenheim 1 Plate). We can see how well the tube signal from support plates is eliminated, but in this example we used only two frequencies and one combination and this is not sufficient to eliminate spurious signals inside the tube sheet.

To do so, we used three frequencies and the combination between f1 and f2 to eliminate support plates. We obtained a first result, C_1.

The combination between f3 and C_1 is then used to eliminate internal noise due to internal geometrical variations or tube sheet deformation due to suspension process. We then obtain the second results, C_2.

Figure 4 presents a strip chart record at 240 kHz and with C2 on a real tube in the Bugey 2 plant. This shows that with three frequencies and two combinations it is possible to eliminate, at the same time, internal noise, support plates, spurious signals in the bended part and in the tube sheet.

III. SPECIAL FEATURES OF THE MULTIFREQUENCY APPARATUS

The first equipment made only had two frequencies. The preservice inspection of Fessenheim 1 was made using this first piece of equipment, and during field utilization we encountered three types of problems:

- Due to the differential mode used for the coil, the apparatus was not able to detect change in thickness.

- Due to the process of fabrication (Pilgering pass) the internal noise was very high and gave a deformation on the signal.

- Due to magnetic variations on the internal surface of the tube, there were very high spurious signals on 10% of the tubes.

All these problems are solved in the IC3FA by special features:

- Complex bridge arrangement including differential and absolute mode. This allows us to use one of the three frequencies at the same time in the absolute and differential modes (see Figure 5).

- The use of three frequencies allows us to use the mixing as explained in the previous chapter and, in this condition, eliminates internal noise (see Figure 6).

- Development of special coils (Figure 7) able to go through the bend and including a magnetic circuit. A double recording is done without magnetic saturation when going into the tube and with saturation when returning. Figure 8 shows the effectiveness of the magnetic saturation.

IV. DESCRIPTION OF THE COMPLETE SYSTEM

The in-service inspection of steam generator tubing is considered as a complete examination and, for this reason, the system was divided in four parts:

- electronic eddy current systems
- recording of the data
- automated pusher/puller for the probes insertion and withdrawal
- finger walker inside the water box

This description is shown in Figure 9.

IV-1 Electronic Eddy Current System (Figure 10)

This system has already been described as the IC3FA. The major asset of this equipment is its ease of use. To be of interest for field application, it was made as simple as possible to avoid loss of time during adjustments at each change of the probes. A qualified operator can change a probe and make the new adjustment in 10 minutes. The data registered on the magnetic tape can be displayed with the mixing, if needed.

IV-2 Recording of the Data

All of the data is recorded on magnetic tape (7 to 14 tracks) and a strip chart recorder (6 to 12 tracks). The bandwidth of these two apparatus is 800 Hz minimum. This is of interest because it allows an increase in speed of inspection.

IV-3 Automatic Pusher/Puller

Figure 11 is the pusher/puller. This is a fully automatic device. When the finger walker is in position to begin the examination of a tube, the operator just pushes a button and the process is as follows:

- switching on the magnetic tape and strip chart recorder
- switching on after a little delay the pusher and compressed air if needed
- the probe goes through the complete tube from one side to the other under speed control (if the probe is stopped for any reason the pusher will stop immediately)
- when the probe goes outside of the tube sheet on the other side of the water box the information is given to the pusher/puller by the IC3FA
- the pusher stops
- magnetization is switched on, if needed
- switching on to puller, the probe is retracted
- when the probe goes outside of the tube sheet in the guiding tube of the finger walker the information is given by the IC3FA and the process is:
    - switching of puller which stops immediately
    - switching off magnetization

- switching off strip chart and magnetic tape recorder
- signal given to the finger walker to allow a change in position

The speed of inspection is 0.5m/s. but there does exist a second speed eight times lower (0.065m/s). The pusher/puller can also be used in the manual mode and the speed of inspection can be easily changed.

The control panel of the pusher/puller also gives other information like speed of inspection, broken probe, etc. The coil is stored on cartridge (like films) and the change can be made rapidly.

This automatic pusher/puller is very useful in field application by considerably decreasing the job to be done by the operator.

IV-4 Finger Walker

The finger walker is shown in Figure 12 with its control panel. It works as an X and Y displacement of one pitch, and it can be manually or automatically controlled. It requires only two interventions in the water box for putting and withdrawal. It needs approximately three seconds per pitch and weighs 33 pounds. It was designed with a fail safe mode and cannot drop into the water box. It needs only compressed air (7 bars) and electricity.

V. FIELD APPLICATION

At the time of this writing, the multiple frequency apparatus has been used for a complete inspection of nuclear power plants in Fessenheim (two frequencies), Fessenheim 2, Bugey 2, (IC3FA) Bugey 3,4,5 (IC3FA plus finger walker). For partial in-service inspection in Doel 1 and 2 (IC3FA). For full in-service inspection in Tihange (IC3FA and finger walker).

The automatic pusher/puller was used in all these inspections. The time needed for a complete inspection on a 900 Mw PWR plant is 650 hours with one set of equipment. It can be reduced to 220 hours with three sets of equipment, which means approximately 8 to 10 days working 24 hours a day.

The speed of inspection is limited by the little bend in the first 10 rows of the steam generator, but it can be increased for the other rows reducing the time of inspection mainly needed by the length to be inspected.

VI. DATA PROCESSING

Until now the data was processed manually by experienced personnel and this represents a large amount of the job. The job is easier with multiple frequencies because the signals after mixing are very clean and clear and the first developments are in progress on magnetic recording of digitalized information (10 channels with 72 DB dynamic and 400 Hz bandwidth). This equipment can be used in a very short time and the data processed afterwards by a computer.

Processing by computer will be ready for field use in the very near future. Analysis can be made by Level 1 or 2 inspectors without any difficulty. The computer executes or completely postpones analysis of the information and gives a diagnosis on present defects.

VII. CONCLUSION

It has been shown that requirements of periodic in-service inspection of steam generator tubes have led to new developments in the field of eddy currents. New methods and devices were designed and fabricated which represent a step forward in the objective knowledge of the flaws.

The question may arise if these new methods and equipment must also be used during shop testing.

The requirement of in-service inspection of nuclear power plants has forced us to produce simple, powerful and automatic equipment that can be used in other applications such as the petrochemical plants, as we now do in France.

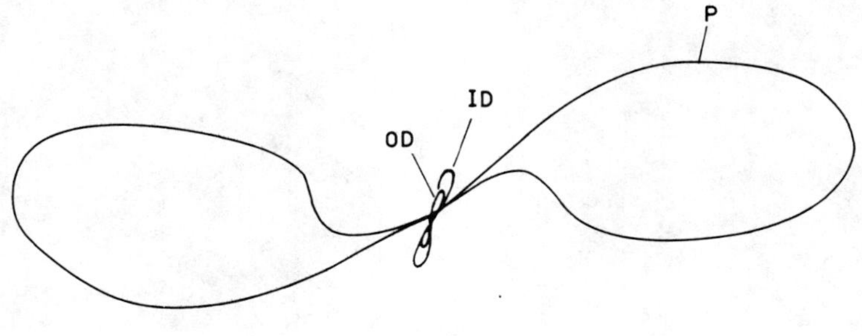

a) Frequency f1 (low)

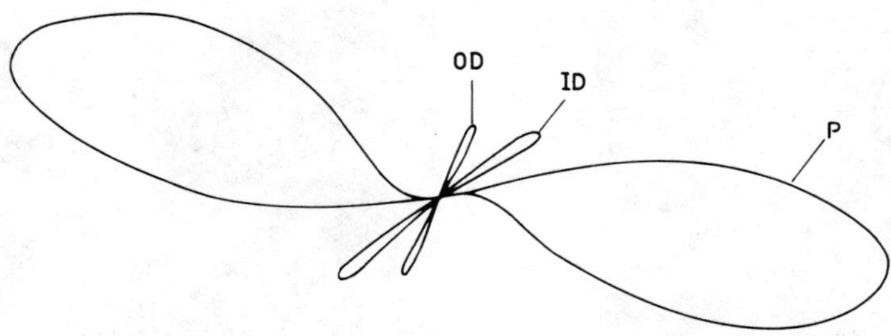

b) Frequency f2 (medium)

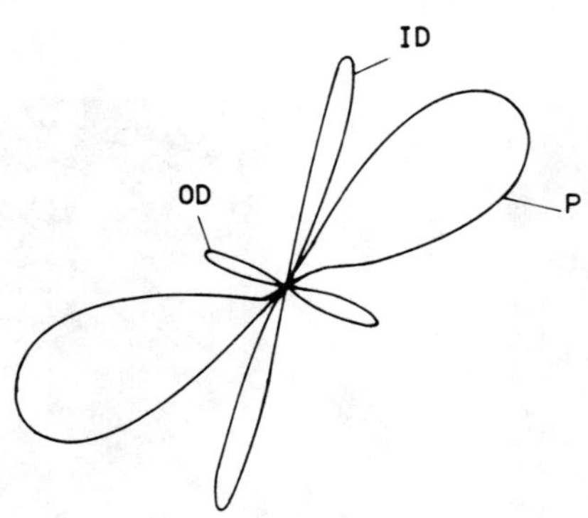

c) Frequency f3 (high)

Fig: 1. SIGNALS AT DIFFERENT FREQUENCIES FROM INTERNAL DEFECT EXTERNAL DEFECT AND SUPPORT PLATE.

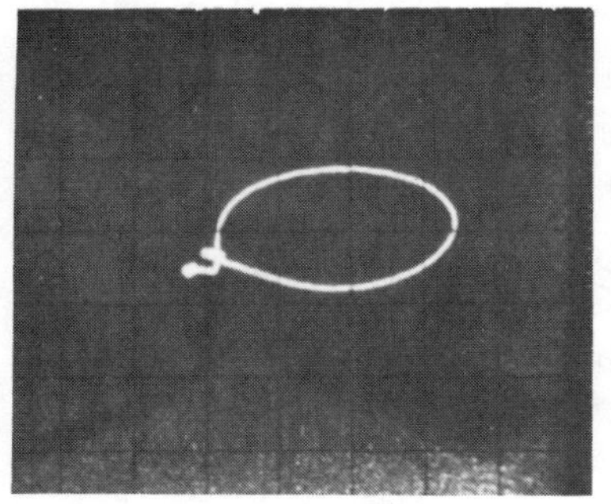

Edge of support plate

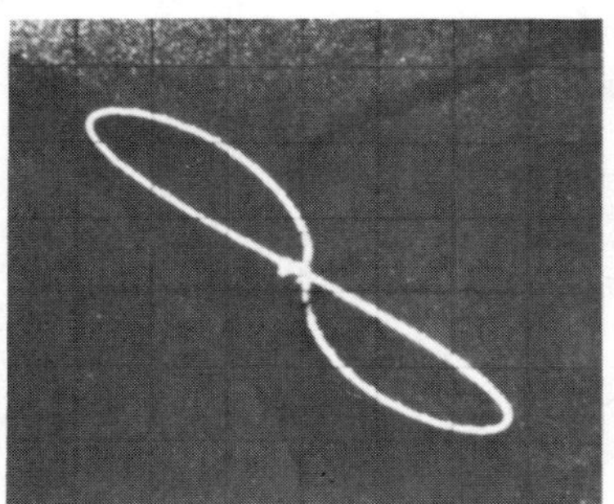

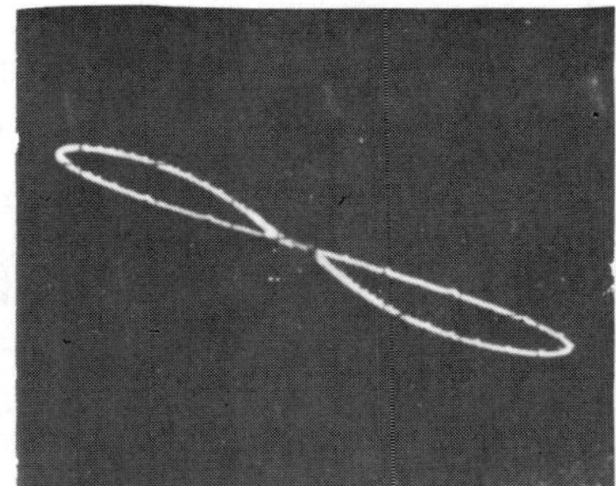

Through hole

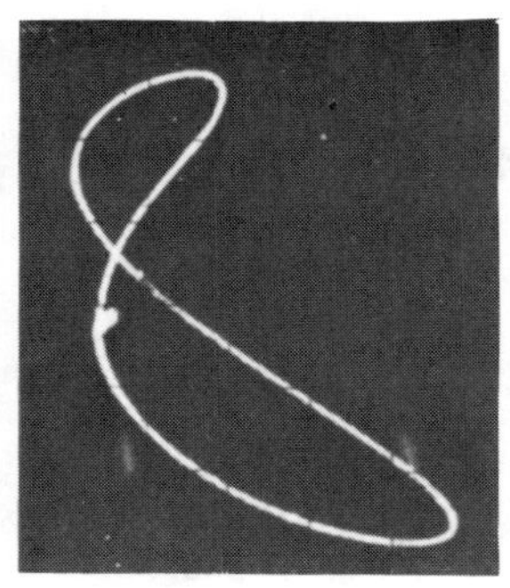

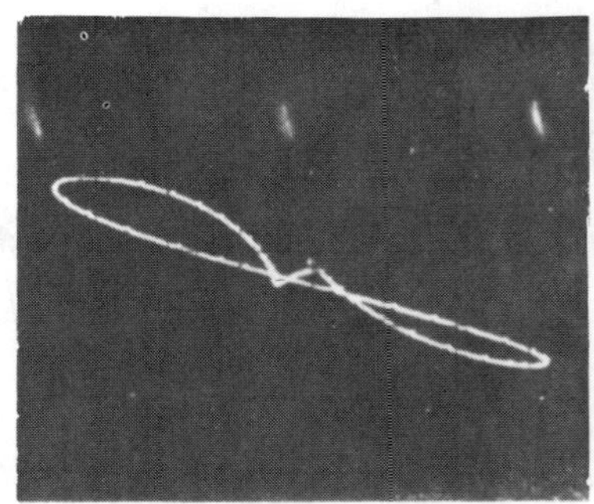

Hole at the edge of support plate

a) Before mixing b) After mixing

Fig: 2. SIGNALS BEFORE AND AFTER MIXING

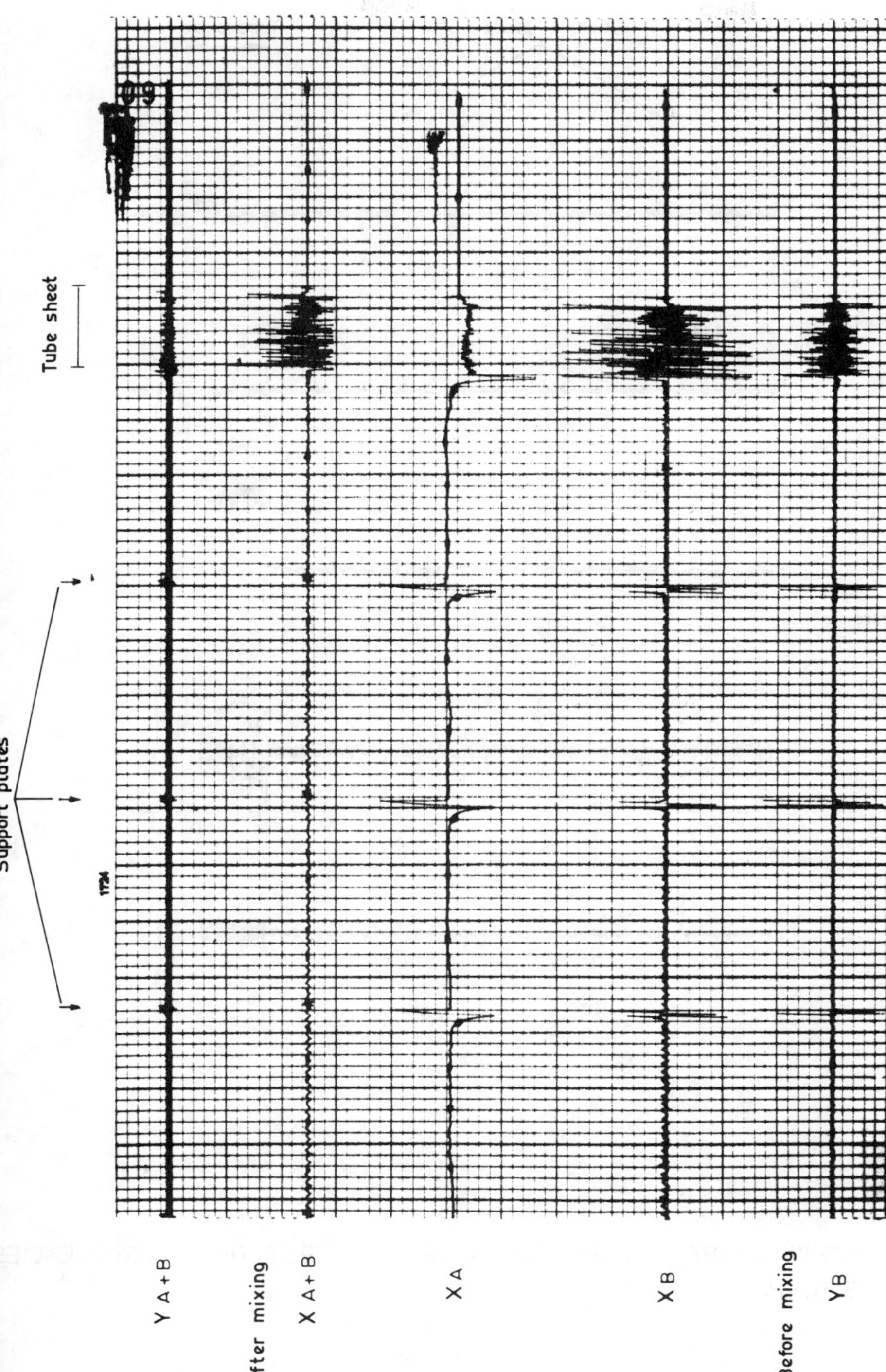

Fig:3. STRIP CHART RECORD OF A FESSENHEIM I TUBE (TWO FREQUENCIES SYSTEM).

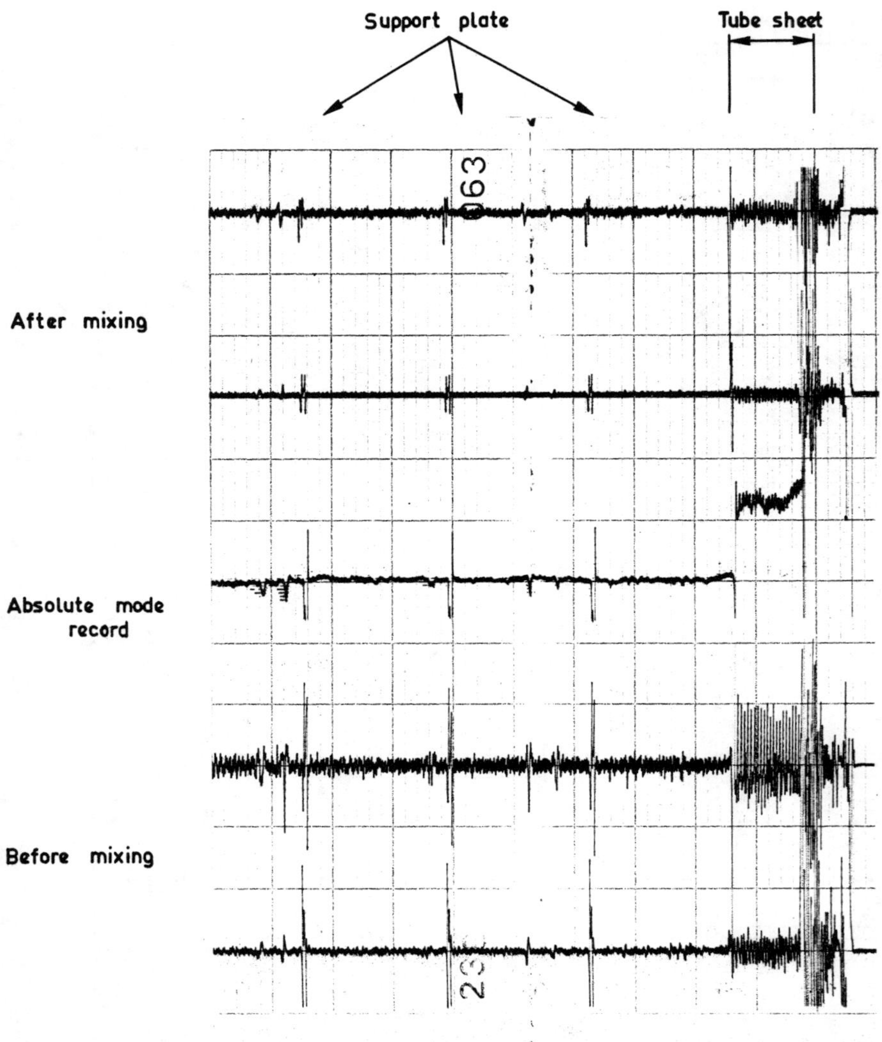

Fig: 4. STRIP CHART RECORD OF BUGEY 2 TUBE (THREE FREQUENCIES SYSTEM).

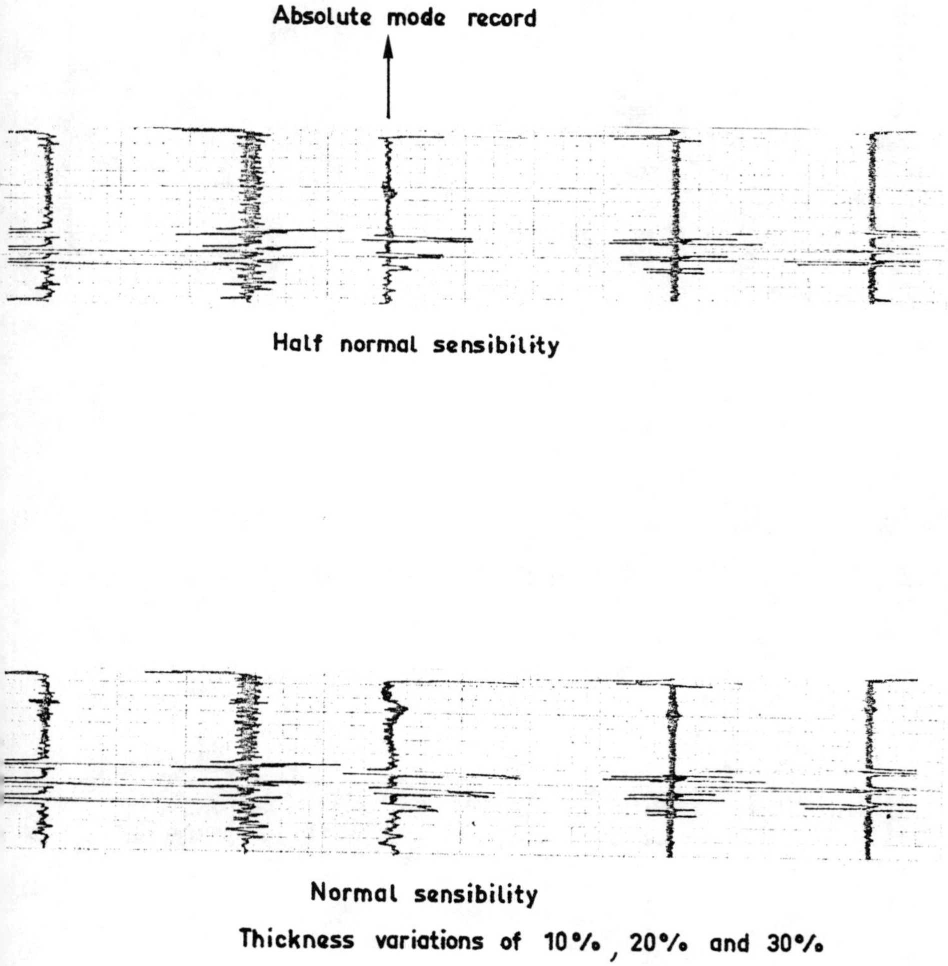

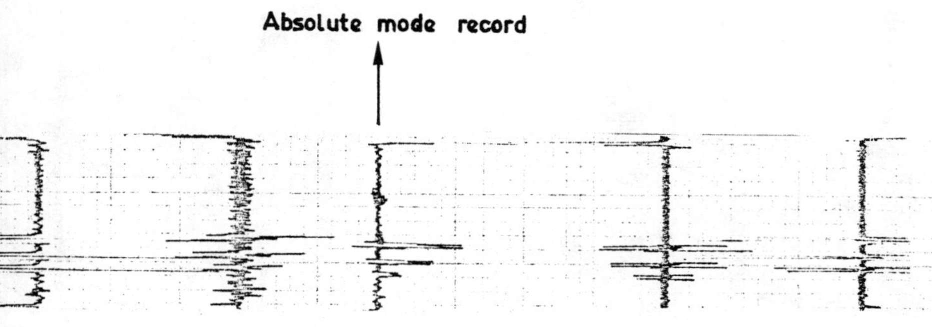

Absolute mode record
Half normal sensibility
Normal sensibility
Thickness variations of 10%, 20% and 30%

Fig: 5. STRIP CHART RECORD OF THICKNESS VARIATIONS

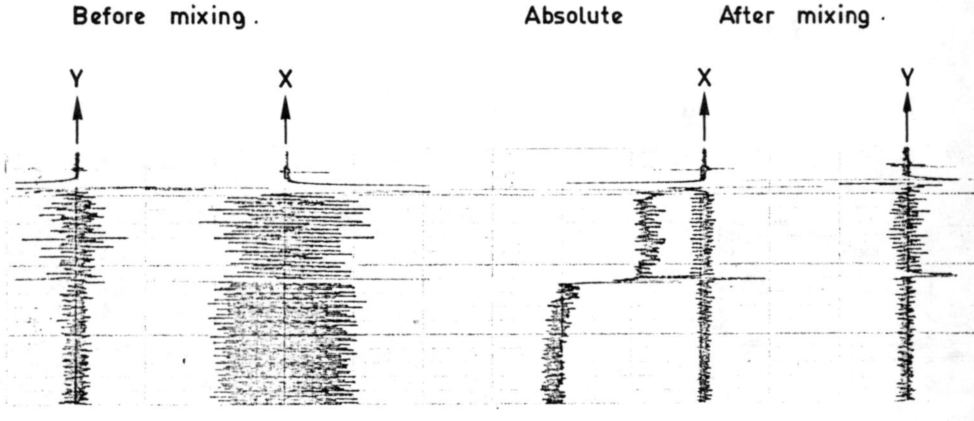

Fig: 6. ELIMINATION OF INTERNAL NOISE ON A REAL TUBE.

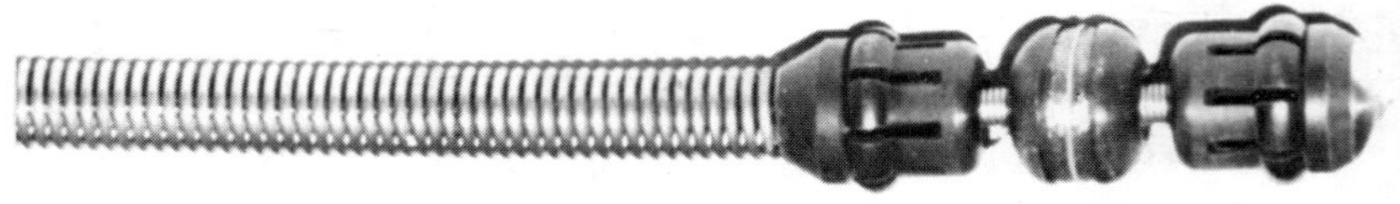

Fig: 7. DIFFERENTIAL EDDY CURRENT PROBE FOR STEAM GENERATOR TUBE EXAMINATION.

Without mixing **With mixing.**

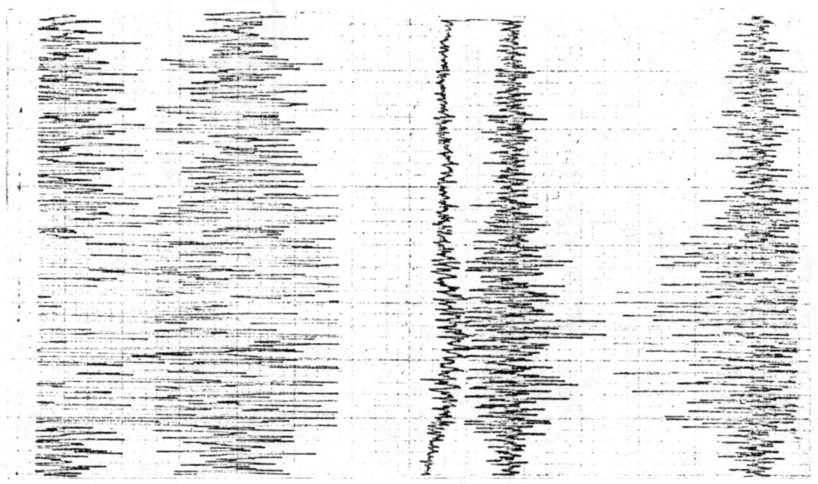

a) Before magnétisation

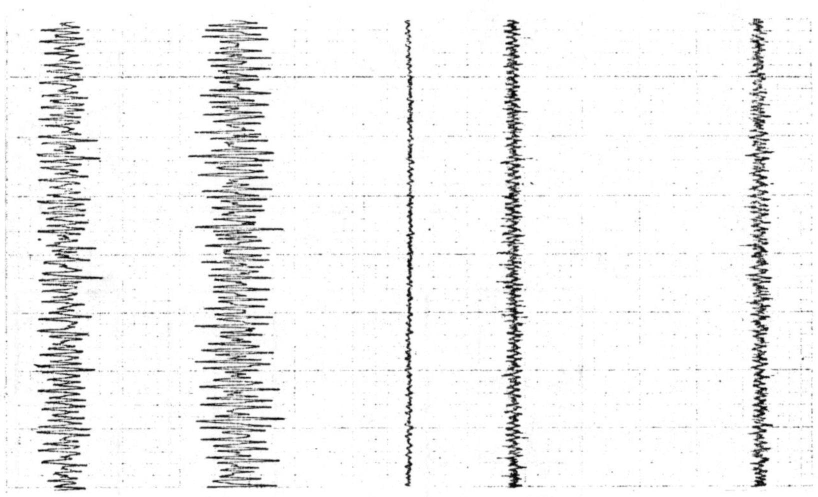

b) After magnétisation

Fig: 8. EFFECTIVENESS OF MAGNETISATION PLUS MIXING

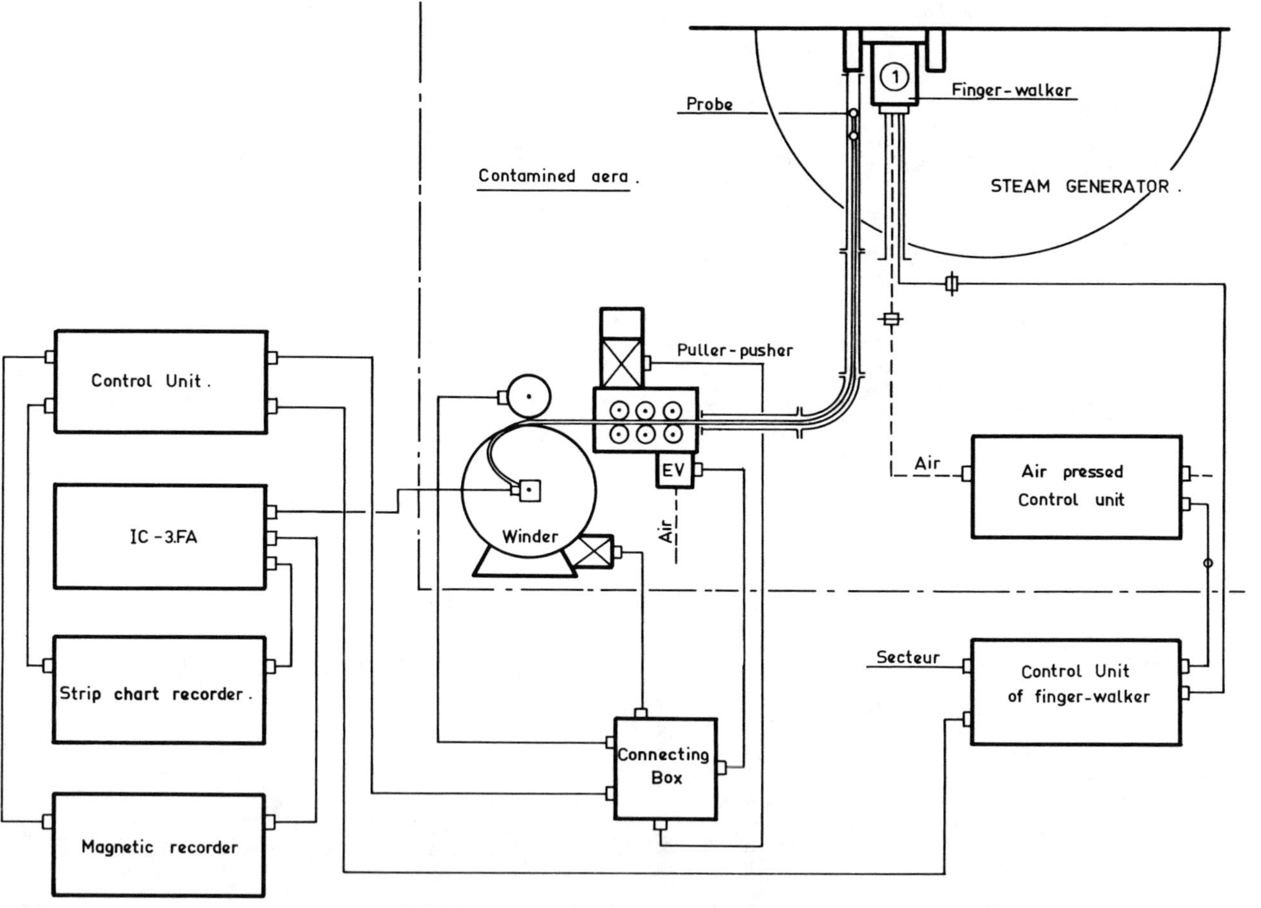

Fig: 9. SCHEMATIC OF THE COMPLETE SYSTEM.

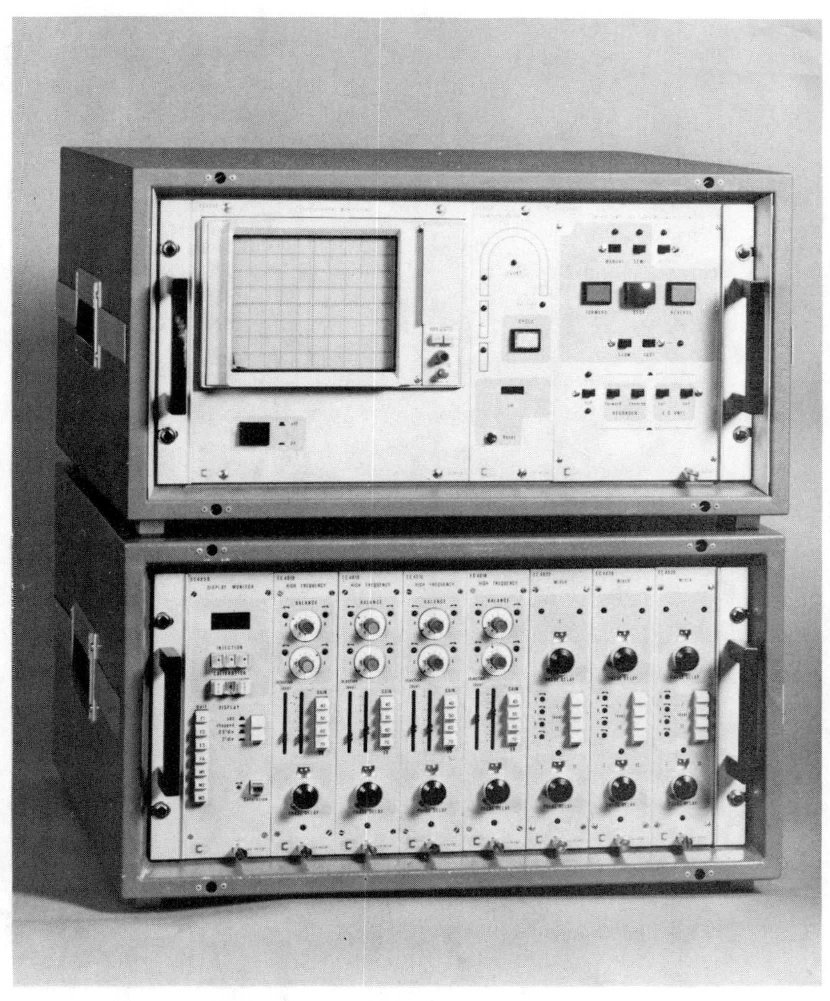

Fig : 10. IC-3FA MULTIFREQUENCY UNIT AND DISPLAY / CONTROL UNIT

Fig: 11 . EDDY CURRENT PROBE PUSHER / PULLER .

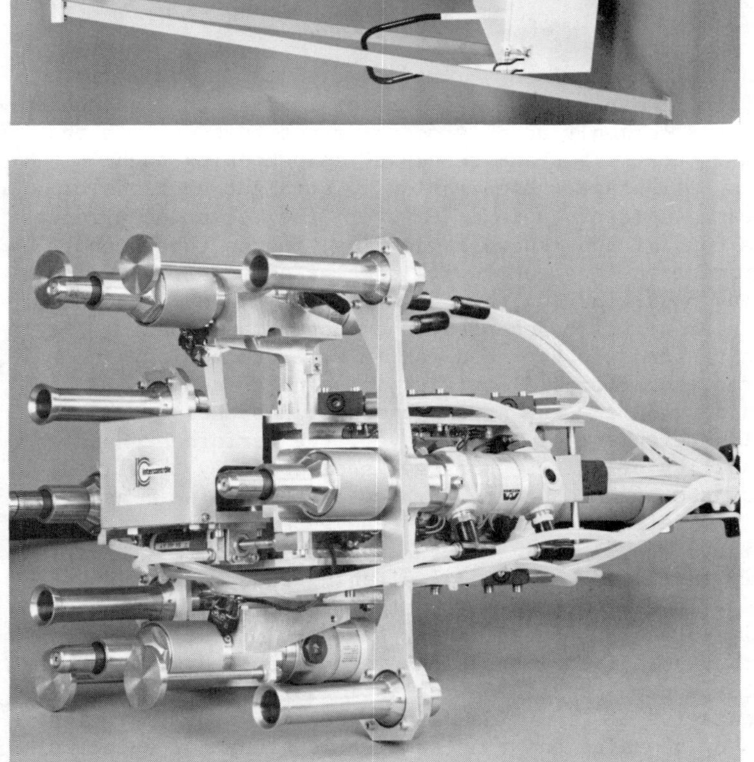

Fig: 12. FINGER-WALKER MANIPULATOR AND CONTROL UNIT.

AN EVALUATION OF PWR STEAM GENERATOR TUBING INSPECTION METHODS

Stephen D. Brown
Battelle-Columbus Laboratories
Columbus, Ohio

Eugene R. Reinhart
Electric Power Research Institute
Palo Alto, California

ABSTRACT

To satisfy regulatory requirements for the in-service inspection of nuclear steam generator tubing, the inspection method most commonly used is eddy current testing. In the past, this test has been successful in detecting such problems as wastage and corrosion in straight sections of the steam generator tubing. However, the recent occurrences of tube deformation, known as denting, in the tube support area provide the inspector with complex eddy current signals which may mask flaw indications. Denting and ovalization of tubing also restrict access by the inspection probe.

In response to the need for improved technology in this area, considerable activity is being funded in NDE systems development for steam generator inspection by several development groups; however, the performance and applicability of this technology is essentially undefined. To study the performance of the new systems as well as define a baseline for existing steam generator inspection capability, the Electric Power Research Institute (EPRI) recently initiated a technical planning study. Four systems were evaluated by a panel of in-service inspection specialists and NDE consultants. As part of the evaluation, simulated in-service inspections were conducted using a steam generator mockup containing realistic defective tubing representative of major industry inspection problems. The details of planning and conducting this study and selected results are presented.

INTRODUCTION

Steam Generator Tubing Degradation

There are forty-six utilities with 125 pressurized water reactor (PWR) units scheduled to be in operation by the end of 1986. Presently there are 37 operating PWR units in the U.S.A., of which 19 have at least one form of tube degradation. No degradation has been found in 8 others, and 10 have only been in operation for a short time. Tube degradation falls generally into two catagories and is summarized below:

Corrosion Attack	Vibration-Induced
• Wastage (thinning)	• Fretting - Antivibration bars - Support plate lands
• Pitting • Erosion-Corrosion • Caustic Cracking	
• Denting at Support Plates - Primary Side Cracking at Drilled Supports and U-Bend Apex	• High Cycle Fatigue Cracks
• Denting Indications at Tube Sheet - Primary side cracking	

Wastage is attributed to design and chemical control problems of the secondary system which result in a concentration of corrosive chemicals at semi-stagnant steam/water interfaces. Wastage tends to predominate on the hot-leg side of the steam generator at low flow areas such as; (a) the tube support region; (b) tube sheet crevice region; (c) within sludge piles just above the tube sheet; (d) isolated straight section tube regions and (e) U-bend area[1].

Pitting has been identified in at least two plants (Palisades and Oconee) and is the result of two different mechanisms. "Teardrop" pitting is the result of erosion/corrosion which occurs because of flow conditions through broached three-point contact tube support plates[2]. Palisades developed pitting after an extended period of dry-layup. Pits were observed both in areas which had exhibited previous wastage, which had occurred during normal plant operation, and in new areas in the upper half of the steam generator[3].

Caustic cracking, i.e., intergranular stress corrosion cracking (IGSCC) results from sodium or potassium hydroxide attack which can be produced and concentrated in tube crevice regions and areas of sludge buildup on the tube sheet on the secondary side of the steam generator system. Primary side SCC, is a denting side effect. Where destructive examinations have been made (Surry), longitudinal cracks initiating from the primary side in the tube support plate region as well as in the inner row U-bends have been identified[4].

Denting is a result of the formation of a non-protective layer of magnetite, Fe_3O_4, in the crevice region between the steam generator tube and tube support plate[5,6]. The magnetite formation is a result of corrosion of the tube support plate. Magnetite exhibits twice the volume as the prexisting carbon steel support plate and hence upon its formation and subsequent growth, rapidly fills the support plate/tube crevice and eventually closes the support plate hole. This latter action results in a circumferentially symmetrical inward radial strain on the steam generator tube causing a constriction of the tube with little reduction in tube wall thickness, i.e., <2 percent.

Denting in its moderate and extensive stages can lead to deformation of the tube support plates leading to the primary side cracking of the steam generator tubes in the U-bend region (Westinghouse only) and at the tube support drill holes. A majority of the denting has occurred at the tube support drill holes and in several cases is suspected or has been identified at the tube sheet area (D. C. Cook 1)[4]. Denting represents the most serious steam generator problem and has severe in-service inspection (ISI) implications.

Fretting has been observed primarily at San Onofre and is a tube mechanical wearing mechanism which results from a rubbing action of the antivibration bars against the steam generator tube on the outer U-bend areas[1]. Fretting of tubes at the support plate lands also has been observed and is a result of flow induced vibration in the once through steam generators (Oconee).

Circumferential fatigue cracks have been identified in the once through steam generators. These cracks occur in the vicinity of the tube sheet and tube support areas and may result from the propagation of a local defect by high cycle fatigue due to vibration. The circumferential extent of these cracks has been observed to extend from 45 degrees to 350 degrees[7].

Steam Generator In-Service Inspection

In-service inspection of steam generator tubing is presently conducted from the primary side using eddy current methods. The basic inspection system has evolved from technology developed during the early 1960's[8,9]. In-service inspection expertise (training of inspectors, analysis of data, etc.) has been derived primarily from the involvement of various groups with the U.S. Nuclear Navy. In the past, eddy current inspection has been successful in detecting tube degradation such as wastage and corrosion in straight sections of steam generator tubing. However, the recent occurrences of denting, the continued occurrence of defects near the tube supports and/or tube sheet (teardrop pitting, HCF cracks) and the need for inspection methods to determine the extent of slow flaw growth to the degree necessary for the assessment of remedial steam generator activities (change to AVT, etc.) have raised questions as to the capability of existing eddy current inspection methods.

Considerable R&D efforts are being funded in steam generator NDE systems development by the Electric Power Research Institute (EPRI), government agency inspection groups, nuclear system steam suppliers (NSSS), and foreign groups. Multiparameter eddy current, new single frequency probe designs, and ultrasonic systems are all in various stages of development. In light of present steam generator problems, the utilities need to sort this NDE activity into the catagories of expected near-term improvement (within six months) and mid-term improvements (within 12 months). The near term improvements should have the potential to improve inspection performance for the next series of major inservice inspections and for units that will be cleaned or where water treatment will be changed. The near-term improvements would therefore reflect system changes that are now ready for field use but require qualification. In the area of mid-term improvements, systems technology should be identified that is amenable to accelerated effort for incorporation into systems that can be used for fall 1978 in-service inspections. In addition, goals for long-term R&D activity should be defined.

To address these needs, EPRI recently initiated a technical planning study.

TECHNICAL PLANNING STUDY BACKGROUND

Program Definition

Although there was considerable information on several types of steam generator tube problems, detailed information on the denting problem was lacking. For a better definition, a steam generator NDE specialists meeting was held on February 24, 1977, at the offices of EPRI. As a result of this meeting a better idea of the nature of the denting problem was obtained, along with considerable information to aid in planning a steam generator inspection performance evaluation study.

From the results of the NDE specialists meeting, a literature survey, and additional communications, an EPRI Technical Planning Study (TPS-77-709) was selected as the vehicle for conducting further effort in this area. Technical planning studies are conducted by EPRI to support research and development planning for the engineering and economic feasibility of proposed technological development and/or hardware options. Such studies permit identification of the most promising options and the major technological issues which must be resolved before the initiation of a comprehensive research program. The technical planning study approach was also selected since this represents one of the most expedient EPRI methods for responding to studying near-term utility problems. Major objectives of this study were defined as:

(a) Establish a baseline performance of present NDE systems (including the operators) in response to a variety of defect types. This baseline would establish the nature and extent of future R&D activities.

(b) The performance of several new inspection methods, techniques, and equipment, should be evaluated to determine their potential for solving present steam generator NDE problems. Both field prototype as well as laboratory methods should be evaluated.

(c) The study should be initiated and completed as soon as possible in order to transmit the information to the EPRI Steam Generator Project Office and other interested EPRI Nuclear Departments for use in planning comprehensive R&D programs.

Since the nature of the steam generator inspection problem was recognized as being very complex, and since EPRI needed to rapidly obtain as much comprehensive information as possible, a technical round-robin program, aided by NDE specialists was selected as the basis for the study. It was felt that the data from simulated in-service steam generator inspections, when combined with the analysis and observations of an expert review panel, would provide considerable insight into the various parameters affecting inspection system performance.

Under the direction of an EPRI Project Manager, a six-man NDE technology evaluation panel was used to assess performance of the various NDE systems. The panel was composed of one NDE inspection specialist from the following NSSS vendors; (1) Babcock & Wilcox, (2) Combustion Engineering, and (3) Westinghouse. The remaining members were selected from the following independent groups; (1) Battelle Columbus, (2) EPRI and, (3) Southwest Research Institute. The above panel was formed to lend perspective and objectivity to the project results. The above groups also supplied examples of simulated in-service defective tubing and aided in the preparation of a realistic test program.

Each steam generator nondestructive testing system was evaluated by the panel in the following manner:

(1) General Impressions - Prior to laboratory tests, details of the system were described by the system supplier.

(2) Scan of known Defects - The panel was allowed to review the system in operation and review data analysis methods.

(3) Scan of Unknown Defects - Data was taken using a mockup containing a series of simulated defective tubing.

(4) Summary of Results - Based on the results of 1, 2, and 3 above, each panel member submitted his conclusions to EPRI regarding the performance of the NDE system under evaluation.

Mockup Configuration

A mockup containing examples of defective tubing was essential for conducting the study. An air transportable mockup was designed since several of the systems that were to be evaluated were in the laboratory or prototype stage of development and transport of these systems from the laboratory was not feasible. Transporting the mockup to the various NDE development laboratories was also optimal from a scheduling and economics standpoint.

Essential features of the mockup are shown in Figure 1. The mockup proved easy to transport, assemble quickly at the test site, and provided a large number of steam generator tube configurations, including tube supports and U-bends.

Figure 1 illustrates the rear of the mockup showing a length of straight section steam generator tubing topped with a U-bend. Lengths of tubing containing simulated defects as well as good sections could be intermixed in any prescribed order. The particular NDE system and the system operators being evaluated were in the front of the mockup and had no prior knowledge of the tube mix. Tube configurations could be readily changed to sequence through the entire set of simulated defects.

Although the study addressed NDE problems associated with all existing NSSS steam generator designs, one tubing size for all the test specimens was selected to simplify the logistics of the program. Tube samples were all 7/8-inch OD (0.050" wall) Inconel 600 steam generator tubing typical of the Westinghouse Series 51 PWR steam generators. The tubing samples were either supplied by members of the NDE evaluation panel from existing test samples or fabricated specifically for the EPRI study.

Simulated defects included in the overall NDE systems evaluational test set consisted of (a) wastage, (b) EDM notches at various orientations relative to the longitudinal axis of the tube, primary or secondary side, (c) dents 5 mils and 100 mils diametral containing primary or secondary side notches transverse or parallel to the tube axis, (d) ovalized dents - various degress of ovalization, (e) pitting, (f) intergranular stress corrosion cracking - laboratory induced, and (g) U-bends containing primary side notches at the bend tangent and apex region. Drilled carbon steel 3/4-inch plate was used to simulate the tube support area. The supports could be placed near, at the edge, and directly over the flaws. Placement was based on field experience with actual flaws.

Systems Evaluated

The following four steam generator NDE systems were evaluated:

(1) Zetec - the basic Zetec single frequency system represents present industry state-of-the-art equipment and techniques. The system can be operated over a range of test frequencies in conjunction with probes designed for specific steam generator test conditions. Frequencies used during the round robin were 400, 225, and 100 kHz. A prototype system

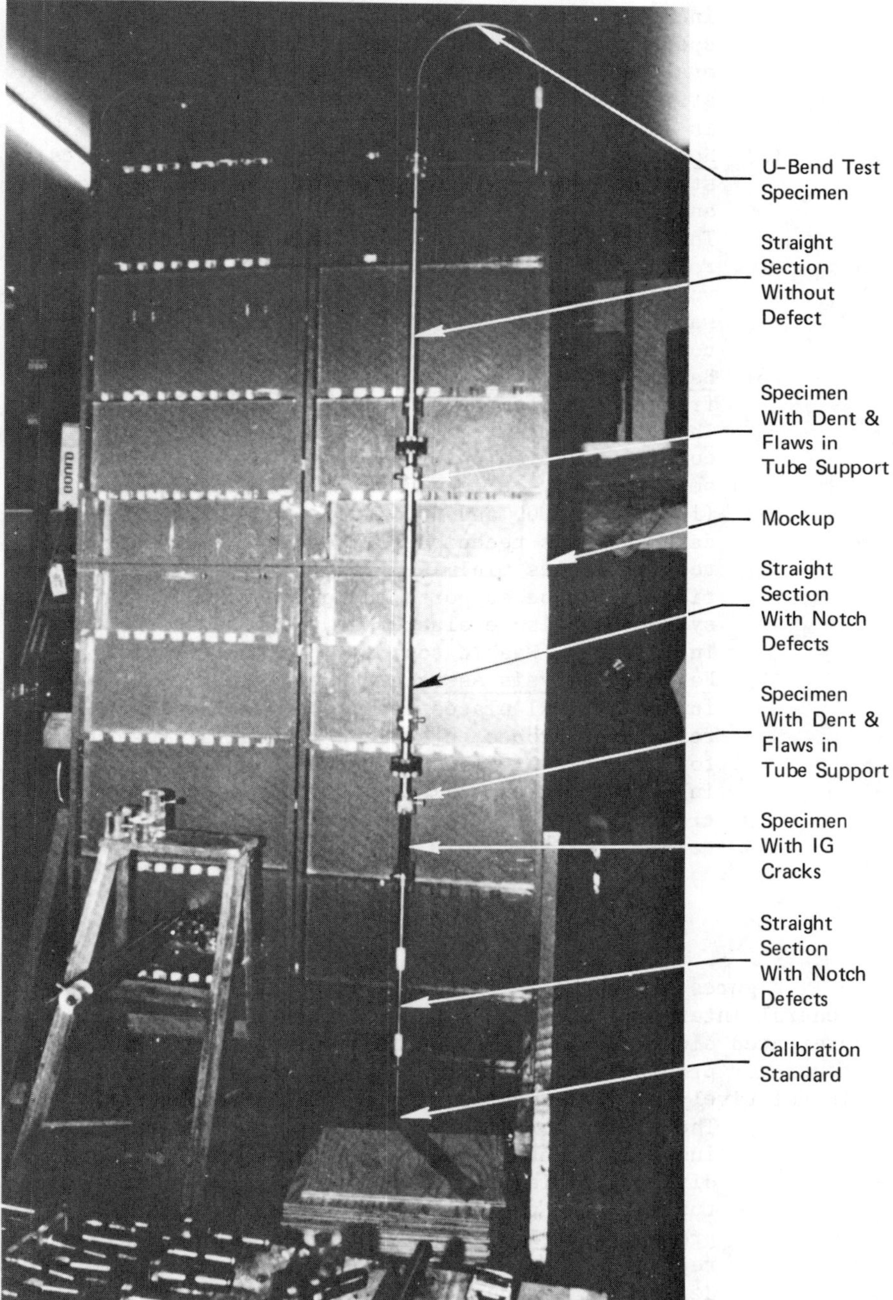

Fig. 1. Air Transportable Mockup

incorporating a rotating pancake coil designed specifically for the inspection of dented and ovalized tubes was also evaluated. Tests of these systems were performed at the Zetec laboratories in Issaquah, Washington.

(2) Holosonics/Intercontrole - this system represents state-of-the-art French field inspection technology and uses a three-frequency eddy current approach. The extraneous variables minimized for the round robin evaluation were tube supports and tubing ID variations. Data analysis is manual. This system was evaluated at the offices of Holosonics/Intercontrole, Richland, Washington.

(3) Battelle-Northwest - this is field prototype multifrequency system being developed by EPRI funding. The system utilized a modified Zetec probe coil combined with an instrumentation system that acquires eddy current data at four frequencies (100, 200, 300, and 400 kHz). Automated and manual data analysis techniques were used. The extraneous test variables minimized for the round robin evaluation were tube supports and probe wobble. This system was also evaluated at the Zetec laboratory in Issaquah, Washington.

(4) Failure Analysis Associates/Reluxtrol - System funded by EPRI places heavy emphasis on controlled reluxtance probes to achieve high axial resolution for analysis of flaws near tube supports and flaws in dented regions and uses single frequency electronics. This system was evaluated at EPRI in Palo Alto, California

RESULTS AND DISCUSSION

Figures presented in this section summarize results of general interest. Not all of the round robin results are presented either because they address very specific steam generator problem areas or because the data base which exists is relatively small. Questions which are addressed include:

(a) The ability of single frequency and multifrequency inspection methods to detect large and small volume discontinuities in isolated straight sections of tubing. The performance parameter is detection probability (P_d) which is defined as the ratio of reported defects to the total number of scanned defects. In general, defect depth was the variable of interest.

(b) The ability of single frequency and multifrequency inspection methods to detect and characterize discontinuities in the proximity of tube supports.

(c) An assessment of single frequency and multifrequency eddy current methods in estimating the depth of detected discontinuities in percent of tube wall. The performance criteria include mean depth measurement accuracy and precision. Mean error is important because it describes the systematic error or bias in measurement of depth and must be considered for the establishment of tube plugging criteria. Precision or scatter in estimating depths is of interest since it is a measure of how reliable changes in flaw growth can be estimated.

(d) The ability of single and multifrequency inspection methods to detect and characterize discontinuities in small dents (5 mils diametral) and an assessment of a single frequency system using a prototype rotating probe in inspecting large dents (100 mils diametral).

Figure 2 illustrates the detection probability of single frequency eddy current inspection methods for large volume wastage defects and small volume axial EDM notches. Notches at depths of 20%, 30%, 80%, and 100% of tube wall were 0.125 inches in length. Notches at 40% and 60% depths varied from 0.125 inch to 0.5 inch in length.

As Figure 2 indicates, not all notches at the 40% depth were detected using the 400 kHz, 86% fill factor probe, ASME Code Section XI setup. The undetected notches were all greater than 0.125 inches (0.160 - 0.5 inch) in length suggesting that a variable other than notch volume is affecting detectability. The low detection probability for axial notches at a depth of 40% and the successful detection of wastage defects to depths of 20% suggests that the minimum eddy current instrumentation gain settings specified using the ASME Section XI procedure are based on the detection of larger volume defects, i.e., wastage.

Small volume detection probability can be improved at 400 kHz by using a larger fill factor probe, i.e., 94%, and operating the eddy current instrumentation at maximum sensitivity. The resultant detection probability for axial notches is shown in Figure 2.

Multifrequency system detection probability results for large volume wastage and small volume axial notches are shown in Figure 3. As Figure 3 indicates, both multifrequency systems (3f and 2f) give identical results in detecting wastage to depths of 10% and axial notches to depths of 30%. In further analysis of the Holosonics/Intercontrole strip chart data, the single frequency 240 kHz uncombined channel gives identical

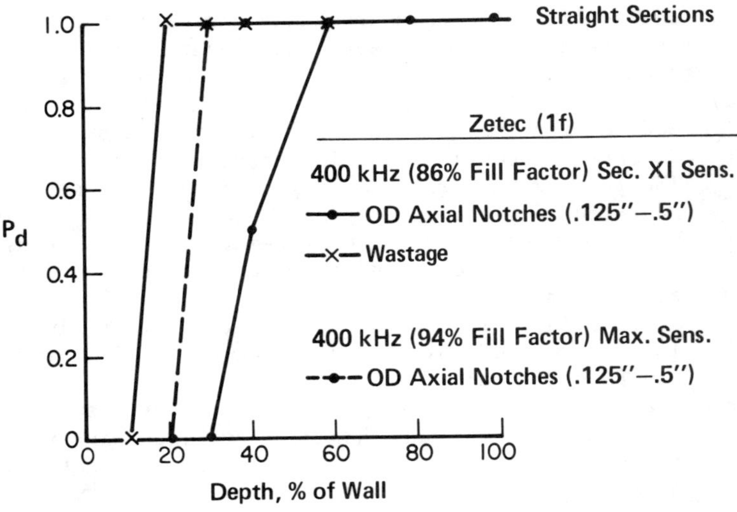

Fig. 2. Single Frequency Detection Probability-Large and Small Volume Defects.

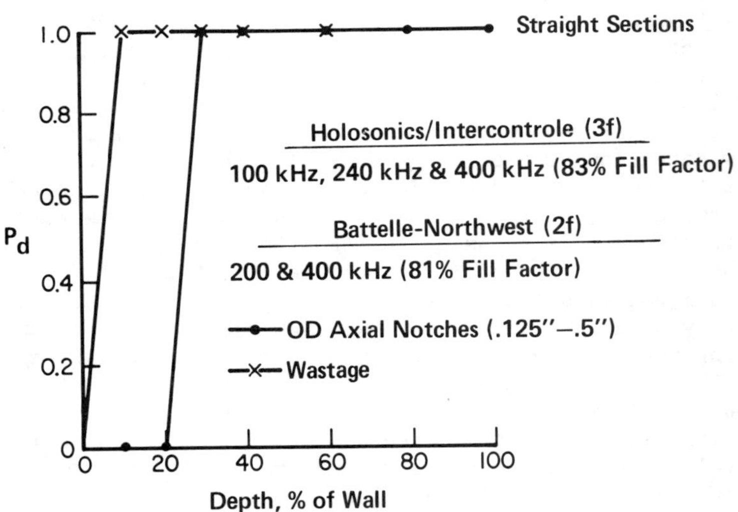

Fig. 3. Multifrequency Detection Probability-Large and Small Volume Defects.

detection probability results as the multiparameter or three-frequency composite channel (3f) for the discontinuities considered in the Figure 3 data set. Thus, although the multi-frequency detection probability results in straight sections of tubing were better than the original single frequency 400 kHz, 86% probe fill factor results (see Figure 2) it was suspected that the difference could be attributed to the low frequency, i.e., 240 kHz or 200 kHz, in the multifrequency composite channel output.

Figure 4 shows a comparison of single frequency detection probability as measured using EDM axial notches at depths of 20%, 40%, and 60% of tube wall. At a given depth, notch lengths were 0.020", 0.040", 0.080", 0.160" and 0.320". Notice that the 240 kHz data were taken using a much smaller fill factor probe (83%) than the 400 kHz data (94% fill factor). The 400 kHz data was also taken at the maximum instrumentation sensitivity. The lower frequency 240 kHz detection probability results are better suggesting that for inspection situations in which small volume discontinuities are expected, reliance on the single frequency 400 kHz only may not be desirable.

The ability of single and multifrequency eddy current systems to detect and characterize flaws in the vicinity of tube supports, which represents an extraneous test variable, was assessed using the defect geometries shown in the upper part of Figure 5.

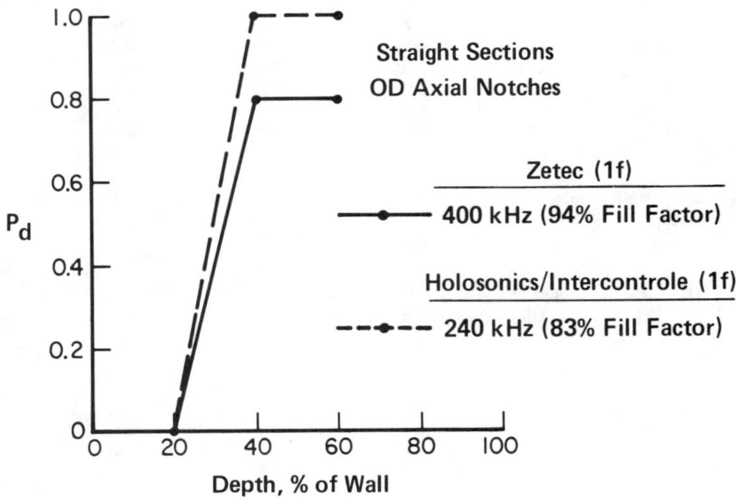

Fig. 4. Small Volume Notch Detection Probability - Low and High Frequency.

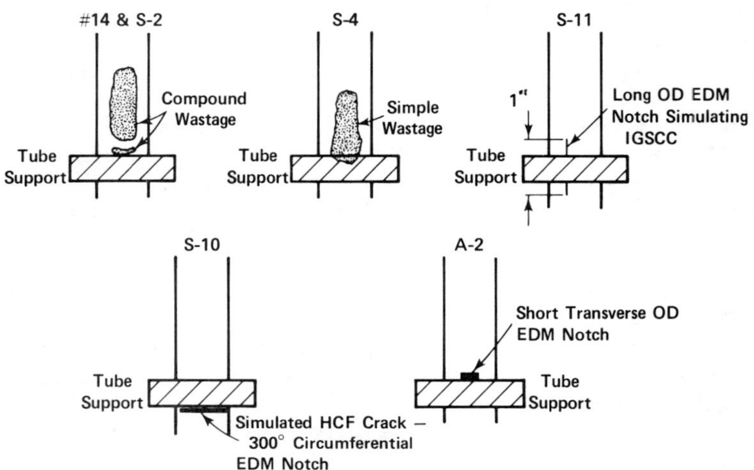

Test Specimen	Defect Depth, % of Tube Wall	Eddy Current Estimated Depth, % of Tube Wall			
		Zetec (1f)	Holosonics/ Intercontrole (3f)	Battelle-Northwest (2f)	Reluxtrol (1f)
#14	20%	34%	60%	20%	No Data
S-2	60%	42%	65%	56%	53%
S-4	60%	82%	70%	20%	45%
S-11	60%	ND	D	ND	30%
S-10	60%	ND	50%	ND	ND
A-2	60%	ND	ND	ND	50%

D – Detected, no depth reported
ND – Not detected

Fig. 5. Summary of Tube Support Defect Detectability and Measurement Data.

Specimen 14 was 20% in maximum depth whereas all other specimens were 60% in depth. The eddy current estimated depths for four systems are shown in the vertical columns of Figure 5. The conventional single frequency system (Zetec 1f) detects the larger volume wastage defects but misses the smaller volume discontinuities. The Holosonics/Intercontrole multifrequency system detected all of the defects with the exception of the short transverse EDM notch. The Battelle-Northwest multi-frequency system results were the same as the conventional single frequency results. The Reluxtrol single frequency results on a percentage basis are equivalent to the Holosonics/ Intercontrole multifrequency results. This is attributed to the high axial resolution (0.050") of the probe with its ability to discriminate defects close to the support region, and the crosswound coil configuration which tends to subtract circular symmetric extraneous signals.

The ability of eddy current inspection methods to accurately and precisely estimate defect depth is of interest as it is relatable to the establishment of plugging criteria and may provide information pertaining to slow flaw growth studies for the monitoring of steam generator chemistry changes. Investigators at Battelle-Columbus have shown that existing single frequency eddy current methods have a tendency to overestimate wastage or large volume defects and under-estimate small volume defects[10]. This has been confirmed by indirect measurements for actual wastage defects in steam generator tubing and destructive measurements of IGSCC from in-service steam generator tubing[11,12].

Figure 6 presents plots of eddy current estimated depth versus actual defect depth for simulated wastage defects and EDM axial notches. Data for three of the systems evaluated is shown. As the figure indicates, there is a tendency for the wastage defects to be overestimated providing a conservative estimate of depth whereas the notches are underestimated. The overestimation of wastage and underestimation of notches is a volume dependent phenomena and is related to the choice of calibration standard used to establish an eddy current phase angle versus depth transfer function.

The eddy current estimated depths for wastage in Figure 6(a) and (c) tend to show less bias about the line relating estimated depth to true depth. The wastage data shown in Figure 6(b) exhibits a larger overestimation of true depth. Both Zetec and Battelle-Northwest used an ASME Section XI eddy current standard for the establishment of the phase angle versus depth calibration curve. The Holosonics/Intercontrole group relied on the use of a smaller volume standard in establishing their calibration curve which would account for the tendency to more severly overestimate wastage depths.

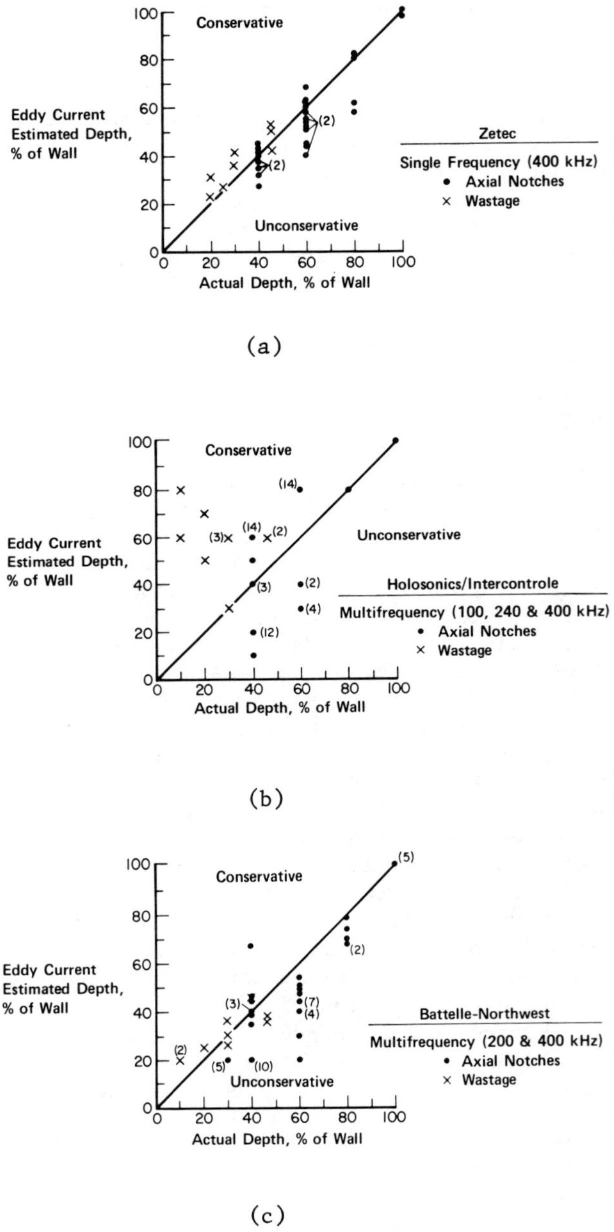

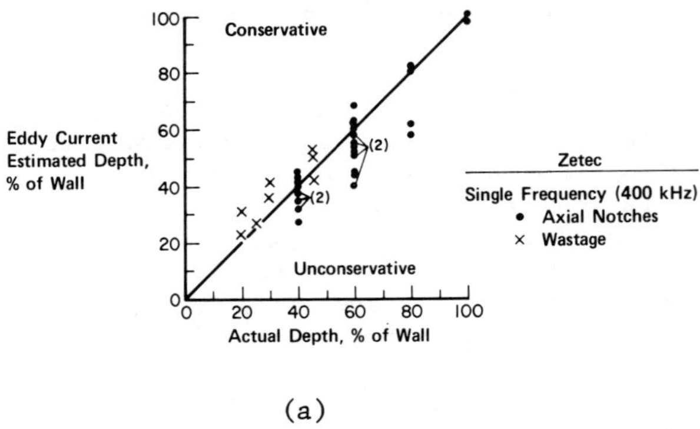

(a)

Conservative

Unconservative

Eddy Current
Estimated Depth,
% of Wall

Actual Depth, % of Wall

(14) (14) (3) (2) (3) (2) (4) (12)

Holosonics/Intercontrole

Multifrequency (100, 240 & 400 kHz)

• Axial Notches

× Wastage

(b)

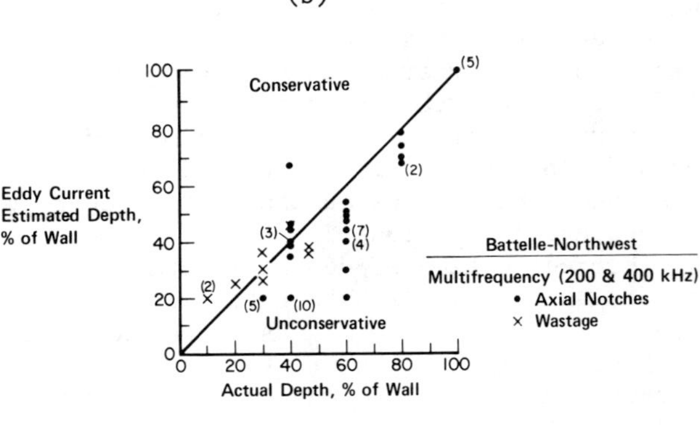

(c)

Fig. 6. Eddy Current Estimated Depth versus True Depth- Wastage and Axial Notches.

Examining the eddy current estimated depths for the notches at the 40% and 60% actual depths for three of the systems in more detail results in distributions shown in Figure 7. In addition to the 400 kHz single frequency data, single frequency 225 kHz data is also considered.

For the 40% notches, the single frequency data exhibits the least scatter (smallest range). The 400 kHz single frequency data shows the smallest mean error (-1%) with the single frequency 225 kHz and the Battelle-Northwest data showing a comparable tendency to underestimate the notch on the average of 8%.

For the 60% notches, the single frequency 225 kHz data and the Battelle-Northwest data have comparable means. The fact that this is true at 40% also suggests that the combined data channel (200 kHz and 400 kHz) of the Battelle-Northwest multifrequency system is dominated by the lower frequency. Both the single frequency Zetec system and the Battelle-Northwest system tend to underestimate notch depth. The single frequency 400 kHz data exhibits the least mean error with the most uniform distribution. It is interesting to note that the scatter in the 400 kHz data is comparable to the 225 kHz scatter.

Figure 8 illustrates repeat data for the same OD axial notch 60% deep. The data forms a subset of data included in Figure 7. Again the single frequency 400 kHz results show the least mean error, but the scatter or range is twice that of the 225 kHz data.

If a comparison of the mean error and range is made for the 400 kHz repeat data for the single 60% notch of Figure 8 and the 60% notch data of Figure 7 from which the data of Figure 8 has been excluded, the results below are arrived at:

	Mean Eddy Current Indicated Depth	Range
(a) Single 60% Notch	58%	51-68% (17%)
(b) All Other Notches	48%	40-58% (18%)

The means of the single notch repeat data and the data of (b) above differ by 10% but the range differances are comparable (17-18%). The range of the 60% notch data of Figure 7(b), which is a composite of the data (a) and (b) above is approximately twice that of the single notch data whereas the mean is 7% less. Thus the best estimate available for the accuracy and precision for eddy current depth measurements is an under-estimation on the average of 2% with a one-hundred percent confidence interval of +10%, -7% about the mean (58%). The volume of the single 60% notch, i.e., (a) above, was greater than the notch volumes of (b) above. This would account for the disparity in mean estimated depth.

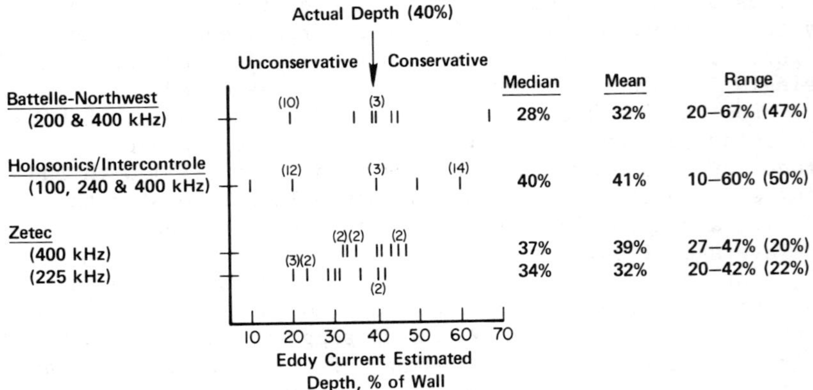

DISTRIBUTION OF EDDY CURRENT ESTIMATED DEPTHS FOR 40% AXIAL EDM NOTCHES (OD)

(a)

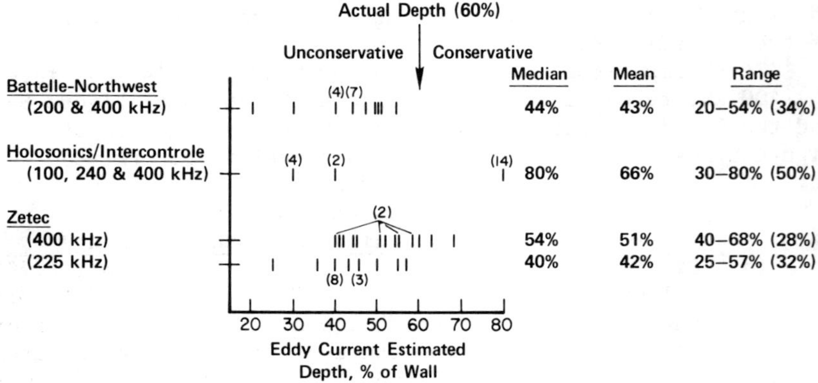

DISTRIBUTION OF EDDY CURRENT ESTIMATED DEPTHS FOR 60% AXIAL EDM NOTCHES (OD)

(b)

Fig. 7. Eddy Current Depth Distribution Data.

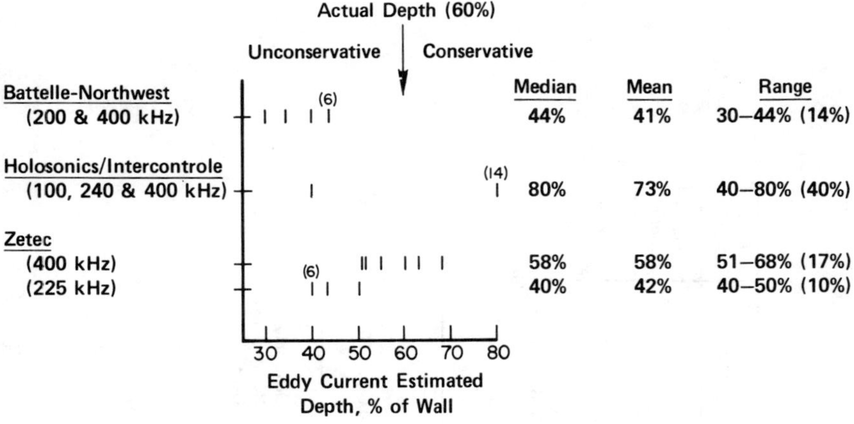

Fig. 8. Distribution of Eddy Current Estimated Depth Repeatability Data for a Particular 60% OD Axial EDM Notch.

The inspection of dented regions using eddy current methods represents a complex inspection situation[13]. Of the systems evaluated, only Zetec had any experience in inspecting dented regions as a result of their extensive field activities. The two multifrequency systems, i.e., Holosonics/Intercontrole and Battelle-Northwest, although generally familiar with the dent, did not make use of multi-parameter methods to minimize the dent signal. Small dents, i.e., 5 mils diametral, containing axial EDM notches were scanned by each of the systems evaluated. Zetec, with their rotating probe, was the only group able to inspect large, i.e., 100 mils diametral, dents.

Results for small dents are shown in Table 1. The actual calls made by each of the respective groups are summarized and an explanation of the coding scheme used is shown on the lower right of the table. There were three major test variables; (1) primary or secondary side notches; (2) notches in the dent center or shoulder region; and (3) notch depth. To see data trends it is best to compare the single frequency bobbin coil with the rotating coil[(a) of Table 1], the bobbin coil single frequency and multi-frequency [(b) of Table 1], the rotating probe coil with the multifrequency bobbin coil [(c) in Table 1], and the rotating coil with the high-resolution coil [(d) in Table 1].

System	40%			60%			100%		Flaw Depth
	OD	ID		OD	ID				Flaw Location
Single Frequency	S	C	S		S	C	S	C	
– Differential Bobbin Coil	D&F	ND	(DSP,D, DOF)	No Data	–	D&F	DSP	DSP	(a) (b)
– Rotating Coil	20%	58% ID	20% ID	No Data	80% ID	50% ID	75% ID	70% ID	(a) (c) (d)
– High Resolution Coil	ND	30% ID	20% ID	No Data	80% ID	40% ID	100%	100%	(d)
Multifrequency									
– Differential Bobbin Coil	DO	85% OD	74%	No Data	D	50%	100%	100%	(b) (c)

S – Shoulder region of dent
C – Center of dent

D&F – Dent & flaw
DSP – Distorted support plate
DOF – Dent or flaw

DO – Dent only
ND – Not detected
D – Dent

Table 1. Summary of Small Dent, Eddy Current Flaw Assessment Data.

The single frequency rotating coil gives more reliable results in detecting and estimating depths of discontinuities than the conventional bobbin coil. The Holosonics/ Intercontrole multifrequency system bobbin coil results are in general more reliable than the single frequency bobbin coil results. Although the dent signal was not minimized using multiparameter methods, large tube ID variations were minimized. Thus a portion of the dent signal was reduced enhancing the multifrequency results, as compared to the single frequency, for the same type coil. The Failure Analysis Associates/Reluxtrol results are comparable to both the single frequency rotating probe and Holosonics/Intercontrole multifrequency results. This is attributed to the "crosswound" coil used which essentially subtracts out the dent signal from dent plus defect signal.

Results for the Zetec rotating coil for the large dents containing axial notches are shown in Table 2. The results are in general positive with approximately 85% of the notches detected. Where notch depth information was provided, the measurement error was on the order of 20% conservative.

Flaw Depth	40%				60%			100%	
Flaw Location	OD		ID		OD		ID		
	S	C	S	C	S	C		S	C
	50% ID	D	No Data	60% ID	D	ND	No Data	100%	100%

D – Detected, no depth reported
ND – Not detected

Table 2. Summary of Large Dent, Eddy Current Flaw Assessment Data - Rotating Probe.

CONCLUSIONS

In isolated sections of steam generator tubing free of extraneous test variables the detection capability of both single and multifrequency methods can be equivalent. The use of low inspection frequencies, i.e., 200-225 kHz for 0.050" wall tube, is the significant factor.

Multifrequency methods have not demonstrated an advantage in defect depth measurement accuracy or precision. Measurement precision in the more important and the lack of precision in multifrequency methods may result from a compression of the phase angle versus depth calibration curve. The conventional single frequency 400 kHz curve has a phase spread of approximately 140° between a 100% and 20% through wall OD discontinuity. For the multifrequency systems considered in this evaluation, the phase spread varied between 55° - 90° for comparable ranges in defect depth. Thus, small variations in assigning the appropriate phase angle to the eddy current Lissajous pattern can give rise to large variations in estimated defect depth.

Single frequency eddy current inspection methods provide conservative estimates of larger volume wastage defects and unconservative estimates of small volume crack-like defects when as ASME Section XI standard is used for the establishment of phase angle versus depth calibration curve. This would suggest the existance of two different plugging criteria and also imply that the eddy current data analysist can discriminate between wastage and crack-like defects.

Multifrequency methods demonstrate a better ability to detect defects in the tube support region than conventional single frequency inservice inspection techniques. However, on an overall percentage basis, the single frequency high resolution coil demonstrates equivalent results.

For small dents, existing single frequency bobbin coil inspection methods are inadequate for small volume defect detection. The single frequency high axial resolution coil, the rotating coil, and multifrequency methods offer significant advantages. For large or ovalized dents, the rotating probe is the only demonstrated inspection method.

ACKNOWLEDGMENTS

This paper is the result of contributions from all members of the EPRI TPS77-709 steam generator NDE evaluation panel and their assistance is acknowledged. The loan of steam generator tubing specimens from an on-going Battelle-Columbus/Brookhaven National Laboratory research program through the courtesy of Dr. John Weeks is also greatly appreciated.

REFERENCES

[1] M. G. Hare, "Steam Generator Tube Failures: World Experience in Water Cooled Nuclear Power Reactors in 1974", Nuclear Safety, Vol. 17, No. 2, March-April, 1976, pp. 231-242.

[2] J. R. Weeks, Brookhaven National Laboratory, Private Communications.

[3] J. R. Weeks, "Corrosion of Steam Generator Tubing in Operating Pressurized Water Reactors", Corrosion Problems in Energy Conversion and Generation, C. S. Tedmor Jr., Ed. The Electrochemical Society, Princeton, N.J. (1974) pp. 322-345.

[4] D. Van Rooyen, "PWR Steam Generator Tubing: Corrosion Problems", International Symposium on Application of Reliability Technology to Nuclear Power Plants, IAEA-SM-218/26, Vienna, Austria, October 10-14, 1977.

[5] E. P. Morgan, et. al., "Examination of Denting and Characterization of Associated Materials in the Plate-Tube Intersections of Westinghouse Nuclear Steam Generators", Westinghouse Scientific Paper 76-702-S6EXM-P1. September 27, 1976.

[6] Steam Generator Update 1976. Westinghouse Nuclear Energy Systems Publication.

[7] Duke Power Company, Operating Occurrence, Report No. RO-269/77-16.

[8] L. J. Chockie, "Nondestructive Testing of Nuclear Steam Generators After Installation", Report No. HW-SA-3245, Hanford Atomic Products Operation, Richland, Washington, October 17, 1963.

[9] H. L. Libby, "An Improved Eddy Current Tubing Test", Report NO. HW-81780, GE/AEC Contract No.-AT(45-1)-1350, Hanford Atomic Products Operation, Richland, Washington, May 7, 1964.

[10] J. H. Flora and S. D. Brown, "Evaluation of the Eddy-Current Method of Inspecting Steam Generator Tubing", Report No. BNL-NUREG-50512R, Battelle Columbus Laboratory/ Brookhaven National Laboratory Contract No. EY-76-C-02-0016, September 30, 1976.

[11] W. D. Fletcher and D. D. Malinowski, "Operating Experience with Westinghouse Steam Generators", Nuclear Technology, Vol. 28, March, 1976, pp. 356-373.

[12] W. E. Berry, et. al., "Examination of Inconel 600 Tubing From the A Steam Generator of the Palisades Nuclear Plant", Battelle Columbus Laboratories, July 11, 1974.

[13] S. D. Brown and J. H. Flora, "Evaluation of the Eddy-Current Method for the Inspection of Steam Generator Tubing - Denting", Battelle Columbus Laboratory/ Brookhaven National Laboratory Report, In-Press, September 30, 1977.

ACOUSTIC SURVEILLANCE IN THE NUCLEAR INDUSTRY - A REVIEW

R.W.B. STEPHENS
Chelsea College (University of London)

INTRODUCTION

Acoustics has become increasingly applied during recent years to the diagnosis and monitoring of both biological and inanimate media and in this contribution particular physical aspects of the problems will be discussed. Some suggestions are also made of possible lines of investigation which could merit attention. The urgent need for improved quantifying of N.D.T. techniques is now more widely realised and the full answer to any method is not just where and when but what has taken place. A brief consideration is given to a quantitative approach to acoustic emission studies. The desirability of correlating two different properties of the same event as confirmatory evidence will also be considered with examples.

ACOUSTICAL DIAGNOSIS

Acoustics in its modern concept covers mechanical vibrations of all frequencies from hundredths of a cycle per second to gigahertz frequencies, and a frequency range of tens of MHz could be involved for example in the early detection of coolant disturbances in nuclear reactors. It will include the detection and location of local coolant boiling in the core, the detection of gas bubbles and obstacles, leak detection and the monitoring of pump noise, e.g. for cavitation.

A comparatively small number of sensors suitably placed about the core would permit the early detection of coolant disturbances, e.g. sodium boiling, but it is essential that the observations are not vitiated by the superimposition of external noise sources upon the measured signals. In such situations as this it is always desirable to correlate another type of physical measurement with the acoustic one so as to preclude the possibility of the chance of random disturbances being evaluated as sodium boiling signals. For the application of the acoustical technique therefore the whole background noise of the reactor must be known over the whole thermal and entire power ranges. As typifying the problem observations of a particular reactor showed that the spectral power density of individual, but completely identical in type, pumps showed a great variation and moreover the overall sound intensity increased by 20% within one year with unchanged working parameters, thus emphasising the necessity for repeated assessment of the reactor noise spectrum. In this case the change was attributed to the shift upwards in frequency of the peak sound spectrum of two secondary pumps by 10% during the year. Such problems can be minimised by the careful design of the vibration isolation system of ancillary equipment. Although the low frequency pump noise may be filtered out yet there remains the difficulty of distinguishing between boiling and cavitation noise since both have essentially broadband 'white noise' spectra.

Acoustical diagnosis has become largely synonymous with ultrasonic testing because of the use of the beaming properties of high frequency radiation or to avoid undesired low frequency background noise, but there is a wide area of interest in audio-frequency analysis as applied, for example, in the monitoring of machinery. Acoustical techniques may be divided conveniently into 'passive' and 'energetic' divisions, (Fig.1), implying respectively a technique involving 'listening' to the information provided by the sound source, and one in which a system is 'probed' by means of an ultrasonic beam.

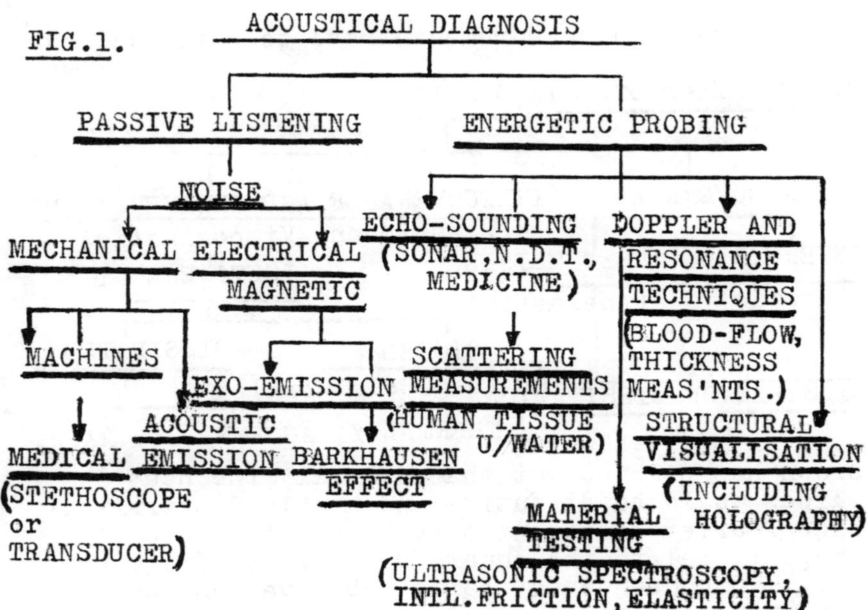

Our interest today is in acoustic emission, of long application by the doctor in his stethoscopic listening to body sounds but actually it was anticipated by Robert Hooke in his Posthumous Works of 1681, to quote 'There may also be the possibility of discovering the internal motions and action of bodies ... whether animal, vegetable, or mineral, by the sound they make; ...' Various parameters have been used to describe acoustic emission investigations but there is increasing support for taking the energy or power within a given frequency bandwidth as the measure (1). This has led to a more quantitative approach concerning, for example, the relation between acoustic emission and the energy dissipated in thermodynamically irreversible processes as indicated in Fig. 2, which follows after the work of Brown (2).

Modern theories of crack propagation are evolved around the strain energy release rate of a material, a fraction only of which is detected as stress waves, i.e. acoustic emission. It would be useful information to know the relative division between the mechanical wave energy and the heat produced at source, but the temperature rises at ordinary temperatures are extremely small and only a few attempted measurements have been made.

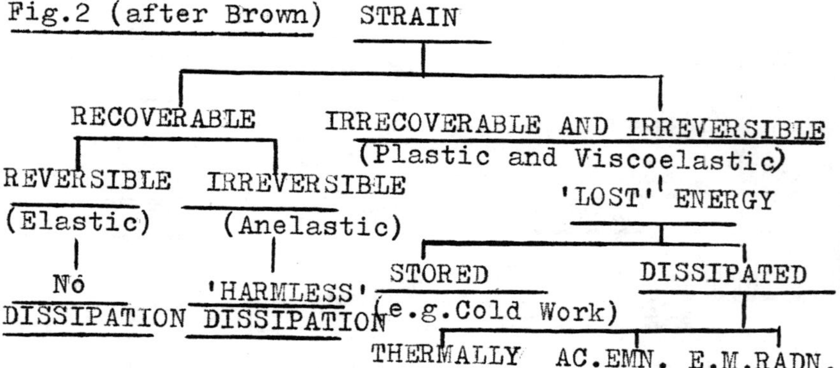

Brown has pointed out that normally the heat pulses would be diffusive and highly dispersive, each Fourier component travelling at a velocity proportional to the square root of its frequency (3). In consequence thermal observations would not provide a measurement of an event complementary to that of acoustic emission, but he has suggested the possibility of making experiments at liquid helium temperatures, with a suitable single crystal of low electrical conductivity so that the phonon mean free path is increased to the order of mm. There are many sensitive infra-red types of detector available to make such an experiment a distinct possibility. A technique which might lead to useful quantification is the use of a laser beam focussed at an interior point of the specimen, the advantage here being flexibility in varying the known length and intensity of the thermal stressing pulse.

ACOUSTIC EMISSION AND AN ANALOGY

An acoustic emission system may be likened to the room reverberation problem but with a more complicated wave-system, due to mode conversion at the boundary surfaces or discontinuities within the specimen or structure. The nature of the medium external to the specimen will also influence the rate of decay of the emission energy, i.e. the equivalent of the reverberation time of room acoustics, so that when the specimen is immersed in a liquid the energy from an emission pulse will decrease much more rapidly than with an air or gas surround. In course of time the acoustic emission will degrade through attenuation so that the pulse shape and its duration will alter, the attenuation arising from passing through visco-elastic regions varying as the square of the frequency. Furthermore due to non-linearity there will be an energy flow from the fundamental to 'feed' a higher harmonic. In a sufficiently long specimen the observation of this change by say a string of three transducers might give source location information.

Before appreciable degradation has taken place the received spectrum of the emission will have a relevance to the nature of the medium and to the type of deformation. It is noteworthy that in room reverberation measurements a greater emphasis is being given nowadays to the early parts of the sound decay. Hence consideration should be given to the significance in acoustic emission of the ratio of the energy of the early part of the received emission signal to the total emission energy of the event. The degradation time will depend mainly on the attenuation or damping coefficient of the medium and so should have a linkage with Q^{-1}, where Q is the mechanical quality factor of the system as a whole. This correlation has received some experimental verification. The frequency content of the pulse will gradually assume the natural modes of the specimen which lie within the emission bandwidth. The position of the source will influence the number of modes excited and for a specimen or structure with parallel walls there is the possibility of a 'flutter' echo being produced.

Another factor which will influence the nature of the emission signal is that concerned with specimen shape and, for a given volume of material, there will be difference in the emission signal depending on whether the cross-section say is square or rectangular.

When the boundary surface is uneven with respect to the component acoustic wavelength, then specular reflection will give way to scattering and the reverberation conditions are more appropriate to those within the ocean. Scattering will of course also occue at 'foreign' regions of suitable dimensions within a specimen, just as artificial scattering elements are introduced in concert halls to create a more uniform sound intensity. In the ocean too this 'volume' scattering occurs from algae and fish shoals and is considered separately from the scattering at the sea-air surface. In the ocean environment some useful information is obtained from only a brief inspection of the manner of decay of the reverberation level with time. For example if t is time,then a decay in level of A(log t), serves to separate the scattering from layers or from an infinite distribution depending on the value of A, (4). It is suggested that this procedure might be modified for A.E. observations on large structures.

LEAK DETECTION

A very common problem in any industrial set-up is that of leak detection. There are a number of industrial equipments for the purpose whether for gas, vapour or liquid. Recently some Czech workers, Pokora & Taraba, made an exhaustive investigation on the noise from a small aperture and determined the amplitude-frequency spectrum over a range of 50 Hz - 290 kHz (using a 3-octave analyser) and pressure drops (with a packed nozzle) between 0.4 and 120 atmospheres (5). They sought a relationship between sound pressure and geometrical parameters of the jet. The doubt one has in such experiments is how close the experimental set-up conforms with the industrial conditions, for the surrounds may have some cavity volumes giving rise to resonances as for the Hartmann jet whistle. Such a situation could arise in a system of pipes. The alternative detection system could be by transducers monitoring the stress waves set-up in the pipe walls. These could be surface waves or Lamb waves depending on the geometry of the system and the type of excitation. These could be quite sensitive monitors as the attenuation would not be large. However, there is a problem in relation to any preferential texture of the pipe for a crack along the preferential direction will show little emission if any.

The use of a transducer as a receiver of signals at a distant point, often inaccessible, is dependent on the constancy of performance at a distance. Assuming any change in the behaviour of the material of the transducer would either be slow or catastrophic then its constancy of response will be dependent on the permanence of the coupling to the workpiece. Holroyd has shown that this may be monitored by monitoring the admittance of the transducer in-situ (6). In his procedure the admittance is measured by driving the transducer at constant voltage from a variable frequency oscillator and measuring the supply current by the voltage drop across a series resistance. Since the response of the transducer depends upon its own particular resonances and these are not equally affected by the damping and mass loading incurred by attachment to the work surface it is necessary to carry out an initial experiment to ascertain the most effective frequency of operation.

Alternate to either of the above procedures I suggest that perhaps it might have some advantages, in the case of steam or water pipes, to utilise the phenomenon of reverberation which has not received application in leak problems.

The absorption of sound in a room is compounded of a surface absorption at the walls and boundary surfaces and a volume absorption in the contained gas. When the boundary absorption is poor the gas absorption is of greater importance and becomes very significant at frequencies around 2 kHz. On classical theory the acoustic absorption in a monatomic gas arises mainly from viscosity and thermal conduction but for a polyatomic gas the experimental value can greatly exceed that deduced from classical theory. This molecular absorption is associated with a relaxation process involving the vibrational energy of oxygen molecules and is influenced by the water vapour content of the atmosphere. Knudsen showed that the presence of the vapour, even if partial condensation on the walls has taken place, made no significant difference to the boundary absorption (7).

For small boundary absorption the reverberation time T of an enclosure may be given by

Sabine's formula :-

$T = 55.3/(c.A)$, [s], where c is the velocity of sound in the atmosphere $[ms^{-1}]$ and A is the equivalent sound absorption surface area$[m^2]$. If S is the actual boundary surface area $[m^2]$, V is the air volume of the room $[m^3]$, α is the average surface absorption coefficient and β is the effective air attenuation coefficient $[m^{-1}]$, then $A = \alpha S + 4\beta V$. Benedetto and Spagnolo have made an exhaustive series of measurements in a large reverberation chamber in the frequency range 1 to 5 kHz and expressed their results in terms of the rate of decay of sound i.e. $R = 60/T\ [dBs^{-1}]$ (8).

At 5 kHz the decay rate at $20^\circ C$ is approximately 0.91 dBs^{-1} for a 1.1% change in humidity, which at $10^\circ C$ is approximately 0.93 dBs^{-1}. This suggests a possible monitoring system for leakage of water or steam in a room containing an assembly of pipes carrying the fluid by automatic measurement of reverberation time. Such a space would normally have little surface absorption and from the data of Benedetto and Spagnolo (Fig. 3)

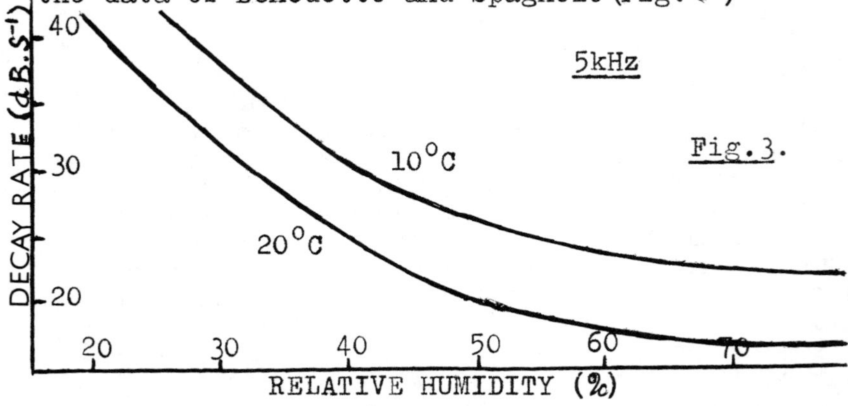

Fig. 3.

it is evident that a strict temperature control is not necessary. Another procedure which might have application is to utilise the change of velocity with humidity as revealed by the disappearance or emergence of particular modes of the enclosure.

In a system where there is for example a network or assembly of pipes the problem of location of faults by audio observation at a distant point becomes more complicated. Effectively it means the identification of the significant paths from the source to the receiver. For non-dispersive systems for which the phase velocity is constant, such as propagation through fluids or for certain modes in solids, then cross-correlation analysis may be used. For dispersive transmission paths where phase velocity is a function of frequency, such as near cut-off frequency in ducts, cross-correlation analysis has been used with the limited application to short-path lengths and relatively low frequency noise. Considerable advances in the technique have been made by Barger but in applying his analysis to a reactor pipe system it would be necessary to make a survey with arbitrary located noise sources (9). The development of the acoustic telescope for shorter range discrimination could provide an answer to the problem. Mungar and Fahy (10) have evaluated the transmission of sound through an array of pipe scatterers using an analysis similar to the determination of the energy states of electrons in the band theory of solids. Their results suggest the existence, in the frequency domain, of various zones of attenuation separated by transmission zones.

NOISE IN FERROMAGNETICS (OR FERROELECTRICS)

As mentioned previously in N.D.T. work there is greater confidence if a different type of test can be used as a confirmatory back-up for the main procedure. Steel as a material looms prominently in construction work but despite the wide application of acoustic emission and ultrasonic methods there can still exist some uncertainty in the assessment observations. Now in considering the energy balance for ferromagnetic or ferroelectric materials, undergoing changes in mechanical stress or polarisation, an extra component of noise is involved. This is associated with an effect known as Barkhausen noise and arises from the discontinuous changes in polarisation accompanying smooth changes in the applied magnetic (or electric for a ferroelectric) and/or mechanical stress fields. The irregularities in domain (or Bloch) wall movements are presumed to arise from crystalline inhomogeneities in the solid which result in local energetic impediments to continuous motion. As might be expected therefore the Barkhousen pulses, recorded by pulse-counting techniques, have been shown to give a distribution in size corresponding to a range of imperfection effects. The Barkhausen 'jumps' in induction are observed as voltage pulses in a search coil surrounding the specimen and may be characterised in terms of a noise spectrum. The associated theory with the experimental observations, indicate that the Barkhausen discontinuities are concerned with much smaller volumes than those of the primary Weiss domains and so it is now considered that the main source of the noise is due to irregularities in the movement of domain walls.

If a sinusoidal slowly varying magnetic field is applied to a ferromagnetic core the Barkhausen 'jumps' may be observed as already mentioned as voltage pulses in a 'pick-up' coil. In general these pulses are random in time sequence and constitute a noise voltage (ΔV) which may be expressed as a noise power spectrum. This may be conveniently expressed as :-

$$\langle \Delta V^2 \rangle = \frac{\left[16n^2 \cdot fo \cdot B_m \cdot V_B\right] \cdot \left(V_p \cdot I_s^2 / B_s\right)}{(1 + \omega^2 \cdot \tau^2)} ,$$

which is a simple relaxation spectrum characterised by the time constant τ and proportional to the size of the elemental volume V_p.

In the expression for the numerator the terms in the 'square' brackets are experimental parameters while those in the 'curved' brackets are material constants.

n = number of turns in the 'pick-up' coil.

f_o = frequency of exciting magnetic field intensity.

B_m = maximum magnetic induction

B_s = saturation magnetic induction.

V_p = total sample volume.
V_B = elemental volume.
I_s = saturation magnetic intensity.
$\omega = 2\pi f_0$

A time constant of 10^{-4} secs. is characteristic of Barkhausen Noise Studies and is a manifestation of magnetic and eddy current losses in the material that limit the rapidity with which magnetization changes can occur. (Fig. 4).

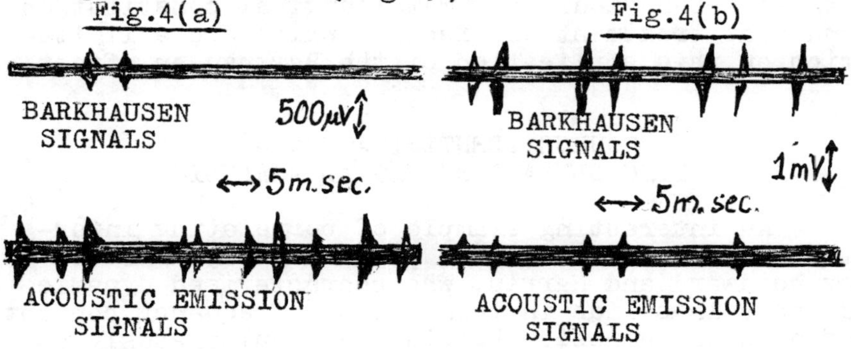

Fig.4(a)

BARKHAUSEN SIGNALS

500μV

↔ 5m.sec.

ACOUSTIC EMISSION SIGNALS

Fig.4(b)

BARKHAUSEN SIGNALS

1mV

↔ 5m.sec.

ACOUSTIC EMISSION SIGNALS

SIMULTANEOUS OSCILLOGRAPHY TRACES OF MAGNETIC AND ACOUSTIC A.E. EMISSION SIGNALS

Fig.4(a) For <u>Moderately Hard Specimen</u> (R_c= 47) Fig.4(b) For a <u>Very Hard Specimen</u> (R_c=61)

When a fatigue crack propagates in a ferromagnetic material there are discontinuous changes in the magnetisation of the medium which are similar to those associated with the Barkhausen effect. Experiments by Mc.Clure et al in which simultaneous measurements were made of acoustic emission and the Barkhausen 'noise' have confirmed this association (11). These workers arranged conditions for a crack to form in a steel sample and it was surrounded by a pick-up coil. With the growth of the crack the volume of magnetised material decreased within the coil giving rise to discontinuous changes in magnetisation and consequently to voltage signals. The specimen was subjected to alternating stresses at 30 Hz at a low enough stress value to permit of several thousand cycles before failure. Simultaneous a.e. measurements were made using a shear transducer of 0.1 MHz resonant frequency. It was found that for a specimen of high hardness the magnetic method was more sensitive for detecting fatigue crack propagation but in a moderate hard material the reverse was the case. By monitoring the Barkhausen discontinuities it is possible to determine quantitatively the state of stress in the surface of a ferromagnetic material. Magnetic recordings, which are proportional to the number and size of Barkhausen discontinuities, increase in magnitude with the applied tensile stress but decrease for an applied compressive stress. This suggests a possible correlation with Rayleigh surface wave measurements. The dependence of the magnetic behaviour upon metallurgical constitution and heat treatment are factors which are a limitation on this application of the Barkhausen effect.

SOURCE IDENTIFICATION - OPTICALLY AND BY SIGNAL ANALYSIS

An interesting example of correlating information from two different procedures is afforded by Woodward and Harris, who characterised dynamic defects in solids in terms of the frequency content of acoustic emission bursts and simultaneously observed the surface deformation events by optical microscopy (12). They computed a new single parameter, the median frequency which divides an energy spectrum into two parts of equal area and hence of equal energy. Now although the bursts are mathematically non-deterministic and non-stationary they have finite energy and can still be characterised statistically and analysed as for stationary random signals.

In this study the single point representing the energy spectrum for each burst is given by the intersection of the median frequency and the mean energy spectral density - the mean of 500 values comprising the energy spectrum. The optical microscopic observation of the surface deformation events were recorded on movie film or videotape and the acoustic signals were detected by a shear-wave transducer. It was found preferable to present the results by an acoustic emission event histogram, i.e. the number of bursts with median frequencies in spectral windows of 1 KHz bandwidth are plotted against the median frequency (Fig. 5). Using zirconium specimens, for an elongation of 0.5mm.,

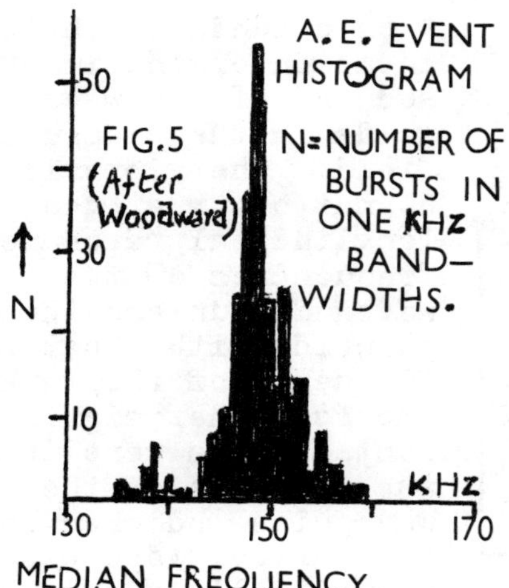

the data fell into two separate groups, the larger of which was assumed to be due to twin initiation as confirmed optically. Furthermore visual observation showed in addition twin broadening and slip. On replotting the data of Fig. (5) by dividing the elongation into two equal ranges of 0.25mm. there was acoustical evidence of three groupings in agreement with the optical observation.

A MECHANICAL WAVEGUIDE AS A SOUND DETECTOR

A possible method of detecting boiling in a liquid-metal reactor is the use of a stainless-steel waveguide as an acoustic transformer which responds to boiling-noise signals. The design of a laboratory experiment (Fig 6) to simulate the problem and the results are briefly discussed (13). The objective is to utilise the waveguide as a warning device should there be a diminution of coolant (liquid sodium) flow or a power surge due to some perturbation of nuclear reactivity. In this way a possible fuel melt-down and subsequent release of fisson products might be averted.

As a simplification to the laboratory experiment water replaced the liquid sodium being justified from their similarity in acoustical behaviour and the fact that the characteristic impedance (ρc) of water at room temperature is approximately the same as that of molten sodium in the range 400-600°C. The mechanical waveguide under test is partially immersed in water contained in a thin-walled tank which is free to vibrate in its natural modes. At these modal frequencies, each being characterised by a discrete pressure configuration, energy is coupled to the waveguide and excites it into vibration. Up to 6 kHz the amplitude of this vibra-

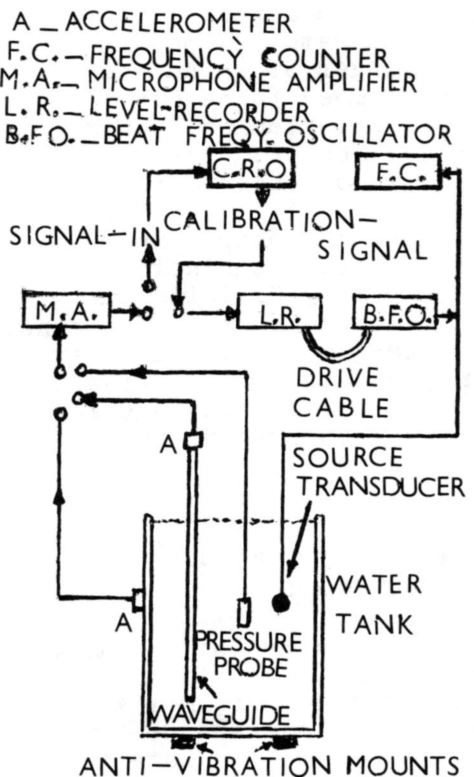

FIG.6 (AFTER WOODWARD)

tion depends on the location of the immersed end of the waveguide in the pressure field. The vibrations do not occur at the longitudinal resonance frequencies of the waveguide unless these coincide with a natural frequency of the tank. Amongst other significant results were that the end-face of the waveguide rod when its cylindrical face was acoustically shielded gave a greater response than when the rod was unshielded and secondly that when the rod was bent sharply through 90° the system became effectively a high-pass filter (13) (14).

USE OF MODELS OF A SYSTEM UNDER INVESTIGATION

The use of visualization techniques to ascertain the 'geometrical acoustical' properties of an industrial plant does not appear to have been greatly explored. It was used first some sixty years ago for the elucidation of problems of room acoustics in two dimensions by means of a ripple tank, or in a three-dimensional model using a light beam to simulate the sound beam and the varying optical reflecting powers of the surfaces chosen to match equivalent acoustic absorption coefficients. Hall has recently described the technique of stroboscopic photoelasticity and discussed the principles involved in the visualization of compressional shear and Rayleigh waves (15). He uses glass as the medium in his experiments, in which the specimen is subjected to an applied stress by simultaneous observations of acoustic emission it might be possible to ascertain the best locations for placing the transducers in a given structural system.

REFERENCES

1) R.W.B.STEPHENS & A.A.POLLOCK 'Waveforms and freqy. spectra of acoustic emissions.' JASA 50 (1971) 904.
2) C.R.BROWN European Working Group on Acoustic Emission 4th Meeting, SACLAY, FRANCE (1975)
3) C.R.BROWN 'On the problem of developing acoustic emission as a quantitative technique'. Proc. Inst. Ac.4-10-1 (1977)
4) R.P.CHAPMAN 'Sound Scattering in the Ocean' Chap.9'U/water Acoustics'Vol.2.Plenum Press(1967)
5) I.N.POKORA & O.TARABA 'Emission of U-sound by gas jet issuing from leaking system' XV Int. Conf. Acoustics Prague 1976.
6) T.J.HOLROYD 'Monitoring A.E.Transducer Couplings' Inst. Acoustics Dec. (1977) Meeting.
7) V.O.KNUDSEN & L.OBERT 'Absorption H.F.Sound in O_2 contng. small amounts H_2O vapor' JASA 7 (1936)
8) G.BENEDELTO & R.SPUGNOLO 'Sound Decay Rate in Rvbn. Chamber, v. Humidity & Temp.' (to be pub'd in ACUSTICA)
9) J.E.BARGER 'Noise Path Diagnostics-using Cross-Correlation Analysis'Noise Control Eng.6,3,1976
10) P.MUNGAR and F.H.FAHY 'Sound Transmission through array of scatterers' J.S.V. 9 (1969)287
11) J.C.MC.CLURE and K.SCHRODER 'Determination of Crack Growth with Barkhausen Effect' Phys. Stat. Sol. (a) 19 K57 (1973)
12) B.WOODWARD & R.W.HARRIS 'Use of signal analysis to identify sources of acoustic emission ' Acustica 37 3 (1977)
13) B.WOODWARD 'Use of mechanical waveguides for sound detection in resonant vessels' Acustica 24 3 (1971)
14) K.F.GRAFF 'Waves and vibrations in curved ultrasonic transmission lines' Ultrasonics March 1972
15) G.HALL 'Ultrasonic Visualisation as Teaching Aid in N.D. Testing Ultrasonics March 1977.

REVIEW OF JAPANESE RESEARCH AND DEVELOPMENT ACTIVITIES ON ACOUSTIC EMISSION APPLICATIONS IN THE NUCLEAR INDUSTRY

Hiroyasu Nakasa, Hideo Kusanagi and Hironori Ohno
Central Research Institute of Electric Power Industry (CRIEPI)
Tokyo, Japan

INTRODUCTION

Japanese activities on acoustic emission (AE) research and development works have recently been more and more active, and AE application fields have become wider and wider, as well as in the United States of America, Europe and elsewhere. Especially, the expectation of the AE technology for assurance of the structural integrity in nuclear components increases due to its high potentiality to detect incipient failure, to monitor crack-growth and leakage and to give warning against catastrophic damage. This review starts with the general survey on Japanese AE research and development activities, and then describes some topics of recent AE applications in the nuclear industry.

GENERAL SURVEY ON JAPANESE AE RESEARCH & DEVELOPMENT ACTIVITIES

Japanese activities on AE research and development works are relatively well-coordinated by the Japanese Committee on Acoustic Emission (JCAE), formed in 1969 in the High Pressure Institute of Japan in co-operation with the Japanese Society for Nondestructive Inspection. JCAE is composed of about 50 members from universities, national and neutral research organizations, electric power companies, plant fabricators, steel makers, inspection and instrument companies, and others. JCAE has held regular meetings bimonthly and a National AE Conference and 3 AE Symposia in alternate years. Extensive and profound research works are found in the Proceedings of the 1st (1972), 2nd (1974) and 3rd (1976) AE Symposium (in English) and in the Proceedings of the 1st National Conference (1977; in Japanese). At present, the 4th AE Symposium is under planning to be held in Tokyo during September 18-20, 1978.

JCAE has the following recent activities:

(1) research and development information activities.
(2) educational and publicity activities.
(3) calibration standards of AE transducers and instruments.
(4) round-robin program.
(5) pile-up of AE characteristics data of various materials.

JCAE has organized 4 tutorial sessions and edited several special issues on the AE technology. The results of the 3rd tutorial session were reported as the "Special Issue on Experiments of Acoustic Emission" in the Journal of Japan High Pressure Institute. The typical example from the Issue is shown in Fig. 1, where signal location and event-count rate displays using 6 transducers are given for the PSI and ISI simulated pressure tests applied to a 3m-diameter spherical pressure vessel in the Ship Research Institute (1).

A round-robin experiment using the standard AE instrument and test specimens is under way by the JCAE members for the purpose of getting the overall calibration of the AE instrumentation of the members with real AE signals during tensile testing. Furthermore, 2 working groups have been formed in JCAE for the pile-up tasks of AE characteristics data. The typical examples of the preliminary activities are shown in Fig. 2, where AE event-rate data in the designated location area were analyzed in a fatigue test using a large WELTEN-80 steel specimen with welding defects (2).

There are 2 other AE committees, the FAE Subcommittee in the Atomic Energy Research Committee of the Japan Welding Engineering Society and the Committee for the TAB-AE Project in the Technical Approval Bureau (TAB) of the Industrial Research Institute, besides JCAE.

The FAE Subcommittee sponsored by PNC, the Power Reactor & Nuclear Fuel Development Corporation, has the future goal for practical use of the AE technology in on-line monitoring of FBR pipe failure through mechanisms such as fatigue, creep, creep-fatigue, creep-buckling, thermal shock and thermal ratcheting. In 1976 and 1977, the FAE Subcommittee did a world-wide survey on AE applications to every type of the nuclear power plant, with stress being laid on the following special items, and has presented a survey report (3):

(1) basic data of AE characteristics from materials used in the reactor primary and secondary systems.
(2) signal attenuation properties on structure components.
(3) background noise spectra.
(4) AE transducers and their coupling.
(5) signal-conditioning techniques.

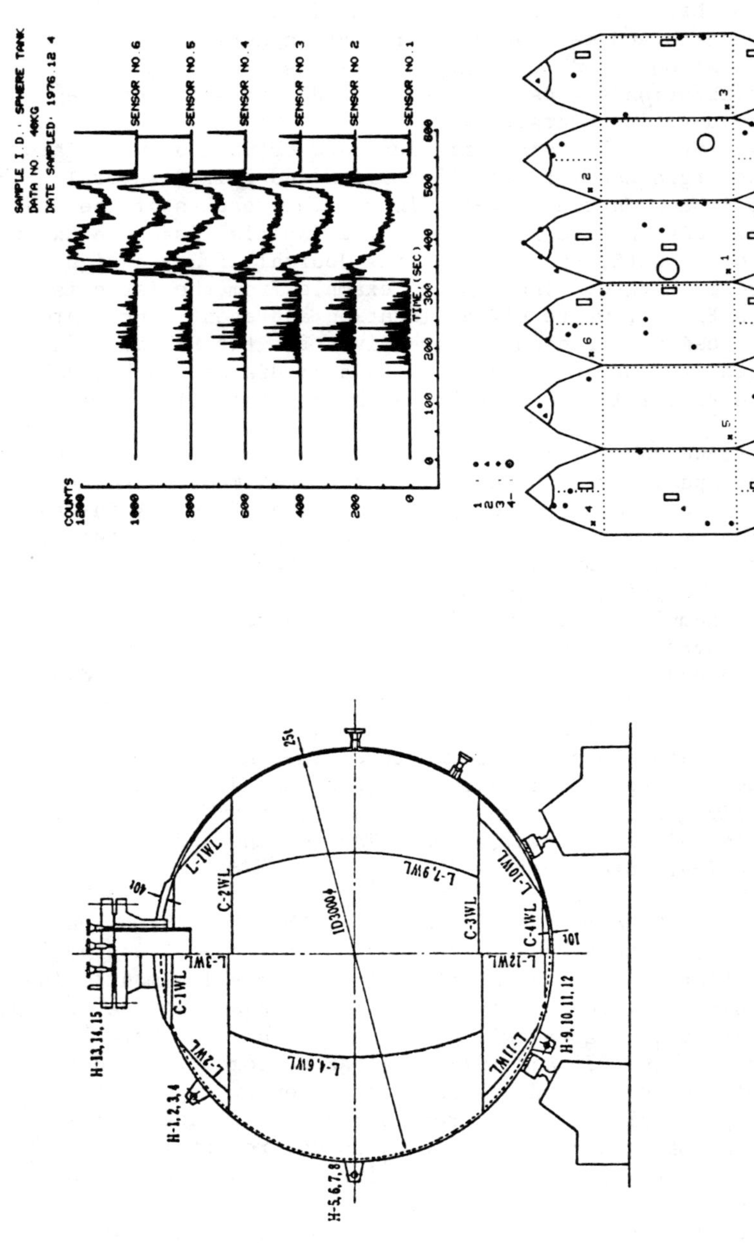

Fig. 1. Typical Example of JCAE Educational Activities.
Pressure Test of 3m-dia. Spherical Vessel.

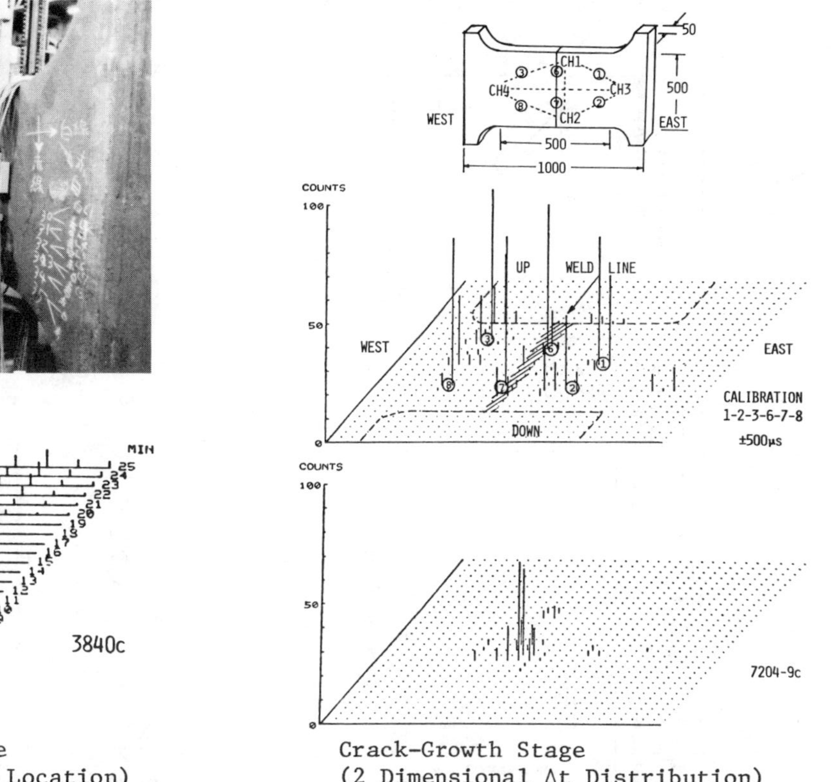

Crack Initiation Stage
(1 Dimensional Source Location)

Crack-Growth Stage
(2 Dimensional Δt Distribution)

Fig. 2. Fatigue Test of a Large WELTEN-80 High-Strength Steel Specimen with Welding Defects.

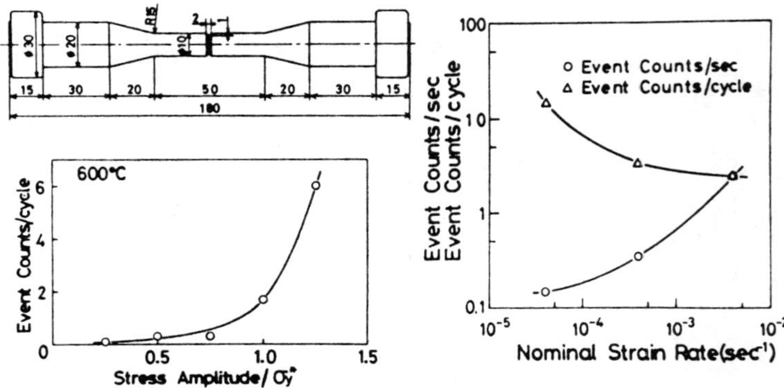

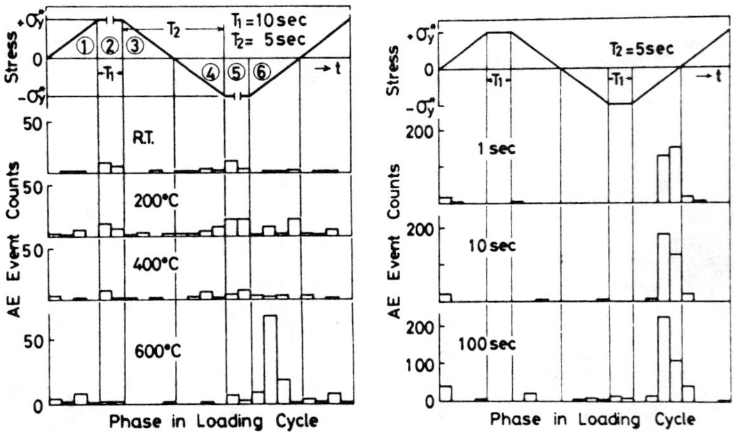

Fig. 3. Fatigue Test of Type 304 Stainless Steel Specimens.

At present, FAE Subcommittee, together with the JCAE working group, has started a basic research effort directed toward understanding the fundamental AE characteristics at various steps of elevated temperatures for steels used in FBR (304SS, 316SS, $2\frac{1}{4}$Cr-Mo). The preliminary results have been presented at the 1st National AE Conference, and Fig. 3 shows typical AE characteristics during 600°C fatigue tests of Type 304 stainless steel specimens (4).

On the other hand, the objectives of the TAB-AE Project are (5):

(1) to evaluate the potential of AE instrumentation systems presently available.
(2) to identify problems and difficulties in the present arts.
(3) to establish more reasonable AE test procedures.
(4) to propose a new code for the AE examination.

In the TAB-AE project, about 20 pressure vessel tests have been performed and monitored by a 40-channel mobile AE system. Services for locating and evaluating AE sources in actual steel structures are offered to the related members. At present, an interesting program is under way, in which a steel pressure vessel shown in Fig. 4 is prepared and AE sources are being investigated during water-pressure tests in various stages from fabrication to operation.

Fig. 4. Model Pressure Vessel for AE Tests in the TAB Project.

AE APPLICATIONS IN THE NUCLEAR INDUSTRY

The assessment of structural integrity of the primary coolant system by inservice inspection and/or on-line monitoring is one of the most important problems in the nuclear power plant, until the truely perfect reactor system will be build up. The following AE research and development programs have been presented at the Governmental survey meeting on the state-of-the-art by the Science and Technology Agency with the intention of systematic research orientation in the nuclear industry (6):

(1) Safety Assessment of Reactor Vessels by AE Techniques. (by Ship Research Institute)
(2) Basic Research on AE Applicability to On-line Monitoring of FBR Piping (7). (by PNC)
(3) Development of the AE Leak-Detection Method for On-line Monitoring of ATR "FUGEN" Piping. (by PNC)
(4) AE Techniques for Integrity Assessment of Reactor Pressure Boundaries (8). (by CRIEPI)
(5) Research on Inservice Inspection Techniques. (by Tokyo Electric Power Co.)
(6) Research on AE On-line Monitoring Techniques. (by Tokyo Shibaura Electric Co.)
(7) Development of High-Temperature Transducers (9). (by Tokyo Shibaura Electric Co.)
(8) Development of the AE Technology for Evaluation of Defects Detected in the Nozzle Region. (by Ishikawazima-Harima Heavy Ind. Co.)
(9) Research on Structural Safety of Nuclear Power Plants. (by Hitachi, Ltd.)

As seen in the above programs, the utility industry and the plant fabricators take strong interest in availability of the AE technology. The utility industry is, however, highly regulated and regards the use of the AE technology on the conservative basis, and so at present practical AE applications to preservice inspection, inservice inspection and on-line monitoring are limited in number for Japanese commercial nuclear power plants.

Experimental approaches through AE on-line monitoring of specific critical areas in the primary system are under way, together with the back-up fatigue test shown in Fig. 5 (10), in order to get a better basic understanding, to accumulate various experiences and to fully evaluate the performance. At present, the following views have been obtained through the preliminary analysis on AE data of a BWR on-line monitoring in various stages of reactor start-up, steady state operation and reactor shutdown:

(1) AE activities are very weak in comparison with the environmental noise, and detectability of AE signals may depend on both the location monitored and the environmental conditions associated with reactor operation.
(2) AE peak-amplitude distributions and the derivative parameters such as total event number and energy, may be useful for the data evaluation.

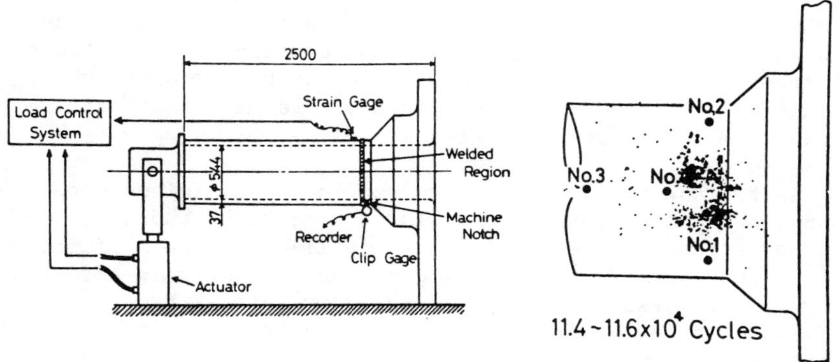

Fig. 5. Fatigue Test of a Model Nozzle Junction with Notch.

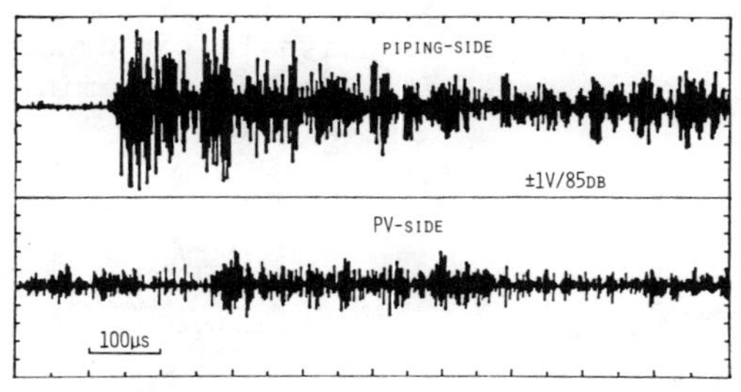

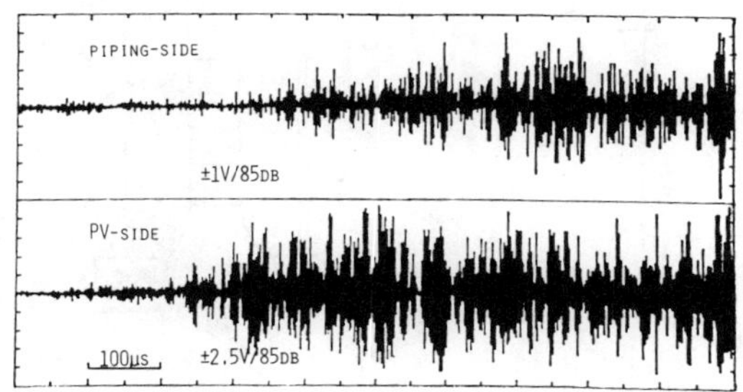

Fig. 6. Signal Waveforms Detected During On-line Monitoring of a BWR Nozzle Region.

(3) a lot of basic research works and data accumulation are needed for the interpretation of the signature of acoustic signals detected, of which typical waveforms are shown in Fig. 6.

Another research program has been performed for the AE application to material property evaluation in a PWR pressure vessel steel surveillance test. In the program, AE characteristics have been analyzed using neutron-irradiated tensile and WOL test pieces after 1 year operation, and some results have been gained that the AE peak-amplitude distribution is strongly affected by irradiation hardening. An advanced program using specimens irradiated to higher degrees in a materials testing reactor (JMTR) is now in progress.

As for the SCC problems occurred in BWR stainless steel piping, there is a Governmental large project by the Ministry of International Trade and Industry (MITI). In the MITI project, a test program to prove the reliability of weld-affected-zones in stainless steel piping under simulated BWR operating conditions is now under way. The related organizations and companies are constructing test facilities including a BWR model plant and accelerated pipe-failure test equipments, as shown in Fig. 7, and they expect the high potentiality of the AE technology to detect stress corrosion cracking early and firmly. At present, AE systems suitable for crack-growth monitoring and leakage detection are being examined.

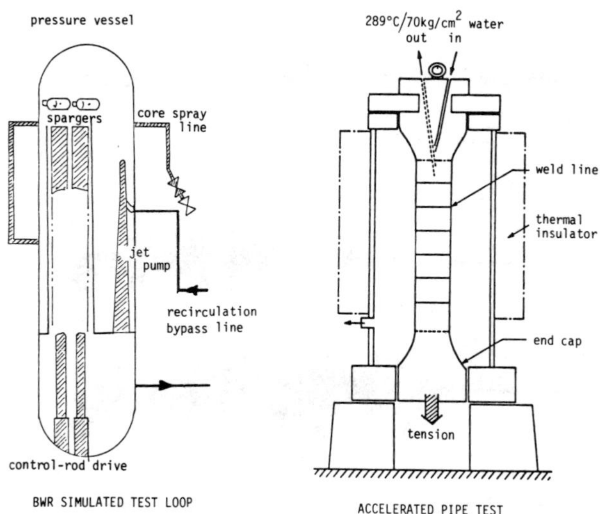

Fig. 7. MITI SCC Project.

TYPICAL RESEARCH & DEVELOPMENT WORKS IN THE NUCLEAR APPLICATION FIELD

Fatigue tests of model reactor pressure vessels have been conducted as the co-operative work of the Japan Atomic Energy Research Institute and CRIEPI in order to study the relationship between AE characteristics and behaviors of crack propagation around nozzle corners. As shown in Fig. 8, AE systems used have successfully traced the crack propagation (11), but there appeared such troublesome problems that AE signals from fatigue crack growth were very feeble and that the signal-to-noise ratio was very low.

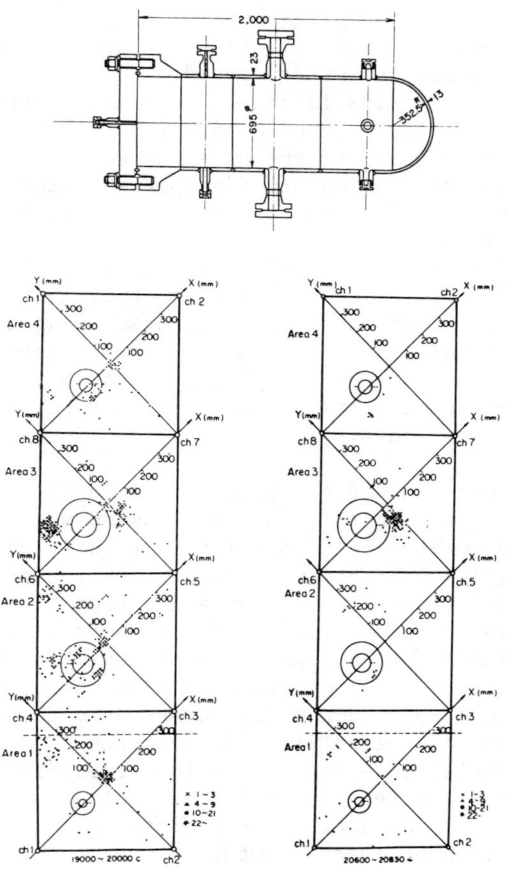

Fig. 8. Fatigue Test of a Model Reactor Vessel.

A great number of fatigue tests have been monitored by AE techniques as the co-operative research project by PNC and CRIEPI since 1972 for the FBR piping components such as elbow, branch, reducer and nozzle components, and the results demonstrated the usefulness of the AE techniques not only in fatigue tests but also toward in-service monitoring of piping components in practical use (12). In the PNC/CRIEPI project during 1974-1976, several tests have been performed under conditions as near to FBR operation as possible. Figure 9 shows a typical result of AE monitoring in a 600°C creep-fatigue test carried out on 12 inch diameter large elbow pipe (13), and also Fig. 10 shows another typical result of AE monitoring in a thermal ratcheting test using a liquid-sodium piping test loop (14). In the latter sodium-loop test, the following results were gained:

(1) background noise originated from sodium flow, etc. was very low, and high AE activities due to thermal expansion and contraction were observed.
(2) the AE activities increased when axial stress rose and then gradually decreased to a finite emission rate, which was proportional to the axial stress.
(3) the proportional nature disappeared after changing of failure mode from thermal ratcheting to occurrance of crack growth.

At present, a new PNC/CRIEPI co-operative work is under way, where the applicability of the AE technology to the detection of fatigue crack growth is investigated using 12 inch diameter pipes of both stainless and carbon steel. According to operating conditions of FBR and LWR, parameters to be varied during repeated loading are crack size, temperature (RT-600°C), stress level and strain rate.

AE research and development works in the nuclear application fields are primarily directed toward water-pressure tests and on-line monitoring of steel structures. However, some interesting AE applications have been performed outside the area of steel structures. Figure 11 shows a typical result in AE monitoring of an external water pressure test of a model concrete vessel for radioactive waste sea disposal (8). Various failure modes of model concrete vessels can be investigated using such a simple AE monitor as shown in the figure, where the following AE characteristics are seen:

(1) AE signals are emitted from low-level pressure.
(2) the Kaiser effect, that is, lack of AE signals for repeated loading of the same stress, plays an important role as the indication of crack-growth.
(3) the emission rate decreases exponentially under constant pressure loading, whereas it increases exponentially as pressure rises up to failure.

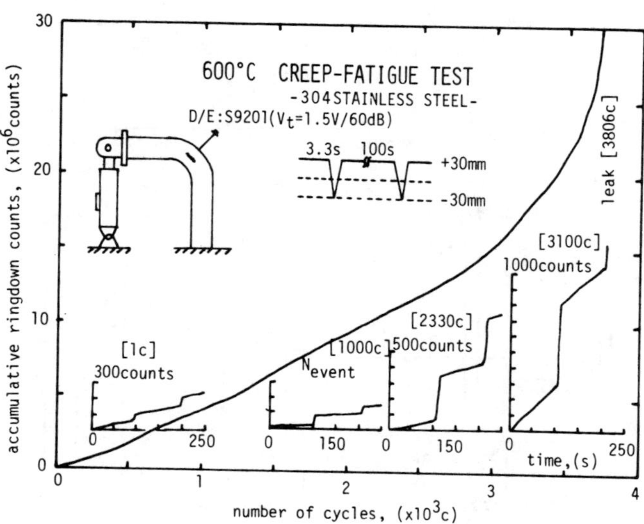

Fig. 9. Creep Fatigue Test of a Type 304 Stainless Steel Piping Component.

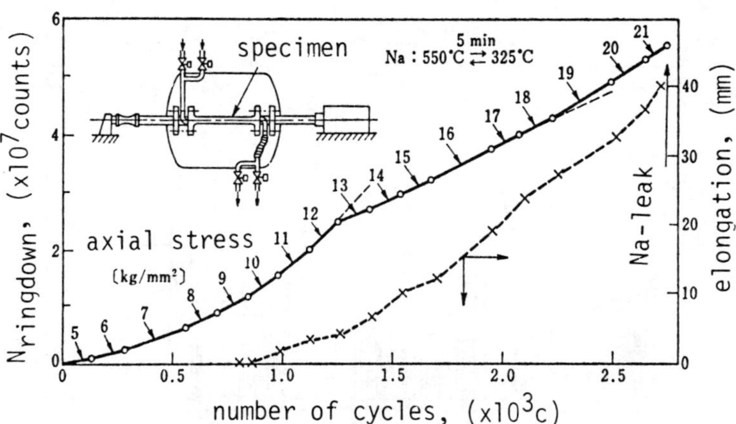

Fig. 10. Thermal Ratcheting Test of Model LMFBR Primary Coolant Piping.

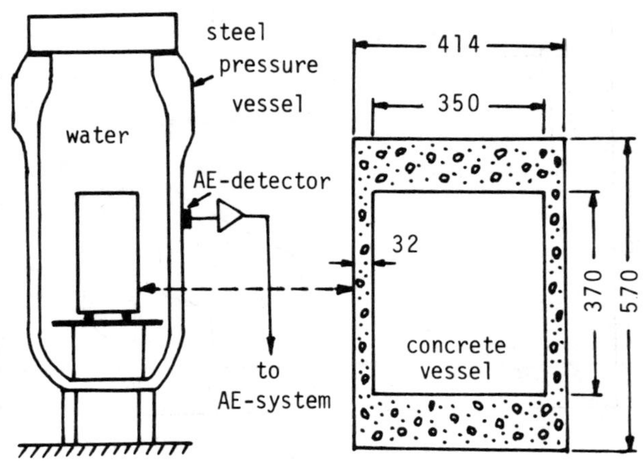

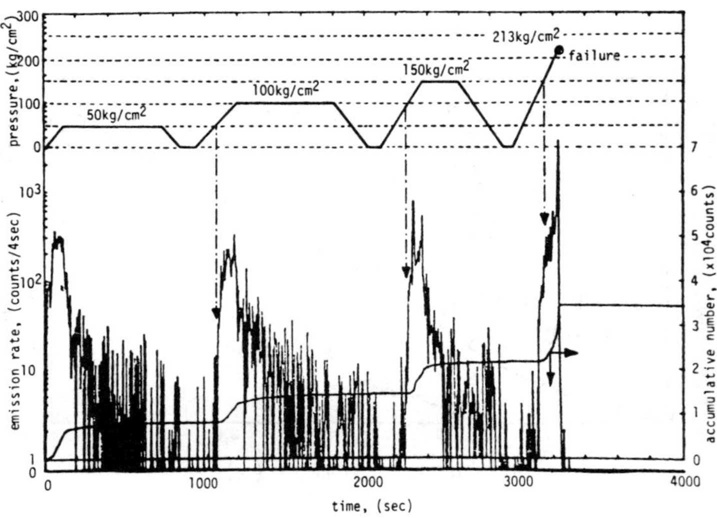

Fig. 11. Hydrostatic External-Pressure Test of a Model Concrete Vessel for Radioactive Waste Sea Disposal.

The Kaiser effect is also useful to get informations on geo-stress (15), which might evaluate the strain energy level of rock foundation. Figure 12 shows an example to be applying the method to evaluation of a fault located beneath a nuclear power plant under construction.

SUMMARY & CONCLUSION

In this review, it has been demonstrated with various typical examples that Japanese AE research and development activities are very active, but also that the AE technology begins to face practical problems to be solved as early as possible.

As almost all techniques have traced the history from birth to practical use that they might never be able to fill their initial promise perfectly, so the AE technology, though it has the unique potentiality, might hereafter have to seek for the viable fields. To nurture the AE technology to practical use in the nuclear industry, it is desirable to establish the international close contact and co-operation.

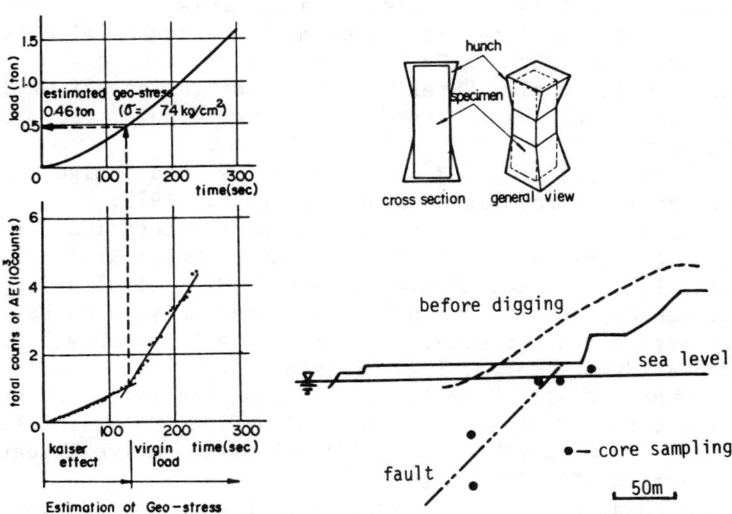

Fig. 12. Geo-Stress Estimation Using the Kaiser Effect and Its Application to a Nuclear Power Plant Construction.

REFERENCES

(1) H. Nakasa, et al., "Analysis of AE Characteristics by Magnetic Tape Recording and Its Application to AE Test of a Pressure Vessel," JHPI, Vol. 15, No. 3, 1977, p.137.

(2) H. Nakasa, et al., "AE Characteristics During a Tensile Fatigue Test of a High-Strength Steel Specimen with Weld Defects Using a 3000 Ton Fatigue Testing Machine," Proc. 1st National AE Conf., Tokyo, 1977, p. 43.

(3) FAE Subcommittee, "Study of Acoustic Emission Technique for Fast Breeder Reactors," J225-77-03, 1977, Atomic Energy Research Committee, Japan Welding Eng. Soc.

(4) T. Nagata, et al., "AE of Fatigue Crack Growth in 304 Stainless-Steel," Proc. 1st National AE Conf., Tokyo, 1977, p. 49.

(5) M. Onoe, "Development of Acoustic Emission Techniques for Pressure Vessel Inspection--TAB-AE(IRI) Project," Proc. 3rd AE Symp., Tokyo, 1976, 2-4, p. 118.

(6) K. Iida, et al., "Application of Acoustic Emission Techniques in the Nuclear Industry," J. Atomic Energy Soc. Japan, Vol. 19, No. 7, 1977, p.429.

(7) T. Nagata, et al., "An Application of the AE Technique to Low-Cycle Fatigue Test of Piping Components," Proc. 3rd AE Symp., Tokyo, 1976, 1-1, p. 11.

(8) H. Nakasa, et al., "Instrumentation System for Parametric Analysis of Acoustic Emission Characteristics and Its Application to Integrity Assessment of Structural Materials," Proc. 3rd AE Symp., Tokyo, 1976, 2-5, p.131.

(9) N. Uesugi, et al., "Preliminary Investigations of On-line AE Monitoring System for Nuclear Power Plant," Proc. 3rd AE Symp., Tokyo, 1976, 2-7, p. 167.

(10) K. Uchida, et al., "AE Characteristics in a Fatigue Test of Nozzle," Proc. 1st National AE Conf., 1977, p.37.

(11) M. Onoe, et al., "Multichannel AE Source Location System and Its Application to Fatigue Test of Model Reactor Vessel," Proc. 2nd AE Symp., Tokyo, 1974, Session 2, p.82.

(12) H. Nakasa, et al., "Acoustic Emission Analysis on Fatigue Failure of Fast Breeder Reactor Piping Components," Proc. 2nd AE Symp., Tokyo, 1974, Session 9, p. 21.

(13) H. Ohno, et al., "Application of Acoustic Emission Techniques to a Creep-Fatigue Test of a Type 304 Stainless Steel Elbow Component," PNC/CRIEPI Co-operative Research Report, CRIEPI 77002, 1977.

(14) H. Nakasa, et al., "AE Monitoring of a Thermal Ratcheting Test in the Sodium Piping Loop," J. NDI, Vol. 24, No. 2, 1975, p. 82.

(15) T. Kanagawa, et al., "Estimation of Spatial Geo-Stress Component in Rock Samples Using the Kaiser Effect of Acoustic Emission," Proc. 3rd AE Symp., Tokyo, 1976, 3-4, p. 229.

PROGRESS IN ACOUSTIC EMISSION MONITORING OF NUCLEAR PLANT - A REVIEW

DR. ADRIAN A. POLLOCK
DUNEGAN/ENDEVCO, SAN JUAN CAPISTRANO, CALIFORNIA 92675

ABSTRACT

The motivations for using acoustic emission in nuclear plants are identified and a chronological review of work conducted in the United States and Europe to date is presented. This perspective shows major progress from early pioneering efforts through the EBOR program to successful primary hydrotests, leak detection and limited-area on-line monitoring accomplished today. Comments are made on some specific technical issues. A list of commercial reactors on which AE work has been performed is included.

INTRODUCTION

The motivations to apply acoustic emission (AE) to nuclear plant inspection are very strong, and substantial progress has been made during the last twelve years despite major technical obstacles.

Three approaches, each with a different motivation, may be distinguished. First, much attention has been paid to the prospect of 100% volumetric inspection of major plant components - for example, the reactor pressure vessel or the entire primary loop. Here the motivation has become closely related to regulatory requirements and to the thrust to minimise failure probabilities and maximise safeguards. Second, AE may be called into use where recognised operational problems exist. Here the technical problems of AE monitoring are not so great, and the motivation is related primarily to the potential savings of down-time and improved maintenance efficiency through advance awareness of incipient problems.

Third, AE may be considered at the plant design stage. For some secondary components, there is a prospect that inspectability requirements can be fulfilled much more economically by AE than by other available techniques.

With respect to crack detection, the issue of feasibility is still very complex. Many factors are involved. Monitoring may be conducted on-line, and/or during periodic proof tests. Both material behaviour and instrumentation requirements depend on the temperature at which testing is conducted.Material behaviour alone is a complex issue since the concern is generally with weldments rather than parent plate, and since secondary effects, such as crack face interference and the cracking and crushing of oxides and slag inclusions, may make a major contribution to the observable AE.

A key question has long been asked, whether AE will be detectable above background noise. This question cannot be answered in a word. It must be extended to cover not only detectability but also interpretability, in the case where superficially similar signals caused by many different mechanisms may be present. Next, the extended question must be made more specific, since a useful answer can only be produced in the context of defined material circumstances and a defined monitoring system. This done, one may look for a meaningful and practical answer.

A significant broadening of the scope of AE on-line monitoring stems from the recognition that an AE system may serve as an effective detector of leaks and loose parts.

Projects conducted in this field have ranged in style from exhaustive conceptual analysis to empirical troubleshooting. The main object of this report is to provide by way of an essentially chronological review, a broad perspective and at least a reference tracer to all significant work to date in this area in the U.S.A. and Europe. In this effort apologies are offered to any whose work may have been inadvertently overlooked. Some technical details of current interest are addressed in the appendices.

REVIEW

As early as 1965, when only a handful of people were studying the phenomenon of acoustic emission, P. H. Hutton and J. C. Spanner at Battelle-Northwest recognized the possibility of using it to enhance safety and improve economy in nuclear plants and facilities (1). Their planned development anticipated the technical approaches that have been followed repeatedly in many subsequent programs: material characterization under static and cyclic stress; detection and location of overstressed areas in progressively more complex systems; development of sensors to withstand high temperatures and irradiation, and equipment to operate in a high background noise; area of interest inspection. In the following years through 1971, Battelle-Northwest pursued an AEC-funded program that is best summarized by Spanner (2). This program included noise studies at Dresden I reactor which steered future efforts towards high-frequency monitoring (typically above 500 kHz), hydraulic noise studies (3) and burst tests on piping, and studies on zircalloy and 304-L stainless steel as well as carbon steel.

D. L. Parry, then with Phillips Petroleum Company at NRTS, was also pursuing the application of acoustic emission to nuclear reactors in those early years. His work included monitoring of the PM-2A half-scale reactor pressure vessel undergoing embrittlement studies to failure (4), and monitoring of hydrotests of the SM-1A reactor undergoing an anneal program at Fort Greely, Alaska in 1967 (5). Results were reported at the Gatlinburg, Tennessee conference in 1967 which was also the birthplace of the Acoustic Emission Working Group. Parry pioneered in source location (independently of Green's work elsewhere (6)) and was probably the first to use simulated emission sources to check system calibration.

By the end of 1970 further noise studies had been conducted by both Hutton (7) and Parry at San Onofre, and by Parry also at Yankee Rowe (8). The difficulty of detecting AE from defect growth in the presence of hydraulic noise was well recognized. Parry and Robinson successfully located a leak at Elk River plant in 1968 (8).

It was in connection with leaks that acoustic emission technology first began to repay the substantial research and development investments made within the nuclear industry. In about 1969, a two-channel AE system at Dounreay successfully located a leak which had been impeding progress for a number of months and could not be located by other techniques.

By 1970, also, interest in acoustic emission was revived in Europe and preliminary work directed at nuclear applications was under way at Risley Engineering and Materials Laboratory (from about 1966), at Battelle-Frankfurt (from 1969) and at CEA Saclay (from about 1969). Personnel from both Risley and Saclay participated in the EBOR program in 1972.

The EBOR program, more properly known as EEI-TVA RP79, was initiated in 1968 and completed in 1973. While other NDT techniques were also given attention, the program is known primarily as a focal point for development and assessment of the potential of AE in the nuclear context. The original reports on EEI-TVA RP79 (9, 10) are very long and a valuable condensation to forty pages has been provided by Spanner (2). The program was sponsored primarily by the Edison Electric Institute and the Tennessee Valley Authority, with support from the AEC who made the EBOR vessel available, and from a number of equipment suppliers: Atomics International (10), Dunegan/Endevco (11), Holosonics, Southwest Research Institute (12) and Westinghouse Electric Corporation. Battelle-Northwest (13), Battelle-Columbus, Jersey Nuclear (14) and Teledyne (15) also participated. The program's place in the historical context of plant inspection has been described by E. C. Bailey (16).

Briefly, the EEI-TVA RP79 program highlighted tests on artificial defects grown by cyclic and static loading on the wall of an expendable reactor vessel (EBOR) in the presence of background noise which simulated normal reactor operating conditions. A number of different tests were monitored by various participants. In general the results were encouraging, with most participants reporting success in detecting and locating cracks under various test conditions. Some participants drew the very positive conclusion that small amounts of crack growth would indeed be detectable above background noise in an operating reactor (11). Other participants, however, drew the opposite conclusion. So controversy may begin, in a situation where people differ both in their technological capabilities and in their psychological attitudes.

EEI-TVA RP79 was not confined to the tests at NRTS on the EBOR vessel. Other tasks included the noise measurements at San Onofre mentioned earlier; shop hydrotests of the Arkansas Nuclear-One (PWR) and Peachbottom-3 (BWR) vessels by D. L. Parry (Jersey Nuclear) in 1971; the development of waveguides and system (CDL) concepts at Battelle Northwest (17); and laboratory tests by Battelle Columbus on surveillance specimens (9) removed from the Connecticut Yankee reactor, which showed that the emissivity of SA-302B is enhanced by exposure to nuclear radiation.

This result on irradiation effects runs parallel to Westinghouse findings that the emissivity of A-533B is enhanced by irradiation (31), but contrary to the experience of H. Nakasa (18) that the emissivity of several pure metals is reduced. Evidently irradiation effects vary according to the material and need to be treated with some care.

It is appropriate at this point to mention that Japanese experience in the application of AE to nuclear plant is very strong and dates back to the early 1970's. The only reason for not including it in the scope of this review is that a complete paper on the subject has just been given at this conference by Dr. Nakasa. Japan's positive commitment to nuclear energy as a major source of power has led to vigorous progress in AE research and application.

Two major research programs that were well advanced by the time EEI-RP79 closed were those of Risley Engineering and Materials Laboratory and of Westinghouse Electric Corporation. REML were developing real-time source location capability at the end of the 1960's and by 1972 had conducted quite extensive laboratory tests, pressure vessel tests and a noise survey on the No. 2 Reactor at Berkeley Nuclear Power Station (19).

P. G. Bentley of REML was among the first to recognise the important fact that a crack in ductile parent metal can propagate almost silently. Finding that many AE protagonists in the U.S.A. were unwilling to accept this point, he emphasized it more strongly and a substantial controversy developed. There followed a series of hydrotests sponsored by the British Navy (circa 1972) in which vessels containing large artificial defects in parent metal were monitored by several AE teams who reported conflicting results (20,21,22). The resulting confusion was later reviewed by Stahlkopf and Dau (23). It is unfortunate for the development of AE that a great deal of experimental pressure vessel work has been oriented by standard practice towards parent metal defects, yet the AE characteristics of parent metal and weldments can differ widely. Therefore, AE work that rides piggy-back on standard tests will not be a useful predictor of the capabilities of AE in real-life monitoring where the majority of practical problems arise in weldments. A related misfortune is the fact that since weldments are more complex, more variable in type, and harder to specify and procure than parent metal, the laboratory characterization of AE from weldments has lagged far behind the characterization of AE from parent metal.

Probably typical of the weldment/parent metal situation in thick sections is the experience of M.P.Kelly (24) in monitoring the burst test of HSST Vessel V-9 (1974), one of a number of HSST vessel tests that have been monitored for AE by various agencies. He found that emission from insignificant defects in the nominally sound welds greatly outweighed emission from the large artificial parent-metal defect that was the center of attention.

Definitive work on the problem of the parent metal defect was to come from the UK Central Electricity Generating Board. The behaviour of a carbon-manganese reactor pressure vessel steel was characterized in the Central Electricity Research Labortories by I.G.Palmer (1973) (25), who with P.T.Heald (26) developed an improvement on H.L.Dunegan's model (27) of AE from plastic zone growth. Plastic zone growth in this steel produced quite plentiful emission while ductile tearing was relatively quiet. These results were applied to a test on a model vessel 1.2m in diameter by A.C.E. Sinclair (1974) (28), using an early version of the ACEMAN source location system developed at Berkeley Nuclear Laboratories. Emissions from the growth of the plastic zone at an artificial defect in this vessel were successfully detected and located, and a very encouraging fit to advanced theoretical models was demonstrated.

Several factors were instrumental in Sinclair's success as compared with the conflicting results at Culcheth. It was probably not a matter of system sensitivity since all the testing agencies had good capability in this respect. However the Berkeley vessel, unlike the Culcheth vessel, was carefully prepared to provide optimum data from the defect with minimum interference from background noise sources. All Sinclair's transducers were within about 70 cm of the defect. Futhermore, the defect on the Berkeley vessel initially penetrated only 65% of the wall thickness whereas the largest defect on the Culcheth vessel penetrated 88% of the wall thickness, thus being so large that through-thickness yielding would have been completed during pre-fatiguing of the vessel before the main AE test began. It is possible, also, that the microstructure of the CEGB steel was more emissive.

The Berkeley test established an important positive datum but it did tend to confirm that the ability of AE to identify plastic zone growth in parent metal in a practical nuclear environment will be severely limited. Even though the amplitude distribution measured by Sinclair had a rather low b-value (1.5) compared with other results for ductile steels (29), the number of emissions detected for a given amount of deformation will fall off quite rapidly as system sensitivity decreases. To locate a hypothetical defect of this type is a 100%-of-volume nuclear vessel hydrotest with a reasonable number of transducers it would be necessary to push to the limits of available sensitivity. This procedure is disliked by some experienced operators because of the large number of small signals that may be present from altogether innocuous sources, especially when previously unpressurized components are involved. During plant operation, the effective system sensitivity is so much reduced by background noise that no emission from plastic zone growth in parent metal would be detectable. The greatest practical value of traditionally conceived AE will surely be found in the context of weldments and cases of environmentally induced cracking where higher amplitude emissions are produced.

Let us return to our chronological review. One of the more significant contributors to the EBOR program was the Westinghouse Electric Corporation which has had a substantial AE development program in-house from 1970 to the present time. Early in their program, R. Gopal and colleagues developed a high-temperature sensor and source location electronics, monitored nuclear plant at Prairie Island and Robinson S.E. Plant during hydrotest and during operation, and conducted some excellent wave propagation studies on a large steel plate (30). Several vessels of the HSST program were also monitored to rupture. By 1976, nine plants had been monitored in various modes (Appendix 1) and two had had AE monitoring equipment permanently installed during construction. Rather extensive information on background noise levels at various locations under various plant operation conditions was published. Fatigue tests on unirradiated and irradiated specimens of A533B steel were undertaken and a model was developed relating AE to microfractures in the plastic zone (31). Increasing emphasis on leak detection has been noticable through the Westinghouse program (32); it is felt that in this area an AE monitoring system can be of immediate demonstrable value, while the interpretation of signals from growing defects is harder so that this traditional aspect of AE technology will take longer to mature (33).

Work in continental Europe progressed in the early 1970's with studies on fast breeder reactors in France and hydrotests of PWR and BWR in Germany. Official agencies also reviewed the state of the art (34). Bret described work in the Rapsodie reactor (1972) on the acoustic detection of gas bubbles and liquid sodium boiling (35), a topic also investigated by T.T. Anderson and colleagues at Argonne National Laboratory (36). E.G. Tomachevsky undertook experimental monitoring of the vessel hangers of the PHENIX fast breeder reactor (1972) (37). N. Chretien of CEA-Vaujours (38) and M. Asty of CEA-Saclay (39) discussed source location technology from the mathematical standpoint. J. Roget and S. Masson described the application of a Dunegan/Endevco 1032 system at the ECAN Indret plant of the French Navy (1976) (40). A. Lucia and colleagues at Euratom Ispra (Italy) pursued a fundamental approach to signal processing (41).

P. Jax and colleagues at Battelle-Frankfurt, equipped with a Dunegan/Endevco 1032 system, reported three hydrotests on nuclear plant in 1975. On the Neckar BWR (42) they monitored the initial hydrotest of piping from the steam generators to the outside of the secondary containment and reported high integrity of these components, while also identifying a leak outside the area of interest. The reactor pressure vessel for Biblis B PWR was monitored during shop hydrotest (43). In this test again there were no indications of significant defects, but an interesting feature was the occurence of a number of emissions at the location of the only weld on the vessel that had been stress relieved by induction heating. A third test was conducted on the pressure vessel and adjoining piping of the Isar reactor (cold hydrotest) (44). In this case signal levels were less than on the Biblis vessel, nevertheless eighteen sources of acoustic emission were located and classified as insignificant. The classification was confirmed by ultrasonic inspection, which found innocuous anomalies at some of the AE sites and nothing at others. The point was made that the less significant the indications, the less one should expect correlation between AE and other NDT techniques. Battelle's test procedures were based on a very extensive program of basic research and development over the previous five years. Careful attention to wave propagation features and a broad-based and systematic approach to signal interpretation are evident in their reports.

Back in the United States, the EBOR program and subsequent reports (45) increased the visibility of AE technology among the utilities and in the nuclear industry in general. Some of the more progressive utilities began to utilize the technique. Baltimore Gas and Electric Company contracted for baseline AE measurements to be made during cold hydrotest of Calvert Cliffs Units I (1973) and II (1976), by Exxon Nuclear (46) and Dunegan/Endevco (47) respectively. In both cases the whole primary loop was monitored, a job that required some 60 channels. Findings were broadly similar on the two units, in a pattern now becoming familiar: some dozens of emitting regions were identified, none were analyzed as significant to structural integrity, and some were confirmed and reanalyzed as insignificant by ultrasonics. Other nuclear components monitored by Kelly include three military reactor pressure vessels and, overseas, piping at Fessenheim (1977) and (in association with Battelle-Frankfurt) the primary loop at Goesgen (1977); and by Parry, nine reactors, some with repeats, in the U.S.A. and Europe in the years 1972-77 as summarized in Appendix 1. Typically Exxon Nuclear's work has been the inspection of complete primary loops during initial hydrotests, but other configurations have also been monitored.

On-line monitoring has also progressed in the last five years. In 1972/73 Commonwealth Edison pioneered in the application of acoustic emission to known problem areas, a major form of utilization as has been mentioned at the beginning in this report. Instrumentation procured from Dunegan/Endevco was installed to monitor potentially loose sparger arms on Dresden II and III and Quad Cities III and IV. Much valuable information was obtained in succeeding years, some of it quite unexpected. Another utility actively developing AE monitoring techniques is Philadelphia Electric Company (48); here valve surveillance and pipe surveillance using an ERDA-supported Trodyne system are in progress at the Peach Bottom BWR plant. As utility interest increased, the impact of code requirements and the practical managerial issues implicit in AE utilization came under discussion (49).

Recent sponsored research efforts in the U.S.A. include the monitoring of the shop hydrotest of the La Salle II BWR pressure vessel, and comparison of the results with ultrasonic inspection, by Acoustic Emission Technology Corporation (50). This project was sponsored by EPRI, who have also funded studies of intergranular stress corrosion cracking in stainless steel (51) and basic work on system calibration and physical fundamentals (52).

In an NRC-sponsored program at Dunegan/Endevco, an area-of-interest monitoring system was installed at Dresden and some of the complex issues of data interpretation that could arise in wide-area monitoring were analyzed in depth (53, 54). The characterization and interpretation aspects of this program are essentially continuing on an enlarged scale at Battelle-Northwest (55).

On-line monitoring for crack growth (assuming that detectability and reliability can be demonstrated) will probably become focussed on the times when the reactor is changing state, i.e. start-ups, shut-downs and scrams. By contrast, the functions of leak detection and loose part detection are more pertinent to steady state operation. It may be through the combination of these functions in an integrated system that AE technology will become a cost effective and practical on-line monitoring tool.

Much simpler than on-line monitoring is a case of trouble-shooting which probably saved enough money to account at one stroke for many of the development efforts previously described. A leak in the CANDU reactor at Pickering was indicated by routine tests during operation in 1974. The faulty tube, and subsequently others, were quickly and efficiently identified by acoustic emission while other techniques proved ineffective (56). A few incidents of this kind have enormous leverage where unscheduled shut-downs of expensive plants are at stake.

Evidently the great majority of AE work in the nuclear field has been directed towards assuring the structural integrity of the primary pressure boundary. But other applications have also been explored, sometimes with very positive results; aside from leak monitoring of valves and boiling detection which have already been mentioned, these include a very successful program on the monitoring of welding during fabrication (57), monitoring of cracking in fuel pellets (58), monitoring of concrete components (59), loose part monitoring (60) and inspection of Class III buried piping.

CONCLUSION

A review of this historical perspective makes it evident that acoustic emission is here to stay and that a ponderous momentum of progress has been developed. The motion is that of a large body moving slowly rather than a small body moving fast. Expectations and hopes naively aroused have often not been matched by performance, and the resulting attitudes of potential users have varied across the spectrum (61). The first pioneers thought at the outset that on-line monitoring could be accomplished within three years and their optimism has been echoed by several others through a period lasting four times that long. From a twelve-year perspective we are perhaps half-way towards delivery on those initial hopes of on-line, wide-area integrity assurance, and final delivery is still not guaranteed. But twenty-five years of development is not disproportionate to the forty-year lifespan of a nuclear reactor, and the first returns of practical value in more limited contexts are already in evidence. In the trouble-shooting mode AE can provide valuable results today. Baseline inspections and other hydrotests lie well within the state of the art and are likely to justify their costs in the long run and also conceivably in the short run. On-line, wide-area surveillance for crack growth remains a sufficiently important prospect to justify continued efforts to make it become a reality in the future.

There has been a strong tendency, world-wide, for much of the first interest in acoustic emission in any country to come from the nuclear community. This tendency has been apparent in the United States, Great Britain, Germany, France, Japan, the Soviet Union (62), Italy, Denmark, Hungary, India and South Africa. This reflects the matching of high technologies, the importance of nuclear power and the potency of acoustic emission. Further major improvements in the technology can be confidently expected.

APPENDIX I

The following table summarizes all the identified AE work that has been conducted to date on commercial power plants in the United States and Europe. Tests on the EBOR vessel and work on military reactors are not included. Here is a key to the abbreviations in the last column for the agencies conducting the work:

BNW, Battelle Northwest. W, Westinghouse. IN, Idaho Nuclear. JN, Jersey Nuclear. REML, Risley Engineering and Materials Laboratory. EN, Exxon Nuclear, CEA, Commissariat pour l'Energie Atomique. CEC, Commonwealth Edison. SWRI, Southwest Research Institute. BF, Battelle Frankfurt. AET, Acoustic Emission Technology. T, Trodyne Corporation. DE, Dunegan/Endevco. PE, Philadelphia Electric Company. TEC, Technology for Energy Corporation.

YEAR	REACTOR(S)	TYPE	COMMENTS	BY
1968	Dresden I	BWR	Noise studies	BNW
1969-70	Saxton	PWR	Noise studies	W
1970	San Onofre I	BWR	Noise studies	BNW/IN
1970	Yankee Rowe	PWR	Noise studies	IN
1971	Arkansas Nuclear One	PWR	Shop hydrotest	JN
1971	Peach Bottom III	BWR	Shop hydrotest	JN
1971	Berkeley II (UK)	Magnox	Noise studies	REML
1971	Turkey Point III & IV	PWR	Attenuation studies	W
1971-73	Robinson II	PWR	On-line monitoring	W
1972+	Prairie Island I & II	PWR	On-line monitoring	W
1972-77	Ginna	PWR	Leak monitoring	W
1972	Calvert Cliffs I	PWR	Primary hydrotest	EN
1972	Zion I & II	PWR	Primary hydrotest	EN
1972	Peach Bottom II	BWR	Primary hydrotest	EN
1972	PHENIX (France)	FBR	Hangers	CEA
1972	Rapsodie (France)	FBR	Boiling detection	CEA
1972+	Dresden II & III	BWR	On-line monitoring	CEC
1972+	Quad Cities III & IV	BWR	On-line monitoring	CEC
1973	Beaver Valley I	PWR	Attenuation studies	W
1973	Peach Bottom III	BWR	Primary hydrotest	EN
1973+	La Crosse	BWR	On-line monitoring	SWRI
1974	KRB (Germany)	BWR	Hydrotest	EN
1974	KWL (Germany)	BWR	Hydrotest	EN
1975	GKN (Germany	PWR	Piping hydrotest	EN
1975	Obrigheim (Germany)	BWR	Piping hydrotest	BF
1975	Neckar (Germany)	BWR	Piping hydrotest	BF
1975	Isar (Germany)	BWR	Vessel hydrotest	BF
1975	Biblis B (Germany)	PWR	Shop hydrotest	BF
1975	La Salle II	BWR	Shop hydrotest	AET
1975	Vermont Yankee	BWR	Noise study	T
1975+	Dresden III	BWR	On-line monitoring	CEC/DE
1976	Calvert Cliffs II	PWR	Primary hydrotest	DE
1976	KWO (Germany)	PWR	Hydrotest	EN
1976	GKT (Austria)	BWR	Hydrotest	EN
1976	KWL (Germany)	BWR	Hydrotest	EN
1976-77	Fessenheim (France)	PWR	Hydrotest	W
1977	Indian Point I	PWR	Hydrotest &c.	W
1977	KRB (Germany)	BWR	Hydrotest	EN
1977	Fessenheim (France)	PWR	Piping hydrotest	DE
1977	Goesgen (Switzld.)	PWR	Primary hydrotest	BF/DE
1977+	Peach Bottom III	BWR	On-line monitoring	PE/T
1977+	Peach Bottom III	BWR	On-line monitoring	PE/TEC

REFERENCES

(1) P. H. Hutton and J. C. Spanner, "Detection Of Metal Overstress By Acoustic Emission," BNWL-SA-438, January 19, 1966.

(2) J. C. Spanner, "Acoustic Emission Techniques And Applications," Intex Publishing Company, 1974. Available from ASNT.

(3) P. H. Hutton, "Detecting Acoustic Emission In The Presence Of Hydraulic Noise," Nondestructive Testing 2, 111-115, May 1969.

(4) D. L. Parry, "Non Destructive Flaw Detection By Use Of Acoustic Emissions," IDO-17230, Phillips Petroleum Company, April 1967.

(5) D. L. Parry, "Nondestructive Flaw Detection In Nuclear Power Installations," Incipient Failure Diagnosis for Assuring Safety and Availability of Nuclear Power Plants, USAEC Report CONF-671011, January 1968.

(6) A. T. Green, "Detection Of Incipient Failures In Pressure Vessels By Stress-Wave Emissions," Nuclear Safety 10, 1, January-February 1969.

(7) P. H. Hutton, "Nuclear Reactor Background Noise Vs. Flaw Detection By Acoustic Emission," BNWL-SA-3820, 1971.

(8) D. L. Parry and D. L. Robinson, "Incipient Failure Detection By Acoustic Emission - A Development And Status Report," IN-1398, Idaho Nuclear Corporation, NRTS (USAEC), August 1970.

(9) E. R. Reinhart (compiler), "In-Service Inspection Program For Nuclear Reactor Vessels," Final Biannual Progress Report No. 8, Southwest Research Institute, NTIS Accession No. PB 222 701, May 21, 1973.

(10) E. R. Reinhart (compiler), "A Joint Industry/Utility Program To Develop NDT Techniques For Nuclear Reactor Inspection," Southwest Research Institute, NTIS Accession No. PB 224 698, May 21, 1973.

(11) M. P. Kelly and R. L. Bell, "Detection And Location Of Flaw Growth In The EBOR Vessel," Technical Report DE 73-4, Dunegan/Endevco, San Juan Capistrano, California, 1973.

(12) S. P. Ying, J. E. Knight and C. C. Scott, "Background Noise For Acoustic Emission In A Boiling Water And A Pressurized Water Nuclear-Power Reactor," Journal of the Acoustical Society of America 53, 6, 1627-1631, 1973.

(13) J. B. Vetrano, W. D. Jolly and P. H. Hutton, "Continuous Monitoring Of Nuclear Reactor Pressure Vessels By Acoustic Emission Techniques," Paper C58/72, pp 221-226, Periodic Inspection of Pressure Vessels, Institution of Mechanical Engineers, London, May 1972.

(14) D. L. Parry, "Acoustic Integrity Analysis Of EBOR Reactor Vessel," Jersey Nuclear Company, Richland, Washington, March 1972.

(15) B. H. Schofield, "Utilization Of Acoustic Emission For In-Service Inspection," Paper C40/72, pp 76-82, Periodic Inspection of Pressure Vessels, Institution of Mechanical Engineers, London, May 1972.

(16) E. C. Bailey, "Inservice Inspection Of A Nuclear Unit - - No Hands," 43rd General Meeting of the National Board of Boiler and Pressure Vessel Inspectors in conjunction with the ASME Boiler and Pressure Vessel Committee, New Orleans, May 6, 1974.

(17) W. D. Jolly, "Field Evaluation Of A Prototype Acoustic Emission Reactor Surveillance System," ASNT Spring Conference, Los Angeles, March 1973.

(18) H. Nakasa, "Application Of Acoustic Emission Techniques To Material Diagnosis," Paper IAEA-SM-168/D-5, Nuclear Power Plant Control and Instrumentation, 1973.

(19) P. G. Bentley, T. E. Burnup, E. J. Burton, A. Cowan, N. Kirby, "Acoustic Emission As An Aid To Pressure Vessel Inspection," Paper C30/72, pp 54-66, Periodic Inspection of Pressure Vessels, Institution of Mechanical Engineers, London, May 1972.

(20) D. L. Parry, "NDT - Acoustics Testing Of Culcheth Pressure Vessel," Exxon Nuclear Company, Report XN-129, May 1973.

(21) D. Birchon, R. Dukes and J. Taylor, "Acoustic Emission Monitoring Of A Pressure Vessel," Periodic Inspection of Pressure Vessels, Institution of Mechanical Engineers, London, June 1974.

(22) P. G. Bentley and D. G. Dawson, "Acoustic Emission Measurements On Thick Walled Ductile Steel Pressure Vessels," Symposium on Reactor Inspection Technology, British Nuclear Energy Society (Western Branch), Bristol, February 1975.

(23) K. E. Stahlkopf and G. J. Dau, "Acoustic Emission, A Critical Assessment," Nuclear Safety 17, 1, 1976.

(24) M. P. Kelly, "Acoustic Emission Monitoring Of Vessel V-9", ORNL-TM-4805 Volume II (G. D. Whitman), pp 60-68, March 1975. Also available as DE V-9 from Dunegan/ Endevco, San Juan Capistrano, California.

(25) I. G. Palmer, "Acoustic Emission Measurements On Reactor Pressure Vessel Steel," Materials Science and Engineering 11, 227-236, 1973.

(26) I. G. Palmer and P. T. Heald, "The Application Of Acoustic Emission Measurements To Fracture Mechanics," Materials Science and Engineering 11, 181-184, 1973.

(27) H. L. Dunegan, D. O. Harris and C. A. Tatro, "Fracture Analysis By Use Of Acoustic Emission," Engineering Fracture Mechanics 1, 105-122, 1968.

(28) A. C. E. Sinclair, C. L. Formby and D. C. Connors, "Acoustic Emission From A Defective C/Mn Steel Pressure Vessel," CEGB Report RD/B/N2976, Berkeley Nuclear Laboratories, April 1974.

(29) A. A. Pollock, "Quantitative Evaluation Of Acoustic Emission From Plastic Zone Growth," Technical Report DE 76-8, Dunegan/Endevco, San Juan Capistrano, California, August 1976.

(30) R. Gopal, "On-Line Acoustic Emission Monitoring Systems For Nuclear Power Plants," US-Japan Joint Symposium on Acoustic Emission, July 1972.

(31) R. Gopal, J. R. Smith and G. V. Rao, "Experience In Acoustic Monitoring Of Pressurized Water Reactors," p 1, Third Conference on Periodic Inspection of Pressurized Components, Institution of Mechanical Engineers, London, September 1976.

(32) R. Gopal, "Acoustic Monitoring Systems To Assure Integrity Of Nuclear Plants," pp 200-220, Monitoring Structural Integrity By Acoustic Emission, ASTM STP 571, American Society for Testing and Materials, 1975.

(33) R. Gopal, J. R. Smith and G. V. Rao, "Acoustic Monitoring Instrumentation For Pressurized Water Reactors," pp 505-514, ISBN 87664-362-4, 23rd International Instrumentation Symposium, Instrumentation Society of America, Las Vegas, May 1977.

(34) A. Nielsen, "Acoustic Emission Surveillance Methods," Reports RISO-M-1491 and RISO-M-1896, Danish Atomic Energy Commission, May 1972 and update October 1976.

(35) A. Bret, "Detection Acoustique Dans Le Réacteur Rapsodie," pp 413-432, Journées d'Etudes sur l'Emission Acoustique (P. Gobin et al), Institut National des Science Appliquées de Lyon, March 1975.

(36) T. T. Anderson, A. P. Gavin, J. R. Karvinen, C. C. Price and K. J. Reimann, "Detecting Acoustic Emission In Large Liquid Metal Cooled Fast Breeder Reactors," pp 250-269, Acoustic Emission, ASTM STP 505, American Society for Testing and Materials, 1972.

(37) E. G. Tomachevsky, "Appliance Of Stress Wave Emission In PHENIX Reactor," pp 48-54, Proceeding of the Third Meeting of the European Working Group on Acoustic Emission, EWGAE-ISPRA, September 1974, Commission of the European Communities EUR 5513e, 1976.

(38) N. Chretien and C. Perennes, "La Localisation Des Sources D'Emission Acoustique," pp 279-322, Journées d'Etudes sur l'Emission Acoustique (P. Gobin et al), Institut National des Sciences Appliquées de Lyon, March 1975.

(39) M. Asty, P. Bernard, J. Monier, "Localisation Des Sources D'Emission Acoustique : Possibilités - Limites," pp 258-278, Journées d'Etudes sur l'Emission Acoustique (P. Gobin et al), Institut National des Sciences Appliquées de Lyon, March 1975.

(40) J. Roget and Sh. Masson, "Application Des Techniques De Détection De L'Emission Acoustique Au Contrôle Des Capacités A Pression," Session 1976, Association Technique Maritime et Aéronautique, 47 rue de Monceau, 75008 Paris.

(41) A. C. Lucia and G. Redondi, "Acoustic Emission Signal Interpretation," Proceeding of the Third Meeting of the European Working Group on Acoustic Emission, EWGAE-ISPRA, September 1974, Commission of the European Communities EUR 5513e, 1976.

(42) P. Jax et al., "Ergebnisse der Schallemissionsanalyse bei der Erstdruckprobe der Frischdampf - und Speisewasserleitung im Gemeinschaftskernkraftwerk Neckar," Report No. BF-V-62.793-2, Battelle-Institut e.V., Frankfurt am Main, September 1975.

(43) P. Jax et al., "Schallemissionsmessungen bei der Druckprobe eines petrochemisches Behaelters und eines Kernreaktordruckbehaelters sowie bei der Systemdruckprobe im Kernkraftwerk," Technical Report on Research Project RS 31/13 (Bundesministerium fuer Forschung und Technologie), Battelle - Institut e.V., Frankfurt am Main, April 1975.

(44) P. Jax et al., "Ergebnisse der Schallemissionsanalyse bei der Erstdruckprobe des Reaktordruckbehaelters KKI," Report No. BF-V-31.010-1, Battelle Institut e.V., Frankfurt am Main, August 1975.

(45) R. L. Bell, "The Use Of Acoustic Emission To Detect Incipient Failure In Pressure Vessels," Kerntechnik 5, 216-222, 1974; also available as DE 74-1 from Dunegan/Endevco, San Juan Capistrano, California.

(46) D. L. Parry, "Industrial Application Of Acoustic Emission Analysis Technology," pp 150-183, Monitoring Structural Integrity by Acoustic Emission, ASTM 5TP571, American Society for Testing and Materials, 1975.

(47) M. P. Kelly and R. J. Schlamp, "Acoustic Emission Pre-Service Inspection Of Calvert Cliffs Unit 2 Nuclear Power Plant," Technical Report DE 76-7, Dunegan/Endevco, San Juan Capistrano, California.

(48) W. F. Hartman and J. W. McElroy, "Acoustic Emission Surveillance For Improving Availability Of Nuclear Power Plants," Progress in Nuclear Energy 1, 2-4, 673-680, New Series, 1977.

(49) H. H. Askwith, "Preservice & Inservice Inspection Planning Using Acoustic Emissions As An NDE Tool," presented to the Acoustic Emission Working Group Meeting in Laguna Hills, California, August 1974.

(50) K. E. Stahlkopf and A. T. Green, "Nondestructive Pressure Vessel Testing By Acoustic Emission," EPRI Journal 1, 1, February 1976.

(51) K. E. Stahlkopf, P. H. Hutton and E. L. Zebroski, "Detection Of Stress Corrosion Cracking In Stainless Steel Piping By Acoustic Emission," 3rd Conference on Periodic Inspection of Pressurized Components, Institution of Mechanical Engineers, London, September 1976.

(52) N. N. Hsu, J. A. Simmonds and S. C. Hardy, "An Approach To Acoustic Emission Signal Analysis - Theory And Experiment," Materials Evaluation 35, 10, 100-106, October 1977.

(53) A. A. Pollock, "Application Of Acoustic Emission As An On-Line Monitoring System For Nuclear Reactors," (Bi-annual Progress Report, July 1-December 31, 1975) NUREG-0159, NRC-5, January 1977. Also available as DE 76-9, Dunegan/Endevco, San Juan Capistrano, California.

(54) A. A. Pollock and J. R. Wadin, "Application Of Acoustic Emission As An On-Line Monitoring System For Nuclear Reactors," (Final Progress Report, January 1-September 30, 1976) Technical Report DE 77-4, Dunegan/Endevco, San Juan Capistrano, California.

(55) P. H. Hutton and E. B. Schwenk, "Program To Develop Acoustic Emission - Flaw Relationship For Inservice Monitoring Of Nuclear Pressure Vessels," (Progress Report, February 1-July 1, 1977) BNWL 2232-2, NUREG-0250-2 R5, November 1977.

(56) J. D. Allinson, "Acoustic Emission As Applied To Leak Detection In CANDU-PHW Reactor Pressure Tubes," Nondestructive Evaluation in the Nuclear Industry, American Society for Metals, Denver, December 1975.

(57) D. Prine, "Inspect As You Go With Acoustic Emission," Welding Design and Fabrication, January 1977.

(58) D. S. Kupperman, C. R. Kennedy and K. J. Reimann, "Acoustic Emission From Fuel Pellets In A Simulated Reactor," Proceedings of the Institute of Acoustics 4-22-1, 1976.

(59) A. T. Green, "Stress-Wave Emission And Fracture Of Prestressed Concrete Reactor Vessel Materials," pp 635-649, Materials Technology: An Inter-American Approach for the Seventies, Volume 1, ASME, 1970.

(60) W. F. Hartman, "Continuous, In-Service Acoustic Emission Monitoring Of Nuclear Reactor Components," Paper 5C11, Eighth World Conference on Nondestructive Testing, ICNDT, Cannes, September 1976.

(61) J. F. Borst, "Acoustic Emission - Is It A Promise Or Mirage?" Nuclear Engineering International, March, 1977.

(62) V. Gorbachev, Y. Zacharov, S. Paraev, Y. Resnikov and V. Rozhnov, "Investigations Of Some Acoustic Emission Parameters Of Metals Under Various Kinds Of Tensile Loading," Meeting of IAEA Technical Committee on use of Non-Destructive Testing Techniques for In-Service Inspection of Reactor Pressure Components, Kobe, Japan, April 1977.

ACOUSTIC EMISSION MONITORING AND ULTRASONIC CORRELATION ON A REACTOR PRESSURE VESSEL

K. Stahlkopf, Electric Power Research Institute
A. Green and C. Morais, Acoustic Emission Technology Corp.

INTRODUCTION

Ultrasonic examinations of nuclear reactor pressure vessels as mandated by Section XI of the ASME Boiler and Pressure Vessel Code has proven to be both reliable and accurate. However, newer techniques, such as acoustic emission (AE) must be evaluated to assess their usefulness in further assessing the safety and integrity of the nuclear pressure boundary. Successful development of AE methods would provide a powerful tool for location and monitoring of defects in engineering structures (1, 2, 3). However, there are still many limitations to the use of the current technology (4). Its acceptance as an engineering tool will depend on the confidence that can be placed in the results of its use. In order to accomplish a realistic evaluation of any new nondestructive testing tool, it is necessary to measure its performance under actual field conditions with the performance of a well understood and established method.

The Electric Power Research Institute initiated such a program with General Electric Company and Acoustic Emission Technology Corporation to compare acoustic emission and ultrasonic inspections of an actual nuclear reactor pressure vessel. The objective of this effort was to establish the correlation between defects identified by acoustic emission monitoring of a BWR pressure vessel during its initial hydrostatic pressure test and ultrasonic examination of the same vessel.

Acoustic Emission Testing

The acoustic emission monitoring was performed on the pressure vessel as it was undergoing shop hydrostatic tests at the manufacturer's site. The vessel was a standard boiling water reactor vessel designed to operate at 8.6 MPa (1250 PSIG) and hydrotested to 11 MPa (1563 PSIG). Following the hydrostatic test, a preservice ultrasonic examination was performed to requirements of Section XI of the ASME Boiler and Pressure Vessel Code. A second ultrasonic examination using higher sensitivity than required by ASME Section XI was performed in the areas identified by acoustic emission. Based upon previous experience with RPV steels and pressure vessels, the structure was instrumented with 18 AE sensors placed approximately 5.2 meters (17 feet) apart. The sensors were mounted on the reactor pressure vessel in the form of diamond shaped rows as shown in Figure 1. During the initial set up tests, it was found that the control rod penetrations in the lower head interfered with the normal coverage sensor 18 would provide, thus, its use was limited.

Unfortunately, because of fabrication (a shop error), several of the sensors were positioned incorrectly causing a skew of the real-time source location with respect to actual position. However, this skew was found and corrected in post-test analyses.

Three straps used to secure scoffalding were tightened around the circumference of the reactor pressure vessel and left in place during the pressure test. Relative motion between the straps and the vessel wall created a number of artificial signals. As shown in Figure 1, one strap was almost directly over a girth weld and as a consequence, data originating from the weld seam region were questionable.

Another problem that arose was the artifact noise generated by the sliding seals used to close off the nozzles during the hydrostatic test. These seals moved as pressure was increased and were the most active source of acoustic noise encountered during the test. Even so, postprocessor techniques enabled AE sites to be located between and immediately adjacent to the nozzles. Postprocessor techniques involve the use of spatial discrimination (in conjunction with amplitude levels, test time, and parametric levels) to limit the influence of nearby noise sources. Postprocessing cannot, of course, extract significant data from an excessive volume of false signals. The welds beneath the scaffolding strap and the safe end welds in the nozzle region, therefore, could not be evaluated.

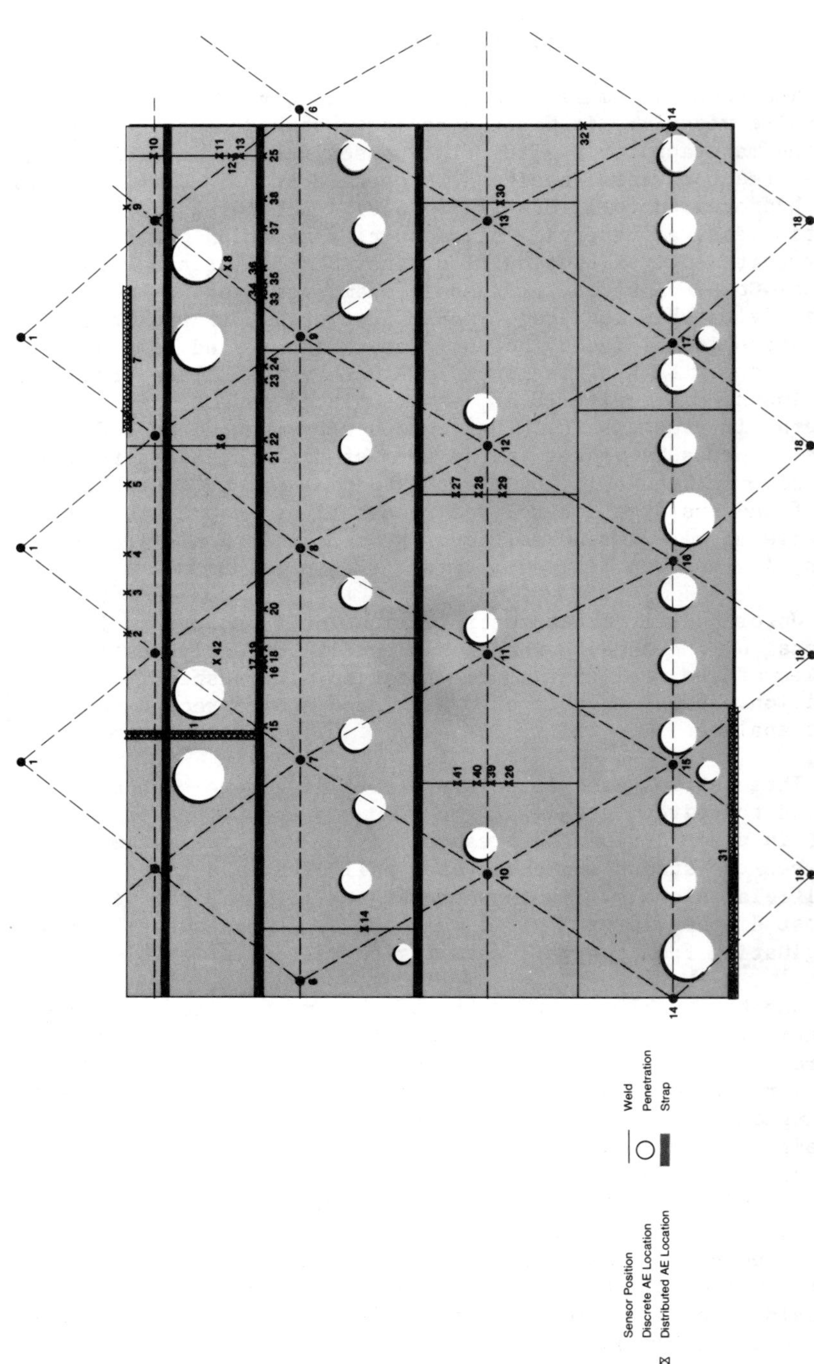

Figure 1 - SENSOR POSITION AND AE LOCATION

The AE monitoring identified 42 sites, 3 of which were along lengths of weld seams rather than discrete locations. These sites are shown in Figure 1. Table I lists the 24 most significant AE locations using three analysis methods. They are: Number of events, summation of amplitude, and individual amplitude. The total number of events shows the most active locations, while the summation of amplitude shows the greatest amount of energy release. Ranking by individual amplitude separates those locations with many small emissions from those with few emissions but of reasonably high amplitude. No attempts have been made to correct the emission amplitude with respect to location distance from the sensor.

As previously noted, emissions were noted along the length of three weld seams, shown on Figure 1 as sites 1, 7, and 31. However, sites 1 and 31 had two distinct emission centers each along its length. These sites have been designated 1a, 1b, 31a, and 31b for ranking purposes.

From the three ranking methods, the top ten locations in each method provide the following information:

a. Locations 2, 5, 6, 7, 9, 26, and 31a rank in the top ten of all three methods.

b. Locations 3, 30, and 31b rank in the top ten of two methods.

c. Locations 13, 36, and 39 rank in the top ten by only one method.

Table II summarizes these data.

Preservice Ultrasonic Examination

A preservice ultrasonic examination was performed with procedures described by ASME Boiler and Pressure Vessel Code Section XI, with one exception. The ultrasonic recording level in the ASME Code is equal to or greater than 50% DAC. In this examination all indications equal to or greater than 25% were recorded. Examinations were made with L wave, 45° shear, and 60° shear. Eight locations identified by AE were found by the preservice examination. All indications were code acceptable and thus, the vessel was acceptable. The sites found by this inspection are listed in Table III. Four of the sites showed multiple ultrasonic indications. An additional 98 code acceptable ultrasonic indications were

Table I. Rank of 24 Most Significant Locations By Three Analysis Methods

Rank	Site No.	Number of Events	Site No.	Summation of Amplitude	Site No.	Recorded Individual Amplitude*
1	7	81	7	376	7	24
2	26	56	26	252	5	12
3	6	39	6	225	31a	12
4	31b	31	31b	182	6	11
5	3	29	2	169	9	10
6	2	27	5	147	30	9
7	9	17	31a	139	26	8
8	5	14	9	127	2	8
9	31a	13	3	106	39	7
10	36	12	30	99	13	5
11	13	10	13	86	31b	4
12	30	9	39	82	40	4
13	39	8	27/28	48	3	3
14	32	8	36	48	27/28	3
15	27/28**	6	40	44	10	2
16	35	6	32	43		1
17	1a	6	10	33		1
18	10	5	35	25		1
19	1b	5	1a	22		1
20	38	5	14	22		1
21	33	5	1b	21		1
22	14	4	37	21		1
23	37	4	17	20		1
24	17	4	34	20		1

*Number of events is equal to or larger than 11, which is an arbitrary maximum recording level.

**Listed as a common site due to lack of localized resolution produced by nozzle geometry.

Table II. Location Occurrence Distribution -- First Ten in Each Ranking Method

Site No.	Three Ranking Methods	Two Ranking Methods	One Ranking Method
2	X		
5	X		
6	X		
7	X		
9	X		
26	X		
31a	X		
3		X	
30		X	
31b		X	
13			X
36			X
39			X

Ranking Methods are: Number of Events
Summation of Amplitude
Individual Amplitude

Table III. AE Sites Confirmed by Preservice Ultrasonic Inspection

AE Site No.	Number of Discrete UT Indications
1	48
10	5
14	1
26	6
29	1
39	1
41	1
42	40

found which were not found during the acoustic emission monitoring of the pressure vessel hydrostatic test.

High Sensitivity Ultrasonic Examination

To try to confirm the existance of defects at the acoustic emission sites where preservice ultrasonic inspection was not sensitive enough to find very small indications, a high sensitivity ultrasonic examination was performed. In this examination after the initial calibration and establishment of the Distance Amplitude Correction (DAC) were completed in accordance with the ASME Code, the gain was increased by a factor of 10 (20db). All indications greater than 2% DAC were recorded. Examinations were made in 45° shear and 60° shear modes. No L wave examination was performed.

Tables IV and V list the ultrasonic examination results in terms of percent DAC for 45° and 60° beam angles respectively. Of the 42 sites identified by AE, 20 gave no discrete indication greater than 2% DAC at either ultrasonic testing angle. Ultrasonic examination ranked sites 1, 10, 11, 13, 24, 30, 38, and 40 as the largest ultrasonic reflectors with 10% or greater DAC.

For comparison, these sites are shown along with their high gain DAC and acoustic emission rankings in Table VI. From Table VI it can be seen that there is little correlation between sites ranked with ultrasonic DAC and those rankings established by AE.

Conclusions

No clear correlation could be established between ultrasonic and AE ranking of sites. However, each method found defects not revealed by the other. Acoustic emission appeared to be more sensitive than ultrasonic testing in locating very small discontinuities, i.e., the 20 sites located by AE that showed less than 2% DAC when examined ultrasonically. However, the presence of these small signals cannot be uniquely interpreted from AE data alone and should be confirmed by other methods to ensure the reliability of the AE siting.

AE site detections are possible in generally high external noise environments such as those encountered in a fabrication shop but detection is limited where interference signals are generated by nozzle plugs and vessel straps.

Table IV. Ultrasonic Examination Ranking (45° Beam Angle)

AE Site	Amplitude (% DAC)	Rank
1	18	1
4	5	11
6	5	11
10	10	2
11	12.5	3
12	8	8
18	2.5	16
21	5	11
23	5	11
24	11	4
25	7.5	9
28	5	11
30	10	5
32	7	10
38	10	5
40	10	5
All Others	2 or less	

TABLE V. Ultrasonic Examination Ranking (60° Beam Angle)

AE Site	Amplitude (% DAC)	Rank
1	11	3
3	6	8
6	6	8
11	12.5	2
13	15	1
14	6	8
23	8	6
24	10	4
28	7.5	7
32	9	5
All Others	2 or less	

Table VI. Comparative Ranking of AE Sites

	Ultrasonic Examination		AE Examination		
AE Site	45° Angle	60° Angle	Event	Summation	Amplitude
1a	1	3	17	19	-
10	2	-	18	17	15
11	3	2	-	-	-
13	-	1	11	11	10
24	4	4	-	-	-
30	5	-	12	10	-
38	5	-	20	-	-
40	5	-	-	15	12

Note: - indicates not ranked in top 24 sites

Further work is needed to establish a sound foundation for ranking the importance of AE sites. The present ampirical approaches, while providing valuable information, are clearly limited but can serve as a guide to more exhaustive examination by other methods.

REFERENCES

(1) A. T. Green, "Detection of Incipient Failures in Pressure Vessels by Stress-Wave Emissions," Nuclear Safety, 1969, 10. 4.

(2) P. H. Hutton and D. L. Parry, "Assessment of Structural Integrity by Acoustic Emission," Materials Research and Standards, 1971, 11, 25.

(3) H. L. Dunegan and D. O. Harris, "Acoustic Emission - A New Nondestructive Testing Tool," Ultrasonics, 1969, 7, 160.

(4) K. E. Stahlkopf and G. J. Dau, "Acoustic Emission, A Critical Assessment," Nuclear Safety, 1976, 17, 1.

ACOUSTIC EMISSION KAISER EFFECT IN ZIRCALOY 4 FUEL RODS

J. R. Birchak
C. W. Chagnon
Babcock & Wilcox Company
Lynchburg Research Center
Lynchburg, Virginia 24505

INTRODUCTION

During fabrication, nuclear fuel rods are pressurized with helium to approximately 385 psig and sealed. Two questions are particularly important: 1) Has helium leaked from the fuel rods? 2) What are the maximum differential pressures encountered during reactor operation? The objective of this investigation was to evaluate the acoustic emission Kaiser effect as a means of nondestructively revealing pressure history and present condition of fuel rods. The phenomenon observed by Kaiser (1), occurs when a material is repeatedly stressed. For example, the first time that a fuel rod is pressurized the material is stressed and therefore produces acoustic emissions. On the second and successive pressurizations, the material is quiet until the previous peak pressure is exceeded.

In these experiments, empty fuel rods were subjected to a variety of pressure cycles, handling conditions and thermal treatments. Each fuel rod had a conventional end cap welded to one end and was pressurized from the other end. During pressurization the acoustic emissions were monitored using a multi-channel system having piezoelectric sensors placed on the fuel rod. The cumulative number of bursts vs increasing pressure were analyzed for evidence that the Kaiser effect gave a reliable indication of pressure conditions.

EXPERIMENTAL PROCEDURE

The specimens were 10 foot sections of Zircaloy-4 cladding with an end cap welded to one end and the other end open.

The end caps were TIG welded to the cladding using conventional manufacturing techniques and inspection procedures. A cross-section of the welded cap end is shown in Figure 1. The tip of the unwelded annulus is of particular importance because it is the location of greatest stress concentration.

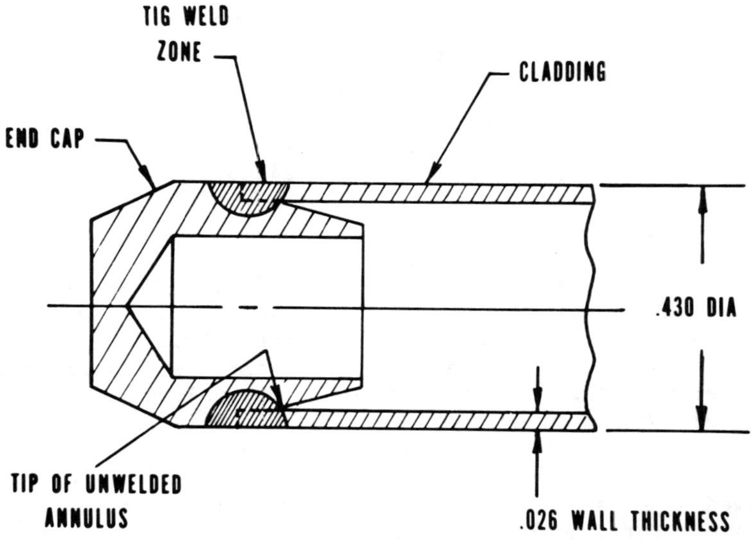

Fig. 1. Cross-Section of Cap End of Specimen.

The specimens were pressurized from the open ends using the system shown schematically in Figure 2. An electronic pressure transducer provided an electrical signal for the parametric input to the acoustic emission system.

The multichannel acoustic emission system was set up to determine the location of sources. During each pressure treatment, the acoustic emission counter was turned on at zero pressure and turned off at the final pressure. Preliminary testing revealed that the welded end cap region had the only recognizable sources. Inasmuch as the remainder of the fuel rod was quiet, the piezoelectric receiver transducers were placed in the vicinity of the end cap as illustrated in Figure 2. Cumulative counts vs pressure and also pressure vs time were recorded for a variety of pressurization conditions.

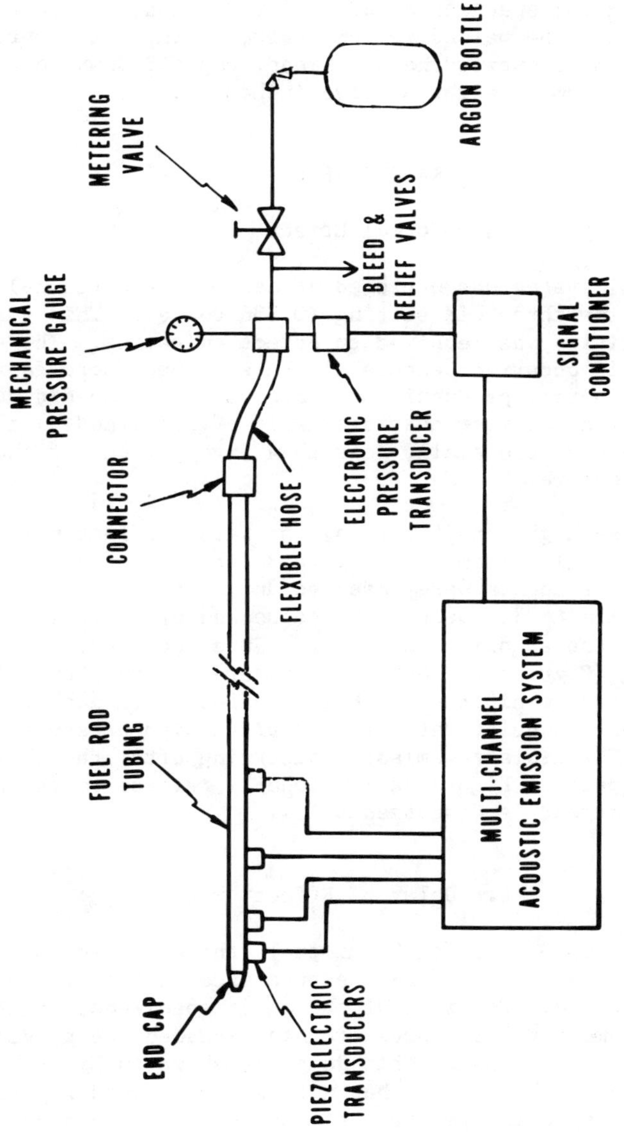

Fig. 2. Experimental Configuration for Monitoring Acoustic Emission During Pressurization of Fuel Rods.

The specimens were subject to treatments to evaluate the effect of heating and handling. The heat treatment consisted of annealing the specimen at 600°F for 30 minutes. In the handling test, the welded region of the tubing was compressively stressed by dropping the specimens, cap end down, onto a steel plate from a height of five inches.

KAISER EFFECT

A. Normal Effect

The AET system was operated at maximum gain (100dB) and relatively low threshold setting (0.336 volts). This very high sensitivity was required to detect emissions. The small signal levels occurred because a) the specimens were fabricated using normal production techniques and hence did not have the strong sources characteristic of flaws and b) the pressure levels were small, less than five percent of the bursting pressure.

At these high sensitivities, the Kaiser effect was observed for all specimens. Figure 3a shows that a significant number of counts were observed during the first pressurization to 385 psig. On the second pressurization (Figure 3b) the specimen was quiet. Subsequent pressurizations to 500 psig (Figure 4a), 800 psig (Figure 4b) and 900 psig (Figure 5) also exhibited the Kaiser effect. In each case the specimen was quiet until the previous high pressure was exceeded. The isolated emission occurring after the third run at 250 psig on Figure 4a is probably related to the effects of heat treatment as discussed below.

B. Delay of Effect

In Figures 3 to 5 for Specimen 1, the acoustic emission counts resume as soon as the pressure exceeds the previous maximum pressure. For most of the other specimens, counts did not resume until the pressure had exceeded the previous maximum by 20 to 70 psi. This delay is shown in Figure 6 for Specimen 2. This delay has not been explained rigorously. A probable origin is premature stress release (analogus to creep) on the first pressurization. Acoustic emissions (which are associated with sudden release of stress energy) would then be delayed upon repressurization until the stress build-up exceeded the levels associated with premature stress release.

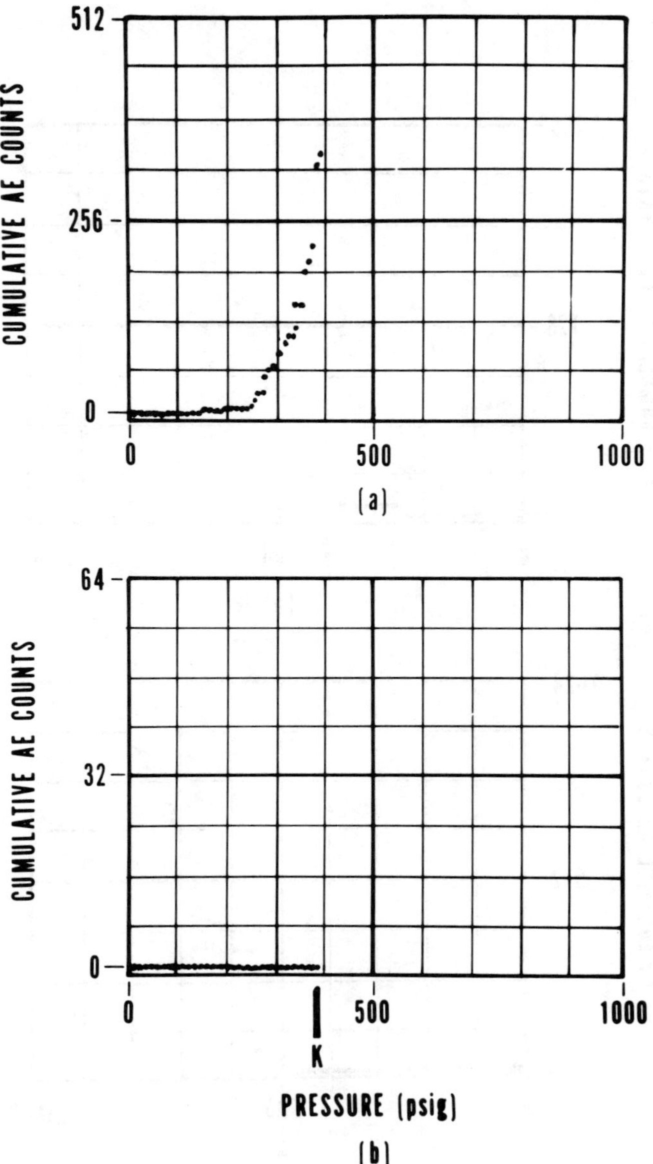

Fig. 3. Cumulative Acoustic Emission Counts for First (a) and Second (b) Pressurization to 385 psig. Fuel Rod Specimen 1. K indicates previous peak pressure.

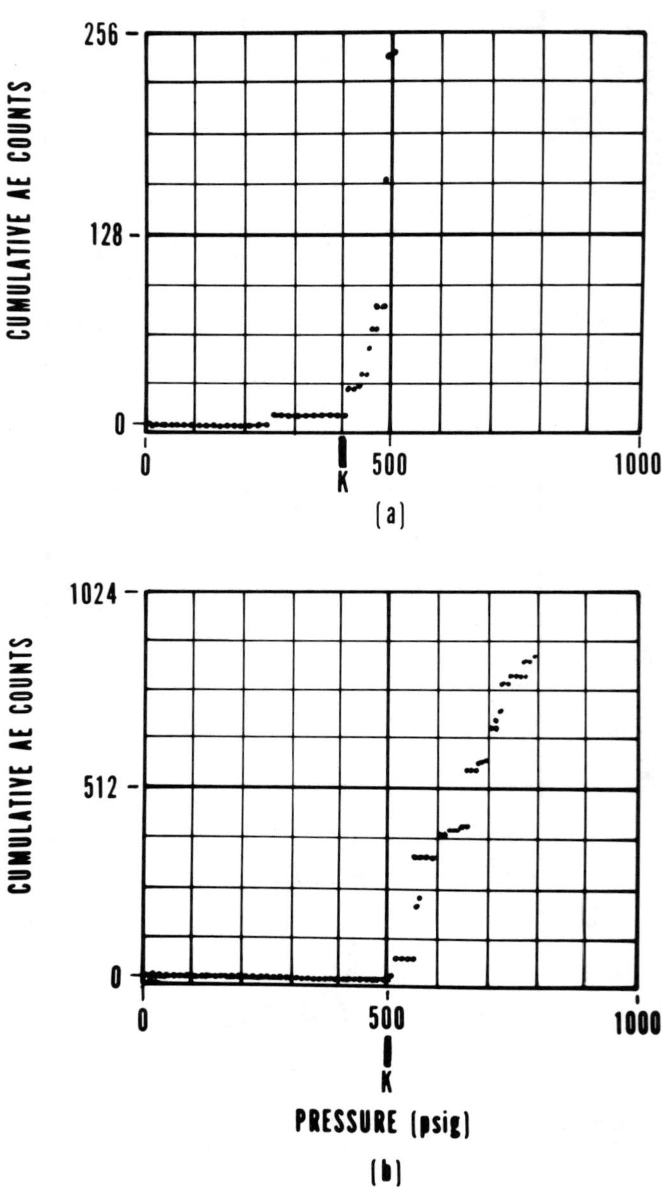

Fig. 4. Cumulative Acoustic Emission Counts for Pressurizations to (a) 500 and (b) 800 psig. Specimen 1.

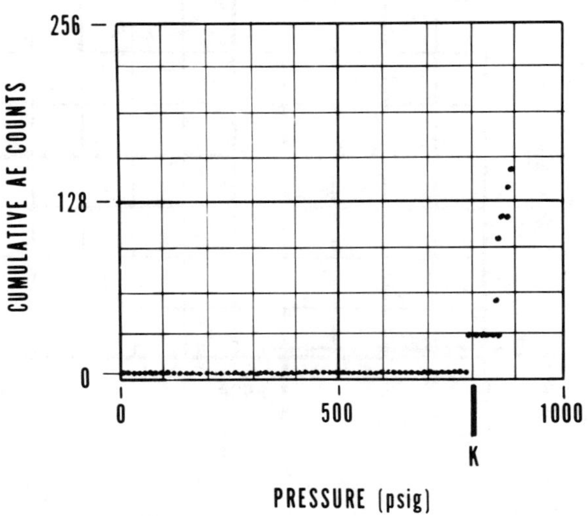

Fig. 5. Cumulative Acoustic Emission Counts for Pressurization from 0 to 900 psig. Specimen 1. K indicates previous peak pressure.

C. Partial Elimination of Kaiser Effect

1) Heat Treatment. Investigations by others (2) have shown that annealing produces recovery; i.e. a partial elimination of the Kaiser effect. This recovery can occur during room temperature aging. Typically, annealed (lower yield) specimens emit more acoustic emissions at small stresses (elastic region) that do cold worked or brittle specimens (3). Heat treatment relieves residual stresses thereby partially eliminating the effects of previous stressing.

This partial elimination is shown in Figure 7a and 7b where cumulative counts are given respectively for Specimens 3 and 4. The solid curves (pre-heat treat) show that the first pressurization from 0 to 900 psig had a Kaiser effect relative to the previous maximum pressure of 800 psig. After depressurization, the specimens were heat treated. Following heat treatment, the specimens were again pressurized to 900 psig.

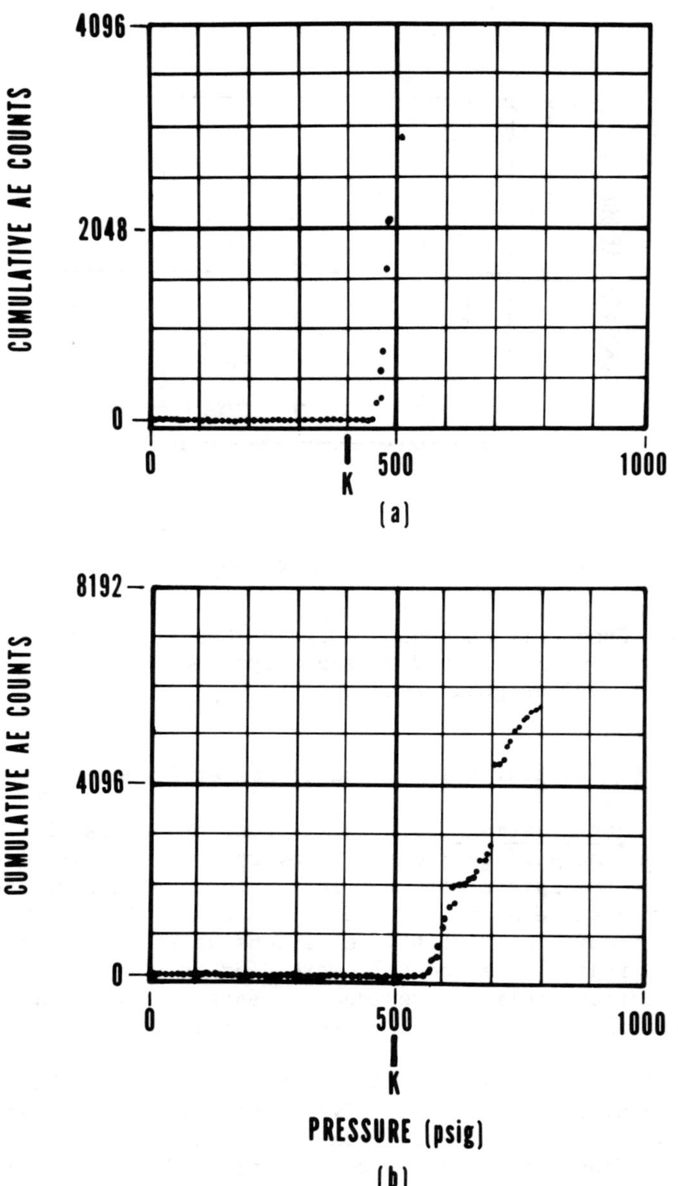

Fig. 6. Delay of Acoustic Emission Counts Until Pressure Exceeds Previous Maximum Pressure by Approximately 60 psig. Specimen 2. K shows prior peak pressure.

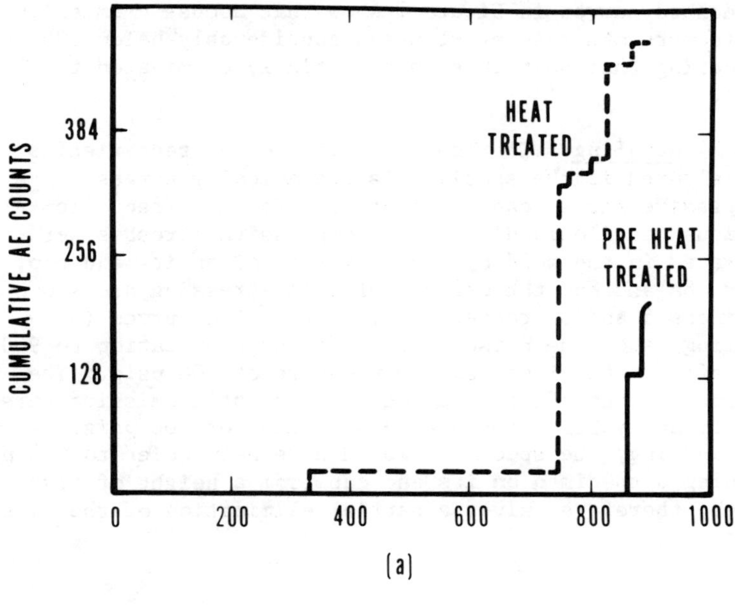

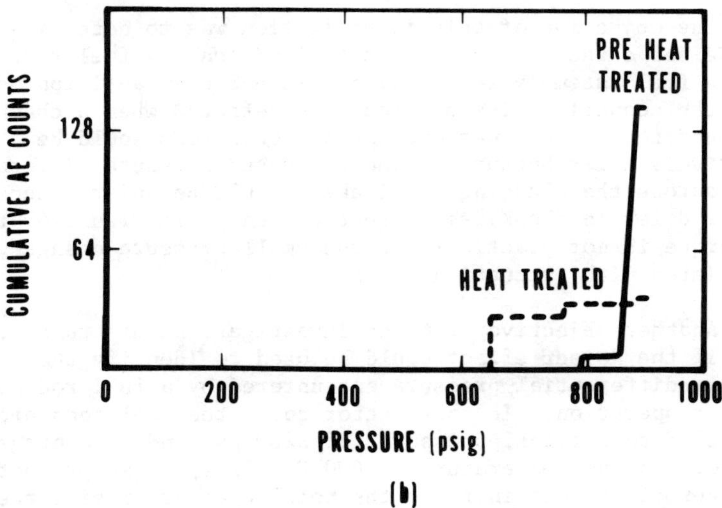

Fig. 7. Cumulative Acoustic Emission Counts vs Pressure for (a) Specimen 3 and (b) Specimen 4. Pre-heat treat pressurization performed after previous peak of 800 psig.

The dashed curves in Figure 7 show that acoustic emission counts were observed at stresses considerably below 900 psig, indicating that heat treatment partially eliminated the Kaiser effect.

2) Handling. The acoustic emission characteristics are also altered if the specimen is mechanically stressed. A high compressive stress can eliminate the Kaiser effect from a previous tensile loading (4). Compressive stresses were generated in the weld by dropping the rod on its end cap. In Figure 8a and 8b, the effects of this stressing are shown for Specimens 5 and 6, respectively. The solid curves (pre-handling) show the Kaiser effect for pressurization to 900 psig after a previous maximum pressure of 800 psig. The pressurizations after handling gave acoustic emission bursts significantly below the previous maximum of 900 psig. Without handling, the specimens would have been quiet to 900 psig. Dropping a specimen on its end cap from a height of five inches, therefore, gives a partial elimination of the Kaiser effect.

DISCUSSION

One objective of this investigation was to determine if the Kaiser effect could be used to find leaking fuel rods during fuel assembly fabrication. It had been anticipated that the acoustic emissions could be detected when a chamber surrounding the rods was evacuated. Good rods would be relatively noisy because of the increased pressure differential across the cladding but leakers would be quiet. Because of the delay in the Kaiser effect evidenced in Figure 6, this technique is not practical for the small pressure change associated with evacuation.

Another objective of these investigations was to determine if the Kaiser effect could be used to identify the maximum differential pressure encountered by a fuel rod during reactor operation. In the reactor core, the fuel rods are subjected to external pressures of 2200 psi and an average reactor coolant temperature of 600°F. Fission gas production and thermal effects increase the total pressure inside the rod to approximately 925 psig near the end of the first fuel cycle. During operation, the internal pressure is lower than the external, resulting in compressive stresses. When the reactor is shut down, the pressure on the exterior of the fuel rod is decreased below the internal pressure, producing tensile stresses in the cladding.

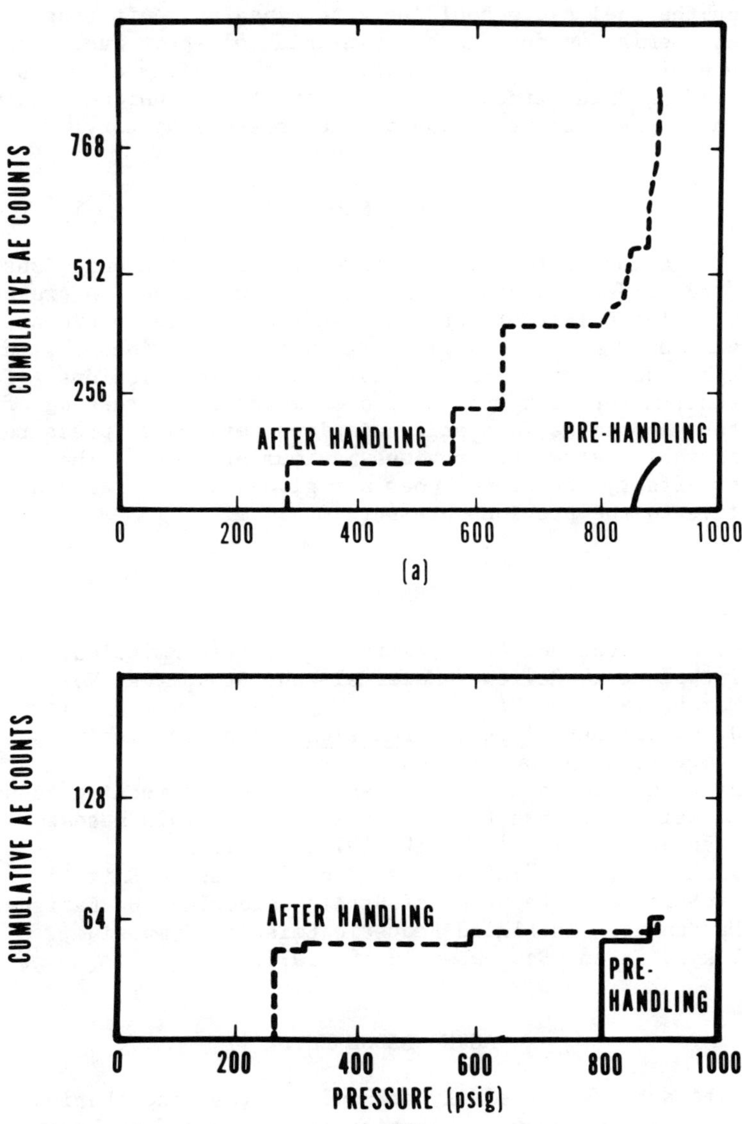

Fig. 8. Cumulative Acoustic Emission Counts vs Pressure for (a) Specimen 5 and (b) Specimen 6. Pre-handling pressurization performed after previous peak of 800 psig.

It was anticipated that the maximum pressure encountered during the fuel cycle could be determined by monitoring acoustic emission during repressurizing of spent fuel rods. Because of the partial elimination of the Kaiser effect by heat and handling treatments, however, the technique does not give a simple interpretation of the pressure history.

CONCLUSIONS

The Kaiser effect was observed for acoustic emissions from the welded end caps of Zircaloy 4 fuel rods. A pressure delay of the Kaiser effect was observed in some specimens. As much as 70 psi excess pressure over the previous maximum pressure was required to reactivate acoustic emissions. In contrast, after heat treatment or compressive stressing of the test specimens, acoustic emissions resumed at pressures considerably below the previous maximum pressure. The Kaiser effect, therefore, does not give a unique response relative to the previous pressure maximum.

REFERENCES

(1) R. G. Liptai and D. O. Harris, "Acoustic Emission, An Introductory Review," Materials and Standards, Vol. 11, March, 1971, p. 8.
(2) J. C. Spanner, Acoustic Emission, Intex Publishing Company, Evanston, Illinois, 1974.
(3) H. L. Dunegan and A. T. Green, "Factors Affecting Acoustic Emission Response from Materials," Materials Research and Standards, Vol. 11, March, 1971, p. 21.
(4) Kishi, et.al., "The Application of Acoustic Emission Technique to the Study of Strain Hardening and Fatigue Hardening," The Third Acoustic Emission Symposium, Tokyo, Japan, September 16-18, 1976.

ACKNOWLEDGEMENTS

The authors gratefully acknowledge the contributions of others to the program. The original concepts for these experiments were suggested by A. E. Wehrmeister and R. E. Womack. P. D. Keith and N. W. White assisted in performing the experiments and the Babcock and Wilcox Company, Commercial Nuclear Fuel Plant, furnished specimens.

INTERPRETATION OF ACOUSTIC EMISSION DATA TO INDICATE FLAW SIGNIFICANCE

P. H. Hutton
E. B. Schwenk
Battelle-Northwest

ABSTRACT

The work discussed is an on-going research program at Battelle-Northwest under sponsorship of the U.S. Nuclear Regulatory Commission. The purpose of this program is to develop an experimental/analytical evaluation of the feasibility of detecting and analyzing flaw growth in reactor pressure boundaries by continuously monitoring for acoustic emission (AE). Major specific program objectives are:

- Develop criteria to distinguish flaw growth AE from other non-significant acoustic signals.

- Develop an AE-flaw growth model as a basis for relating inservice AE to flaw significance.

The investigation is devoted exclusively to ASTM Type A533, Grade B, Class 1 pressure vessel steel. Work to date has included fracture and fatigue crack growth in base metal and weld metal, fatigue crack growth in 3% plastic prestrained base metal, and fracture in a flawed 6 inch (152 mm) wall pressure vessel.

Preliminary results are encouraging to developing a meaningful AE-flaw relationship. AE count and energy both show a consistant relationship to the fatigue crack driving force and fatigue crack growth rate. These offer a potential means for establishing an interpretive relationship.

INTRODUCTION

Although the safety record to date in nuclear facilities has been quite outstanding, the need for continuing effort to provide even greater safety assurance is recognized. The work described here is part of this effort.

This work is sponsored by the U.S. Nuclear Regulatory Commission. The purpose is to develop an experimental determination of the feasibility of detecting and evaluating flaw growth in reactor pressure boundaries on a continuous basis using acoustic emission (AE) methods.

The current primary objectives are:

- Develop a criteria to distinguish flaw growth AE from other nonsignificant acoustic signals including those from innocuous defects.
- Develop an AE-flaw growth model as a basis for relating inservice AE to flaw significance.

The program was initiated July 1, 1976 and is scheduled to continue through FY-83 contingent on intermediate results achieved. The intended end result is a demonstrated method and hardware system for detecting and evaluating flaw growth by continuous AE monitoring.

Primary emphasis has been on developing a model for flaw evaluation by AE data analysis. Without this other program objectives become almost academic. We feel that program results to date allow us to construct a preliminary evaluation model.

Results of two surveys[1,2,3] of defects in pressure vessels, in general, indicated that defects that occurred in manufacturing and fabrication, coupled with operation-induced fatigue crack growth from those defects, was the dominant cause of failure. Thus, to illustrate our approach, we have assumed an application where fatigue is propagating a crack in a vessel. Determination of flaw severity centers on the magnitude of the stress intensity factor (K) and its relationship to flaw growth rate (da/dN).

Our analytical method employs rate of change of AE and fatigue parameters. Laboratory data indicates that crack growth rate (da/dN) and the "fatigue crack driving force" or stress intensity factor range (ΔK) can be obtained from the AE rate (dC/dN). The composite of an AE rate (dC_3/dN) and fatigue crack growth rate from three test specimens is shown in Figure 1. These results were developed from two different heats of A533B Class 1 steel, one weldment, two different thicknesses - 0.5 and 1.75 in. (12.5 & 45mm), and two different geometry specimens (single-edge notch and size two compact tension). Furthermore, one specimen was uniformly prestrained 3% prior to fatigue testing to simulate localized straining in a pressure vessel. The data were characterized using a least squares regression technique assuming a power-law behavior between dC_3/dN and da/dN where,

$$\frac{dC_3}{dN} = 2.529 \times 10^5 \left(\frac{da}{dN}\right)^{(1.182)} \tag{1}$$

The confidence limits on Figure 1 indicate for an average of one AE event count per cycle, there is a 90% confidence that a crack is growing at a rate (da/dN) somewhere between 8 and 90 micro-inches per cycle. From the same three specimens, a Paris-Erdogan fatigue crack growth rate equation was derived,

$$da/dN = 3.657 \times 10^{-9} (\Delta K)^{2.242} \tag{2}$$

This is shown in Figure 2. Substituting the range of da/dN values derived from AE data into Equation 2 yields a crack driving force ΔK, ranging from about 31 to 91 Ksi $\sqrt{\text{in}}$. (or correspondingly a K_{max} level of about 34 to 101 Ksi $\sqrt{\text{in}}$. based on an $R = P_{min.}/P_{max.}$ of 0.1). Thus, knowledge of ΔK or K_{max} based on dC_3/dN (AE rate) alone could permit an estimate of severity of the AE source with respect to a limiting value of K_{max}.

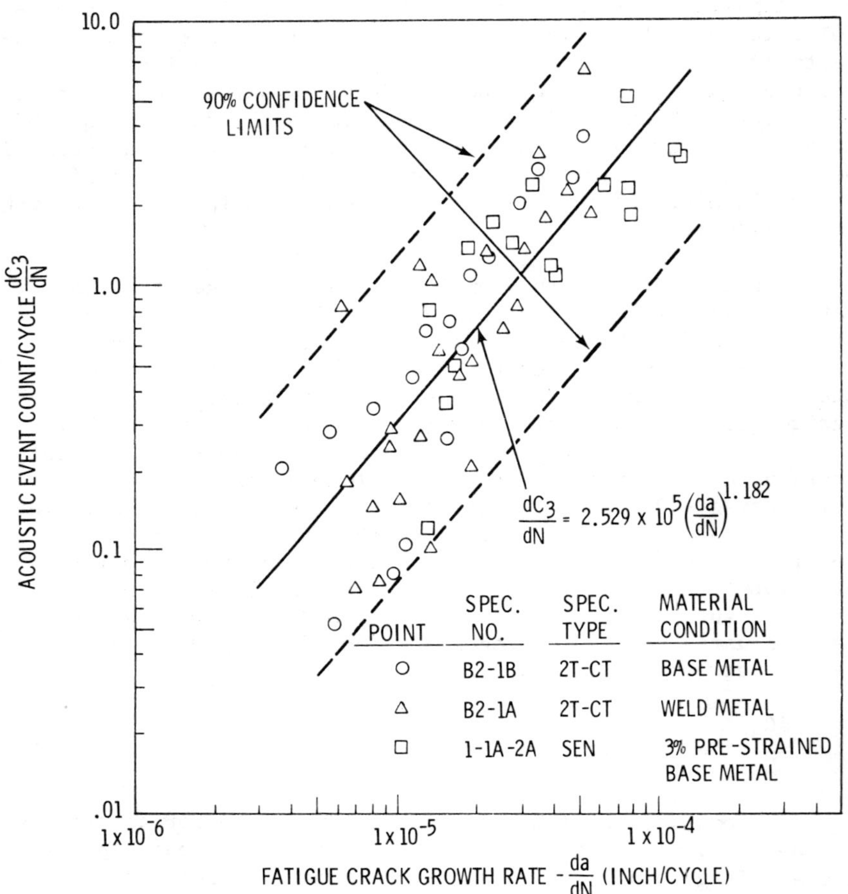

FIGURE 1. AE Count per Cycle dC_3/dN vs Fatigue Crack Growth Rate (da/dN) for Three Tests. R = 0.1, 3Hz

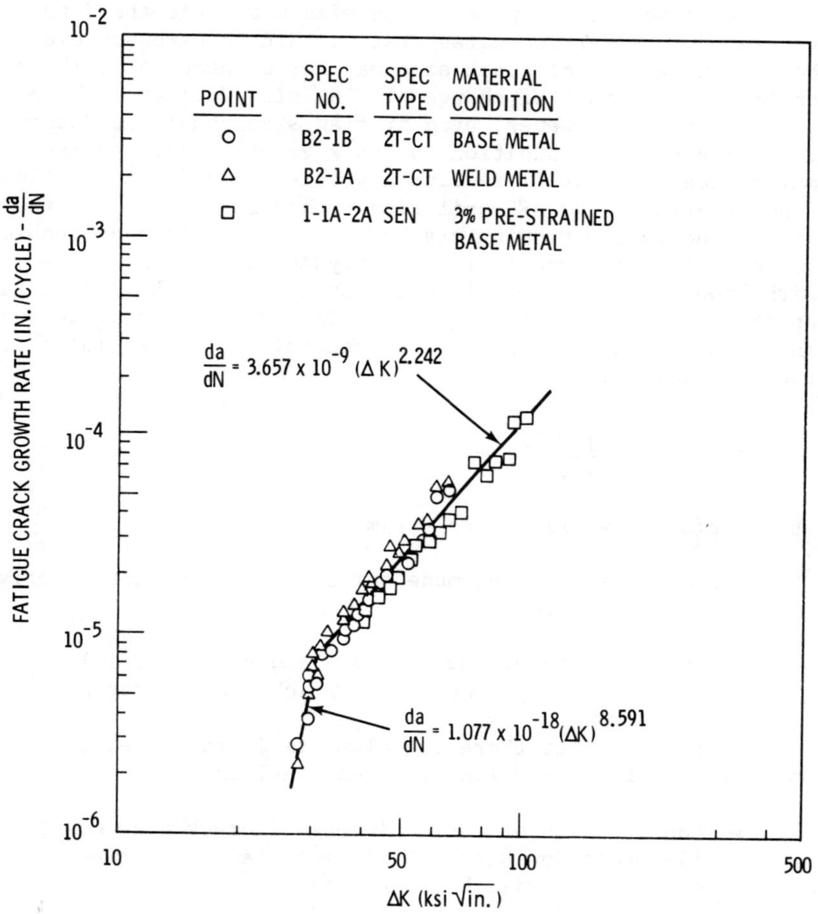

FIGURE 2. Fatigue Crack Growth Rate (da/dN) vs Stress Intensity Factor Range (ΔK) for Three Tests. R = 0.1, 3Hz

The conclusion appears to be viable because the data generated thus far indicates that, within some reasonable bounds, knowledge of crack size may not be necessary. Based on Equation 1, rate of change of flaw size dominates the production of AE. However, even if flaw size should be determined to influence AE in addition to crack growth rate, the basic method described still remains a viable means for evaluating flaw severity. The AE monitoring method should be able to detect and locate growing cracks as being within a definable volume. Finite element stress analysis techniques coupled with known vessel design load ranges in the crack vicinity and an AE estimate of K from Equations (1) and (2) can allow estimation of crack size through the generalized stress intensity factor equation.

$$g(a) = \frac{K_I}{h(\sigma,\phi)}$$

where, $g(a)$ = flaw size parameter

K_I = opening mode stress intensity factor derived from AE

$h(\sigma,\phi)$ = governing stress field parameter (σ) and shape factor (ϕ) which are estimatable values.

We realize that there are elements in the model that must obviously be further defined. These include:

- Confirmation of the influence of crack size on AE-flaw relationships. Different flaw geometries and orientation will be investigated.

- The effect of load-cycle rate on AE generated during fatigue crack growth. This is needed to translate from a load cycle (N) basis to a time (t) basis for reactor application. Definition of load cycles in the sense used here for rate determination would be difficult to establish on a reactor, particularly for variable thermal loadings.

- The effect of R ratio on AE. Data used in this discussion was based on an R ratio of 0.1. On a reactor, outside of startup and shutdown, R ratios may be more in the order of 0.5 to 0.7.

- The effect of flaw geometry on AE. The data herein used was derived from a through-wall crack. Some of our past work has suggested that flaw geometry is not significant. This, however, needs to be confirmed.

- The data scatter needs to be reduced. There is encouragement that this may be improved by more specialized data acquisition methods such as rise time and amplitude discrimination.

- The effect of material thickness on detected AE. The data used was measured for specimens 1/2 to 1-3/4 in. (12.5 to 45mm) thick. One reference point generated to date (AE from a Heavy Section Steel Technology program test[4] vessel with 6 in. (152mm) wall thickness) indicates that material thickness may not be grossly significant except for its effect on source location accuracy.

These factors are all being considered in the continuing program.

In the course of this work, we are taking precautions to standardize AE measurement method from test to test. The same sensors are used in each test and they are calibrated on the specimen both before and after each test. An advanced AE monitor system (Figure 3) is used on all tests. This provides for isolation of only the signals originating from the flaw locations. It produces a measure of six AE parameters (event count of all signals, and event count, energy, peak time and amplitude of isolated signals, plus a combination of peak time-amplitude). Two mechanical parameters (total load cycles and the number of load cycles producing AE) are also measured. These are all recorded on a coordinated basis in solid state digital memories. A tabulated data printout for analysis is obtained directly from the memories.

The various AE parameters are being examined statistically for correlation with fracture mechanics parameters. This serves two purposes:

- It provides an indication of which parameters correlate best with flaw descriptions.

- It provides information on which parameters can be controlled for separation of crack growth AE from other unwanted signals without compromising the flaw evaluation model.

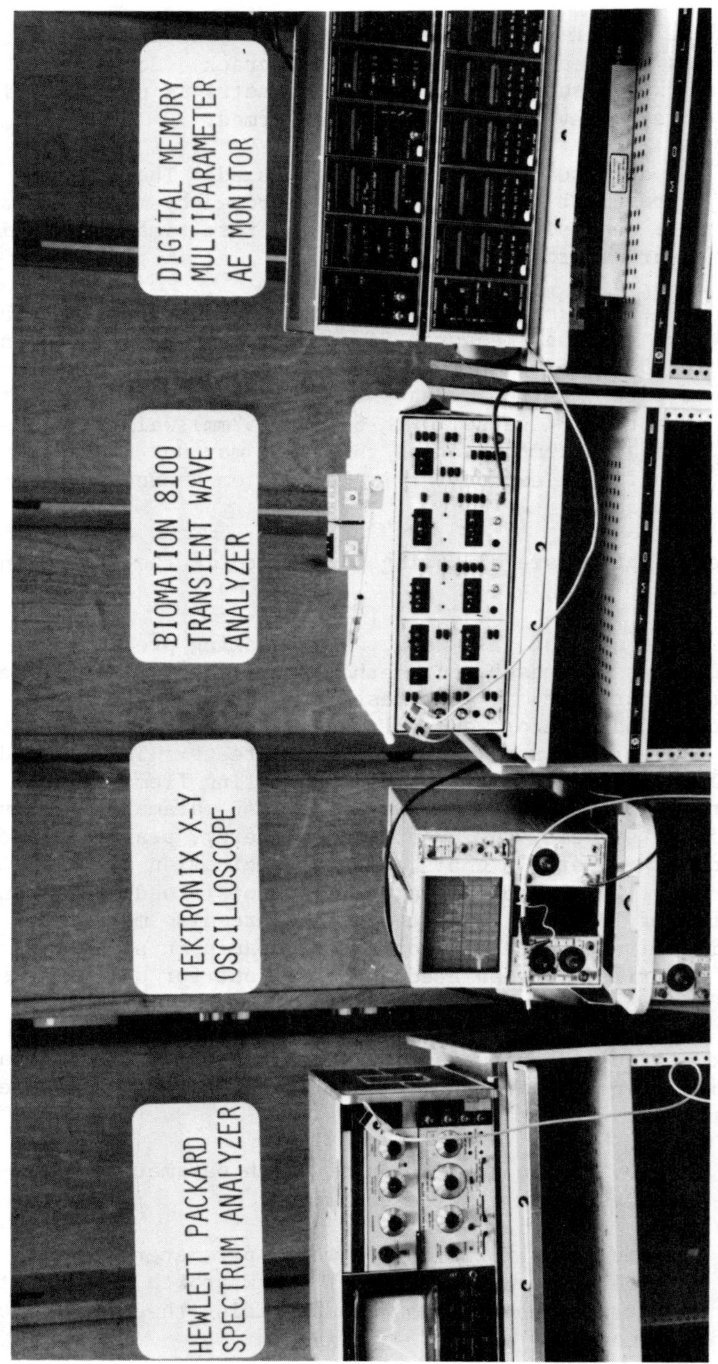

FIGURE 3. Acoustic Emission Analysis System

In summary, we feel that it is possible to make a useful estimate of flaw severity using only the AE generated by that flaw. This is in the context of a tool which can be used by a reactor operating staff.

Periodic reports of the results from this program are and will be contained in a series of Nuclear Regulatory Commission reports under the number NUREG 0250 - (Report No.). These can be obtained from the National Technical Information Service, Springfield, Virginia 22161.

REFERENCES

(1) W. S. Gibbons and B. D. Hackney, "Survey of Piping Failures for the Reactor Primary Coolant Pipe Rupture Study", Geap-4574, General Electric Company, San Jose, CA, May 1964.

(2) E. B. Schwenk and L. A. James, "Estimate of the Fatigue Cracking Behavior of the Fast Test Reactor Primary Piping", Battelle-Northwest, Richland, WA, October 1969, pp. 75-84.

(3) C. A. G. Phillips and R. G. Warwick, "A Survey of Defects in Pressure Vessels Built to High Standards of Construction and Its Relevance to Nuclear Primary Circuit Envelopes", United Kingdom Atomic Energy Authority, AASB(S)R 162, 1968.

(4) R. H. Bryan, "Quick Look Report on Test of Intermediate Test Vessel V-7B, Sustained-Load Test of Flaw Weld Repair", ORNL-SST2, August 3, 1977 (Topical Report forthcoming).

ON-LINE ACOUSTIC EMISSION SURVEILLANCE OF BWR COMPONENTS

William F. Hartman
Technology for Energy Corporation

John W. McElroy
Philadelphia Electric Company

SUMMARY

An experimental study of advanced acoustic emission monitoring of BWR components is being performed at Philadelphia Electric Company's Peach Bottom nucler power station. The monitoring consists of two types: leak noise monitoring of safety/relief valves, and discrete-event monitoring of a recirculation bypass line, core-spray line, and feedwater nozzles. The overall objective of this program is to gather enough on-line data and related information to be able to assess the capabilities of using continuous acoustic emission monitoring for increasing availability of LWR systems.

We here present the measurement criteria of the surveillance system, refined procedural guides for the installation, operation and maintenance of such a system, and some results that encourage further development of data presentation forms.

INTRODUCTION

The efficiency and reliability of nuclear production of energy are commonly related to the capacity and availability of the operating nuclear power stations. Improvements in nuclear unit productivity can produce significant benefits for the utility industry, the nuclear industry and the general public. It has been shown that, because of inflation, the value of a 20% increase in LWR productivity can equal the original cost of the unit (1). In the area of nondestructive evaluation (NDE), advanced surveillance techniques are good candidates for improving plant productivity by improving availability. An objective of on-line surveillance is early detection of degenerative conditions, especially those which are harbingers of eventual malfunction. One NDE technique that can be used in a surveillance mode is acoustic emission (AE) monitoring.

On-line AE monitoring requires detection and interpretation of AE signals which occur during operating conditions. There are two basic types of AE signal, namely, discrete-event signals and steady-leak signals. The former type is more commonly called "burst emission;" while the latter type is not always regarded as AE in a strict sense. Nevertheless, in the case of pressurized components, both types can be detected by the same AE sensors.

Advanced AE monitoring of BWR components is being performed at Philadelphia Electric Company's Peach Bottom Atomic Power Station. The monitoring consists of two types, leak-noise monitoring of safety/relief valves and discrete-event monitoring of selected piping and nozzles. The objective of this experimental study is to assess the feasibility of routine implementation of AE surveillance methods for increasing availability of LWR systems. Eventually, the results of the measurement tasks of the surveillance methods will be compared with the results of inservice examinations occurring during refueling outages.

LEAK NOISE MONITORING

A commonly used safety/relief valve in BWR plants is a pilot-operated type, having a two-stage pilot valve section and a main valve section. A schematic of the valve is shown in Figure 1. In operation, the pilot stage is held closed by the internal pressure acting over the pilot stage seat area. As the pressure increases, bellows expansion reduces the abutment gap between the stem and disc yoke. When the stem abuts against the yoke, further pressure increase lifts the disc from its seat, admitting steam to the operating piston of the second stage, causing it to open. This vents the chamber over the main valve piston to the downstream side of the valve, thus creating a pressure differential across the main valve piston and the valve opens.

The pilot-disc seat is vulnerable to deteriorating processes which lead to increasing leakage across the seat into the chamber of the second stage. A small leak has no immediate major adverse effect, because the pilot valve discharge line can accommodate the flow without substantial pressure buildup. However, as a small leak grows, the second-stage pressure increases until, upon reaching a critical value, the second-stage disc opens, even though the pilot disc was not lifted. Since valve malfunction does not occur until a leak worsens to that critical leak rate, early detection of leakage can allow corrective measures to be scheduled, thereby reducing the loss of plant availability.

Several types of surveillance techniques have been considered for detection of the valve leakage. Sensing the temperature near the discharge line is one possibility. This approach is generally considered to be unreliable and to have poor sensitivity, as we point out below. Direct measurement of the pressure in the second stage chamber could be accomplished by fitting the valve with a pressure transducer. Although this invasive technique can be accommodated in a laboratory experiment, its implementation for plant use would require substantial additional safety analysis and testing of the valve's integrity. The monitoring of the leak noise transmitted to the external wall of the pilot stage is an attractive non-invasive method. A review of on-line leak detection methods for reactor systems indicated that AE monitoring has the greatest potential for providing desirable features not found in other methods (2). Among these are high sensitivity and potential for determining leakage rate.

A valve having a leaking pilot-disc seat was tested under laboratory conditions in order to determine the correlation of the leak noise with leak rate or second-stage pressure. The leak rate, downstream temperature and the RMS voltage of the acoustic signal are plotted against the second-stage pressure in Figure 2. Although not shown in the figure, very soon after exceeding a second stage pressure of 200 psi, the valve "pops." At that time, the second stage pressure is about 400 psi, but, for all practical purposes, the value of 200 psi may be taken as the critical pressure, since the pressure increases rapidly from 200 to 400 psi. The AE sensor was clamped onto the external housing of the pilot works. The signal was band-pass filtered in the range 5-8 kHz. The downstream temperature was measured by a thermocouple in the vicinity of the "pilot valve discharge line." As the second stage pressure increased from 40 to 200 psi, the leak rate increased 59%, the temperature increased 9% and the AE RMS voltage increased 370%. Therefore, the sensitivity of the acoustic detection is excellent. Furthermore, if the leak were intermittent, the acoustic detector would respond accordingly while the thermocouple would not. These measurements provided additional encouragement for utilization of the AE method.

Currently, eleven safety/relief valves at Peach Bottom Unit 3 are instrumented for AE leak detection. The signal from each valve's acoustic sensor is conditioned prior to being channeled to a multiplexer. The RMS values of the signals are sampled at a rate of one per second per channel. Each sample consists of two values, namely, the RMS values in two different bandwidths. In this way, substantial changes in background noise can be distinquished from leak noise. A microprocessor memory stores the last sixty pairs of RMS values for each channel, as well as the last one hundred hourly averages of each, together with the hourly variance. Also, every hourly average value is preserved on a floppy disk data storage device. Figure 3 shows an example of the type of display available from the surveillance system.[1]

DISCRETE-EVENT MONITORING

Intergranular stress-corrosion cracking in 304 stainless steel recirculating by-pass piping and core spray piping has been observed in several BWR nuclear plants (3). Cracks have been found on the inside surface of the pipe, in the heat affected zone adjacent to a weld. Cracks have also been found on feedwater nozzles. Flow-vibration induced cracks on the inner radius weld grow through the thermal recycling of the plant. That area is also adjacent to the feedwater spargers, which have also experienced vibrational fatigue in several plants. Early detection of such cracks, especially before leakage occurs, may not always be provided by standard inservice nondestructive evaluation. In order to minimize loss in plant availability, on-line detection and evaluation of cracking is of ecomonic

[1]Technology for Energy Corporation, Model 1400

interest. If, prior to a scheduled shutdown, surveillance data provide estimates of the probability of cracks in these pipes, then preparations can be made for additional inspections and/or repairs.

During a refueling outage of Peach Bottom BWR Unit 3, AE sensors, calibration devices, vibration transducers and thermocouples were installed on the following components: recirculation by-pass line "B", core spray "B", feedwater nozzles "A" and "E" containing piezoelectric elements of lithium niobate, the sensors have a 500 kHz resonant frequency, and are designed for continuous operation at 315°C. They are mounted "dry" without the use of a "couplant" material. Each sensor is held in place by stainless steel straps. A good contact is enhanced by the thermal expansion of the piping.

This experimental study, although still in progress, has produced some general guidelines for the installation of the sensors and cabling:

1. The location of every sensor should be marked so that later removal and reinstallation can be easily accommodated.

2. The sensor mount should provide electrical isolation of sensor housing from the plant.

3. Quality assurance of high-temperature sensors, preamplifiers, cables and connectors should include durability tests.

4. Routing and securing of cables must be carefully reviewed for vibration effects, proximity to high-energy electrical interference and accessibility within containment.

5. Each cable penetration should be properly labeled and appropriate impedance matching should be performed for each connection to the instrumentation.

6. Electrical power for all instrumentation should be supplied through an isolation transformer.

7. Install remotely-activated AE simulators for checking performance of all sensors.

The AE instrumentation[2] contains 24 channels of amplification with corresponding detectors and processors. The sensors are arranged into four one-dimensional arrays. At the time of the second arrival for each AE event, the following measured parameters are digitally encoded in primary storage registers: event number, arrival time difference, sensor identity codes in the sequence of arrival, the peak voltage of the signal detected at the first-arrival sensor, the real time, temperature at the array, the vibration amplitude and average frequency at that array for the previous one second interval, and a voltage corresponding to the coolant flow. The sensor locations, a description of the system's special features, along with some examples of real-time data displays are given elsewhere (4).

Although the real-time data displays usually depict the AE activity over the entire monitored portion of each component, data may be broken down into several other forms. Using a data-analysis mode of the system, it is possible to display only the AE activity that has occurred in a particular region of a component. This is an approprate way of viewing the source-location data whenever AE activity appears clustered within a region. Examples of such displays are given in Figures 4-6, which also show the time histories of the activity for the same regions, during the period September 14-28, 1977.

For the six-month period ending September 28, 1977, 4,300 discrete AE events had been accepted by the system. Many of these events occurred during operational changes in the plant; that is, while the local temperatures and pressures were varying. For example, in Figure 4, one-half of the activity occurs in less than one day. At that time, a controlled plant run down began a scheduled maintenance outage. The sources appear to be within a 10 cm section of the pipe. However, the center of the cluster is nearly 15 cm from the nearest weld. Therefore, the probability is low that the AE is emanating from cracks in the heat-affected zone of the weld. During routine in-service inspection, the identity of the source will be sought. On the other hand, Figure 5 shows that twenty events occurred very quickly in the core-spray line. The center of the cluster happens to coincide with a weld location. Since this component had little AE activity for three months, repetition of this

---

[2]Trodyne Corporation, MSCD System

behavior during changes in operating conditions will be considered particularly important and could suggest that special scrutiny be given that weld during in-service inspection. Discrete events are frequently recorded for the "E" feedwater nozzle. The typical shape of the source cluster is shown in Figure 6. The "events vs. time" plot shows that the AE activity gradually increased over the period shown.

Thus far, this experimental program has demonstrated that discrete event AE monitoring can be successfully conducted not only on passive components (by-pass and core-spray lines), but also on active components having substantial ambient noise (feedwater nozzles). Results presented here encourage further development of data presentation forms. When the AE results have been compared with the results of scheduled in-service inspection, additional guidelines for interpreting the AE data will be formulated.

REFERENCES

(1) P. M. Lang, S. A. Wisniewski, and D. E. Erb, "Productivity Improvement - Why and How," presented at 1977 Winter Meeting of the American Nuclear Society, San Francisco, California.

(2) G. J. Dau, "A Review of On Line Leak Detection Methods for Reactor Systems," Proceedings of 3rd Conference on Periodic Inspection of Pressurized Components, Institution of Mechanical Engineers, London, 1976, pp. 67-73.

(3) "EXTRA, The BWR Pipe-Crack Situation," Nucleonics Week, Special Issue, January 31, 1975.

(4) W. F. Hartman and J. W. McElroy, "Acoustic Emission Surveillance for Improving Availability of Nuclear Power Plants," "Progress in Nuclear Energy," Volume 1, 1977, pp. 673-680.

ACKNOWLEDGEMENT

This work is supported by the U. S. Department of Energy through a contract with Philadelphia Electric Company.

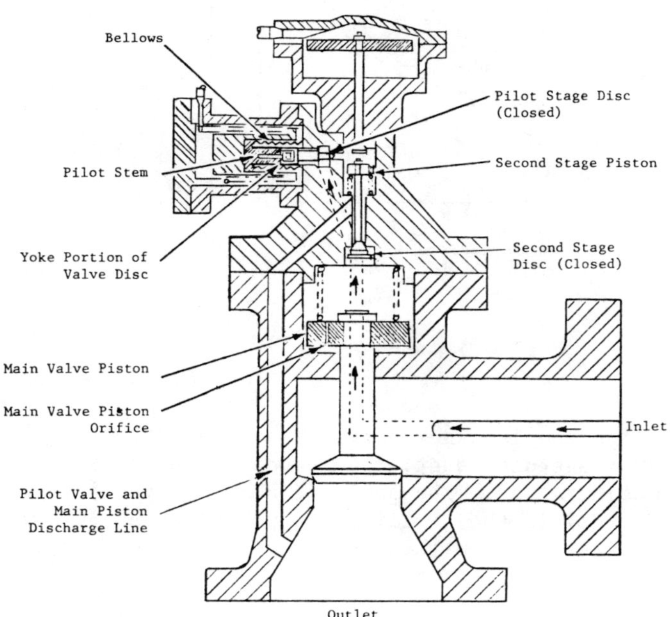

Fig. 1. Schematic of Safety/Relief Valve on BWR Main Steam Lines.

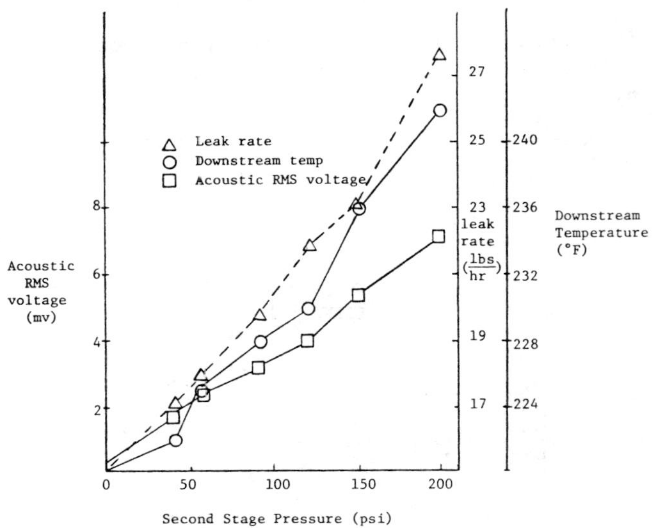

Fig. 2. As the leakage increases across the pilot-stage-disc seat, the second stage pressure increases. The Acoustic RMS Voltage (5-10 kHz) is the amplified output of an AE Sensor installed on the housing of the pilot stage.

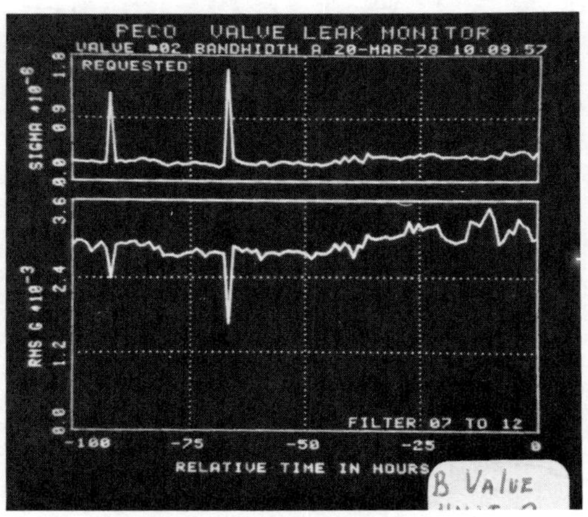

Fig. 3. Example of the display of the past 100 hrs of the acoustic signal from one valve sensor. The ordinate parameter of the lower trace is

$$X = \frac{1}{3600} \sum_{i=1}^{3600} x_i,$$

where x_i is the per-second value of the RMS signal in the bandwidth 8-10 kHz. The ordinate parameter of the upper trace is

$$\sigma^2 = \frac{1}{3600} \sum_{i=1}^{3600} x_i^2 - X^2$$

The spikes correspond to times when the turbine stop valves are exercised.

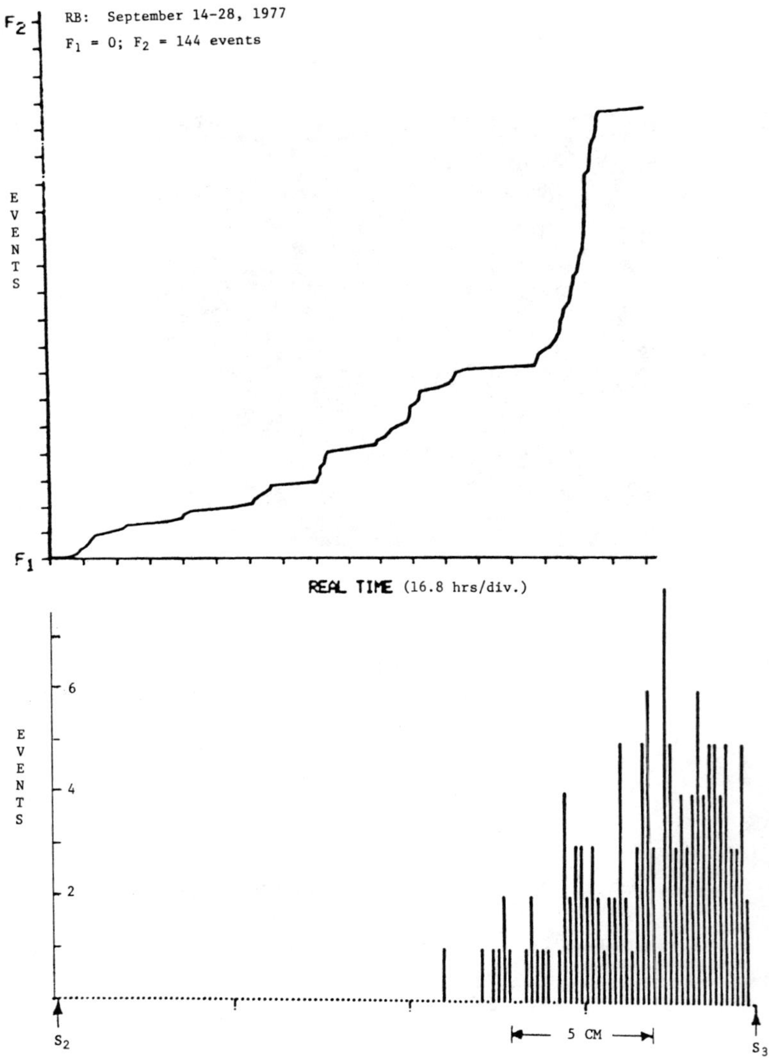

Fig. 4. Activity History (top) and Activity Distribution (bottom) for a portion of the recirculation bypass line. S_2 and S_3 are the positions of the AE sensors.

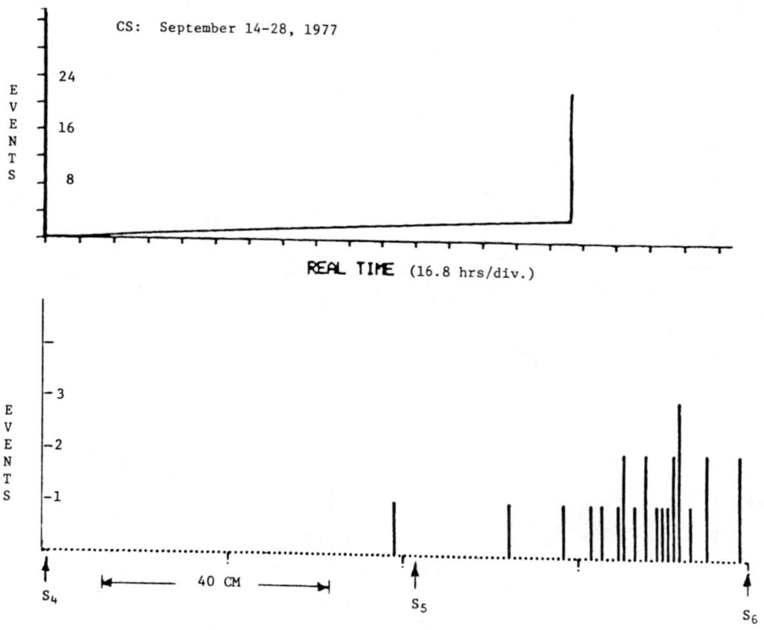

Fig. 5. Activity History (top) and Activity Distribution (bottom) for a portion of the core-spray line. S_5 and S_6 are the positions of AE sensors.

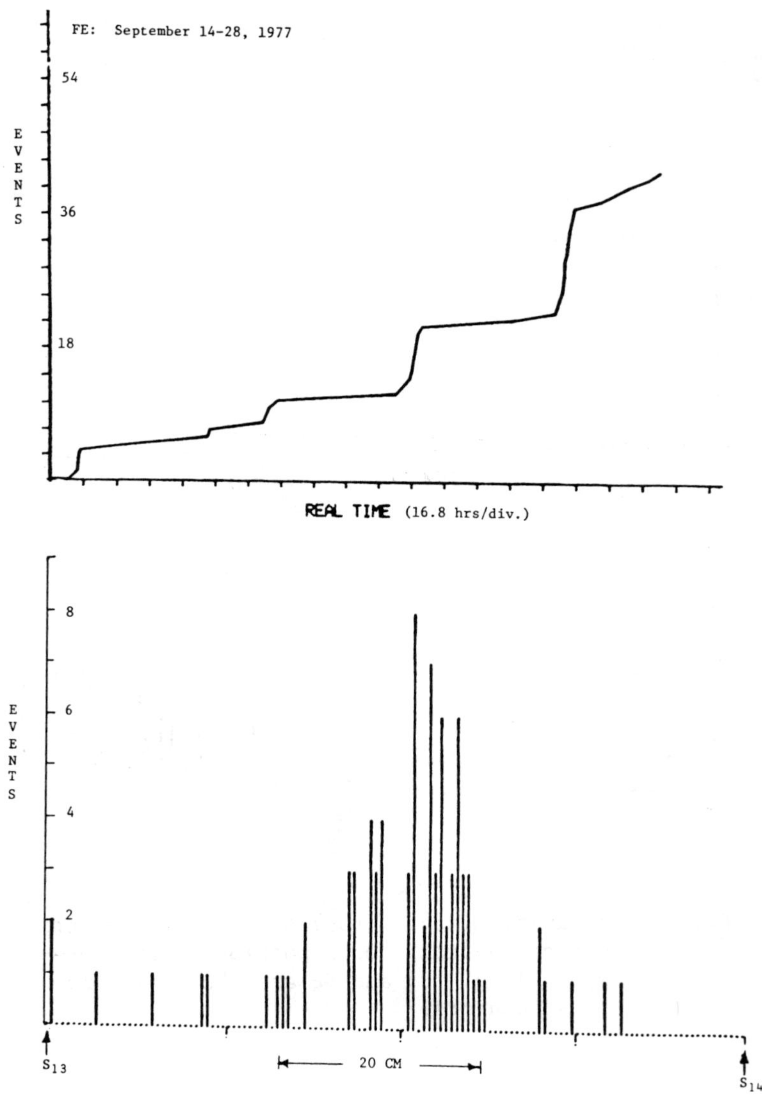

Fig. 6. Activity History (top) and Activity Distribution (bottom) for a portion of the "E" feedwater nozzle.

PRESENT STATUS AND FUTURE OUTLOOK FOR NONDESTRUCTIVE ANALYSIS IN NUCLEAR MATERIALS ACCOUNTANCY IN THE U.K.

A.S. Adamson, A.G. Hamlin
Nuclear Materials Accounting Control Team, U.K.A.E.A.

INTRODUCTION

It is intended in this paper to review applications in the United Kingdom Atomic Energy Authority and British Nuclear Fuels Ltd. of non-destructive assay techniques which are either applied directly to nuclear materials accountancy or which currently provide information primarily for management control, but which also add some assurance to the accountancy. Often in these plant control applications, specific problems have been overcome which may ultimately lead to the technique providing a method for accountancy applications.

The techniques which will be described cover most of the fuel cycle, but it must be remembered that the function of efficient management is not necessarily to apply non-destructive techniques. Many measurements are still best carried out by destructive analysis which, for many purposes, provides more accurate values within the timescale required by the operator. It follows, therefore, that the techniques which are presently applied are considered to be economically justified by plant management, bearing in mind their responsibilities for accountancy, safety or internal auditing of stocks, and therefore represent a technological elite among the choice available.

This paper is divided into two principal parts dealing respectively with existing applications and likely future developments. Within these parts, the applications are grouped according to the basic techniques.

PRESENT APPLICATIONS

Gamma spectrometry

This is widely used in applications ranging from the control of flow of source material for the initial purification, prior to conversion to UF_6, through the fabrication stages of both low-and high-enriched uranium fuel and for various waste measurement streams. Sodium iodide detectors are used for the majority of applications, but where there is a need for better resolution, either high purity germanium or lithium-drifted germanium detectors are used.

In uranium ore processing, a continuous dissolution process is controlled by γ-spectrometry of the 186 KeV uranium peak using a sodium iodide detector. The output is used to control the feed of nitric acid to the dissolver so that the concentration of dissolved uranium remains constant. The system also acts as an additional measurement of the uranium fed to the separation plant.

The enrichment of all UF_6 packed into cylinders is analysed by mass spectrometry on filling but it is considered prudent to confirm the value before the material is used. A portable high-resolution germanium detector gives adequate response to differentiate between, for example, 3.1% and 3.2% enriched material with a five minute count. It is necessary to correct this for thickness of the cylinder and this is measured using an ultrasonic thickness gauge. The measurement has been greatly simplified by the fact that the UF_6 forms an infinitely thick layer uniformly over the cylinder walls and it is not necessary to locate the level of the bulk UF_6 in the cylinder. To avoid the risk of spurious results due to voids, readings are taken at several positions on the cylinders.

The enrichment of UO_2 in drums is measured using a well collimated heavily shielded sodium iodide crystal mounted on a trolley. This measurement is carried out in a store in the presence of a high background from other drums. Results for material of enrichment higher than 1% are within 2% of the mass spectrometry value. This method is, like the previous one, not used directly for accountancy, but rather as a managerial check before the material is used or shipped to another department or factory (1) (2)

The uranium content of freshly manufactured uranium/aluminium fuel plates is assayed for accountancy and process control using a sodium iodide crystal with a two channel analyser. The principal problem in the original commissioning of this equipment was interference from any residual fission products in the recycled uranium. Correction is made by the conventional method of measuring the counts in a channel which covers an adjacent part of the Compton continuum resulting from any fission product present. This method has been in use for many years and has been checked regularly by comparative destructive analysis of complete plates. The accuracy of the measurement is within 0.1% and the precision of the measurement of a single plate is 0.61% (3).

The isotopic composition of plutonium from a number of parts of the fuel cycle is measured by the use of intrinsic and lithium-drifted germanium detectors. On well characterised material from Magnox reactors measurement of the Pu^{239}/Pu^{241} ratio allows the Pu^{240} isotopic content to be deduced to within 5% of the value obtained by mass spectrometry. Normally the 203 and 208 KeV peaks of Pu^{239} and U^{237} are measured but if these are poorly resolved the 129 and 149 KeV peaks of Pu^{239} and Pu^{241} are used instead. The measurement of isotopic composition of material of unknown reactor history utilising Pu^{239}, 240 and 241 and Am^{241} peaks has been successfully applied under laboratory conditions, and attempts are now being made to evolve a simplified procedure for use in the field based on shorter counting times (4).

The assay and control of waste is one of the most important applications of gamma-spectrometry, since destructive techniques are seldom satisfactory for the measurement of waste and very rapid response is frequently required for monitoring systems on effluent lines.

The original method of analysing soft waste involved the measurement of U^{235} or Pu emission with a NaI crystal and a two-channel analyser. The U^{235} was assayed conventionally by measuring the 186 KeV peak but the Pu was assayed by measuring the emission in a window from 355 - 405 KeV, giving a response related to total plutonium rather than Pu^{239} in material of dubious isotopic composition (5) (6).

For the measurement of soft waste containing plutonium of well known isotopic composition, a segmented gamma scanning system incorporating a measurement of the attenuation of Ba^{133} 380 KeV radiation from an external source to provide a correction for attenuation in the matrix is now in use. This material does not contain fission products, and a NaI detector is found to be satisfactory.

For the measurement of soft waste containing plutonium which is possibly contaminated by fission products, a segmented gamma-scanning system incorporating a measurement of the attenuation of Se^{75} 400 KeV radiation from an external source to provide a correction for attenuation in the matrix is being introduced. This utilises a lithium-drifted or Ge detector. At present only Pu^{239} is measured, but it is intended to extend the application of the technique to measure U^{235} and Pu^{241}. The technique is similar to that described by Los Alamos with the difference that the angle of view defined by the collimator is chosen to smooth the dependency of the response on the position of the plutonium in the package to be assayed.

A further technique for measuring the plutonium in waste, not involving measurements of segments, uses two NaI detectors placed near the top and bottom of the waste drum. An absorber whose thickness varies both vertically and horizontally is placed between the detectors and the drum. The varying absorber thickness smoothes the relationship between the height of the plutonium source in the drum and the observed count rates. The drum is rotated to smooth the effect of differences in radial location of the plutonium (Fig.1).(7)

Plutonium in pure raffinates e.g. condensates and cooling water, is monitored by means of the 17 KeV L X-ray. The detector is made just infinitely thick to the 17 KeV rays but thin enough to have very poor detection characteristics for 60 KeV and higher energy emissions. The device is intended to indicate some plant malfunction, therefore accuracy and precision are not important and have not been determined. The lower limit of detection is approximately 0.05 µg Pu/ml of process stream liquor. The upper level has not been established. Count rates equivalent to 1 mg/ml plutonium have been obtained, but not necessarily due to plutonium as the major cause of error has been due to break through of U^{237} or Am^{241} or carry over of solids into the measurement cell.

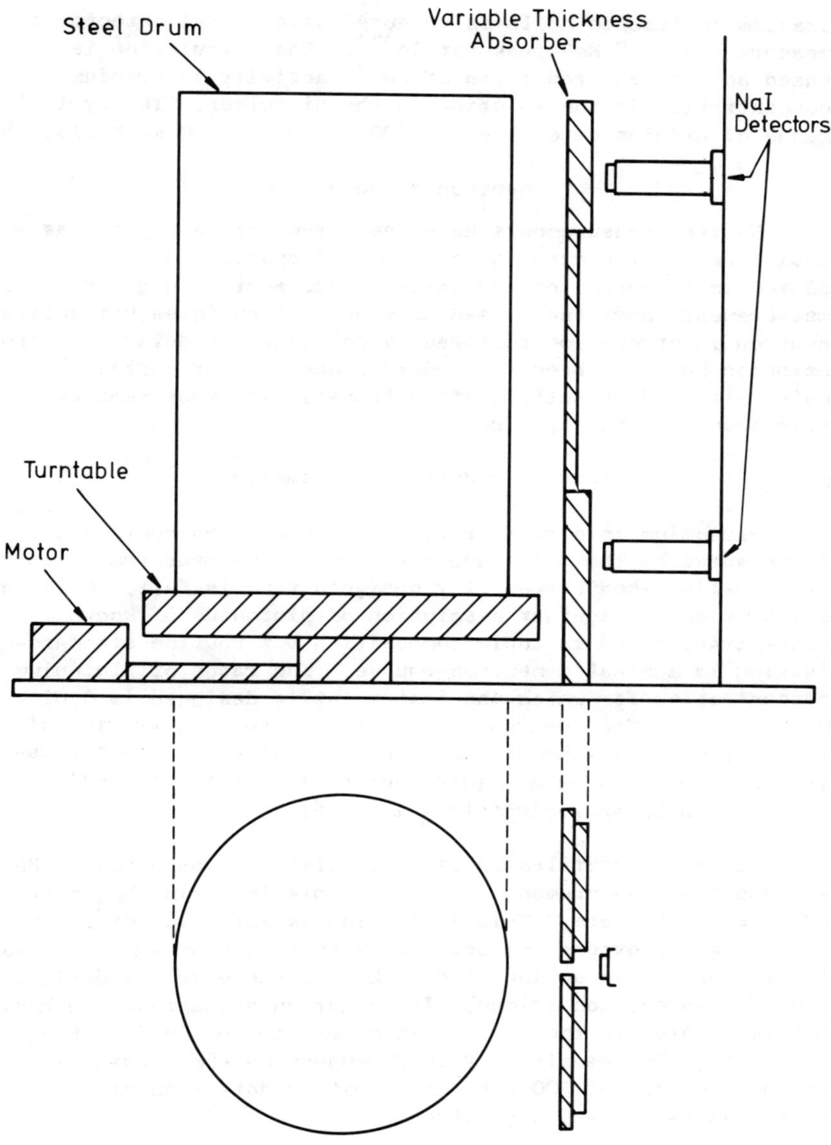

DIAGRAMMATIC REPRESENTATION OF TWO DETECTOR-VARIABLE THICKNESS ABSORBER METHOD FOR THE MEASUREMENT OF PLUTONIUM IN WASTE (SECTION 2.1.6.4. REFERS)

FIGURE 1.

Uranium in leached hulls is measured using a NaI detector to measure the 2.18 MeV peak for Ce^{144}. The calculation is based on the measured ratio of Ce^{144} activity to uranium concentration in the solution in the dissolver. The typical range of uranium determined is 300 - 500g in 100 Kg hulls. (8).

Neutron Methods

Neutron measurements have been used for many years as a basis for plant control of a number of systems involving plutonium reprocessing and waste measurement. In general the measurements have been based on passive techniques, but active neutron interrogation followed by counting the delayed neutron emission has also been a standard procedure for certain materials, and recently neutron transmission measurements have found several applications.

Passive neutron measurements

Plutonium in certain raffinates which also contain U^{237} is measured by the gross neutron count. The measurement is made on-line when a specially designed tank is full. Original calibration was against a solution of plutonium of known concentration and isotopic composition but routine standardisation is against a neutron source. The range of plutonium concentration for which the instrument is designed is 0.02 - 0.1g litre. The instrument is designed to give warning of malfunction of a reprocessing plant and accuracy and precision have not been established, but their combined effect is believed to be approximately $\pm$ 10% (16).

Plutonium profiles in mixer settlers are obtained at BNFL, Windscale by measurement of neutron emission using BF_3 tubes or fission chambers. This indication is intended for plant control and provides the operator with a display which allows timely corrective action to be taken if there is any deviation from flow-sheet conditions. It is not an accountancy method, but is related to the levels ultimately to be found in the raffinate. The sensitivity is dependent on the γ dose rate. In the presence of 500 R/hr the limit of detection of plutonium is 0.2 - 0.3 g/litre.

Neutron measurements have been combined with gamma measurements to assay the plutonium associated with heavy or dense waste. The principal purpose of the measurement is to ensure that disposal conforms to the levels authorised by the Secretary of State for the Environment. Calibration is against standards which are not typical of the material but which will ensure that a low result is not obtained.

An increasing number of measurements are being carried out on waste and also on bulk materials using neutron coincidence counting technique to determine the Pu^{240} content. Equipment has been operated based both on the Birkhoff VDC counter (9) (10) (11) (12) and the Böhnel Shift Register type counter (13) (14) (12). The application includes the measurement of waste in packages varying from approximately 1 litre - 200 litres and to bulk material (usually PuO_2) containing from 1g - 3 Kg plutonium. At present the waste monitoring systems are based on BF_3 counters embedded in polythene giving a comparatively long neutron die-away time ($\sim$100 μsec) and the bulk material assay systems use He-3 counters, in polythene, having a neutron die away time of approximately 30 μsec.

Active neutron measurements

The U^{235} content of soft waste with high associated $\beta\gamma$ activity (up to 750 mR/hr) is measured by neutron irradiation from a Cf^{252} source followed by delayed neutron counting. The lower detection limit of this system is 2g U^{235} per bag of waste and the calibration covers the range from 2 - 20g (15).

Residual undissolved fast reactor fuel in cladding and transport cans is analysed by interrogation with 14 MeV neutrons followed by counting the delayed neutron emission. The lower limit of detection is less than 1g and measurements have been made in excess of 50g. Routine calibration is carried out against a sample of waste to which is added a standard pin containing 3g U^{235}. The accuracy of the measurements is estimated at 15 - 25% (1σ). This technique has been described previously (15) (16).

Neutron transmission measurements

Neutron transmission measurements have been combined with gamma-spectrometry to measure the U^{235} contents of unirradiated fuel bundles for the Steam Generating Heavy Water Reactor (SGHWR). The fuel bundle contains pins whose enrichments vary from 1.2% - 4.5%. Incorrect loading in the inner rings of the cluster can be detected using a Cf^{252} source of approx. 2×10^7 ns^{-1}. This technique is not in routine use but has been thoroughly tested. Further work has been carried out on mock fast reactor assemblies and is discussed in Section 3.

Neutron transmission measurements have been introduced to assay the U^{235} content of hollow U/Al fuel tubes ("inserts") destined, after further extrusion, to be incorporated in MTR fuel elements. This technique was introduced after it was found that the gamma counting technique used for fuel plates could not be applied with sufficient accuracy to these inserts. The method uses a number of Am/Be sources surrounding the alloy tube and a single detector in the centre. Full evaluation of this method is not yet complete but it is expected to become the routine accounting method in the near future.(18)

Calorimetry

Calorimetry is used to measure the plutonium in CaF_2 slags and filters. The calorimeter is of very simple design but has proved effective in the assay of plutonium in these difficult materials. The technique is used to measure from 5 - 300g in samples of slag of total weight of up to 15 Kg. The precision (1σ) is 2.5%, and the accuracy is believed to be about $\pm$ 1-2% in operating conditions, but this has not been fully tested. In laboratory conditions the accuracy was tested by assaying a piece of plutonium metal of known weight and isotopic composition. In these conditions the result was found to be within 0.5% of the true value (16).

Ultrasonic Measurements

A method, relating transmission time to concentration of uranium and plutonium in solvent, has been developed in the UK as on-line instrument in reprocessing plants. Under ideal conditions the coefficient of variation on solutions containing 60g U+Pu/litre is approx. 1%. Calibration and standardisation during operation will be against material which is sampled and analysed chemically. (17).

FUTURE DEVELOPMENTS

The major developments reflect both the increasing requirement for management, either local or central, to obtain assurance that material in stock is in accordance with the records and also the desire to improve measurements in terms of quality, speed and financial cost.

Gamma Spectrometry

There are a number of programmes aimed at developing improved high resolution detectors. Particular promise has been shown in the development of HgI_2 and CdTe crystals at the universities of Wales and Hull respectively. The former have not yet reached the stage of testing as detectors but encouraging results are being obtained from the CdTe.

A further development which seems likely to facilitate work in the field is the production of high purity germanium detectors using Stirling cycle cooling. The advantages are not limited to the elimination of the need for a liquid nitrogen supply, but also includes the increased ability to control the geometric arrangement of sample and detector.

Neutron and combined gamma and neutron techniques

Development is continuing on the construction and understanding of coincidence neutron detection systems. This has involved much computer simulation and it seems likely that quantities of plutonium containing up to 500g Pu^{240} will be able to be assayed to within $\pm$ 3% (1σ) and it should also be possible to combine this with an isotopic composition determined by gamma spectrometry to give an assay of plutonium in a container to $\pm$ 5% (1σ).

Neutron detectors will be installed to measure plutonium in massive objects - e.g. plant equipment, vehicles etc. This will be linked to a recording system to avoid the risk of movements not being registered.

There will be an increased use of neutron measurements in the assay of waste, particularly when the new fast reactor reprocessing plant comes into operation. These will mostly depend on passive measurements, either of total neutron or coincident neutrons, but it is also intended to use the commercial NNC Random Driver interrogation system for one of the waste streams.

The advent of the reprocessing of the plutonium fuel from the fast reactors also requires a more sensitive measurement from the 14 MeV neutron interrogation system, than was satisfactory for the measurements of U^{235}. A programme of work in collaboration with Philips of Eindhoven has shown that the ceramic tube neutron generator can be operated at much higher outputs than were hitherto quoted, and it seems probable that the desired sensitivity will be achieved without a need to resort to longer analysis times.

Combined neutron and gamma measurements have been shown capable of differentiating between the various types of sub-assembly used in fast reactor operations. The tests to date have involved the use of a total neutron count allied with high resolution γ-spectrometry. No final decision has been made on the extension of this work.

It is hoped to install a monitoring device to measure plutonium in the raffinate from the first separation stage of the fast reactor fuel reprocessing plant. Early indication of malfunction of the plant would prevent the need for expensive recycling operations following analysis in the hold-up tank. Passive measurements are unlikely to be satisfactory and possible interrogative measurements are under consideration.

Ultrasonic methods show promise of wide use in concentration measurements and volume measurements. At present the only proven application is to the measurement of heavy metal concentration in the solvent where it is possible to guarantee the composition of the matrix (17).

Thermo-luminescent dosimetry has been shown capable of giving useful results in the verification of inner fuel rods in fuel bundles. The technique is not rapid but the equipment is inexpensive and, since a large number of dosimeters can be employed, a high throughput can be achieved.

Calorimetry may be employed in verification measurements to the extent that it will provide a signature for a particular item. It does not, at present, seem likely that the technique will be applied to the assay of any material.

APPLICABILITY OF TECHNIQUES TO VERIFICATION AND INSPECTION ACTIVITIES

The most pressing requirement of present day nuclear

materials control is the need for large scale verification of quantities of material held at various stages of the fuel cycle or in long-term storage. This requirement is an essential element of international safeguards and is also acknowledged as necessary and desirable by central and local management in the U.K. and by national regulatory bodies elsewhere. It involves different considerations from those which were taken into account when many of the techniques described above were selected for the particular applications. In this section the effect of the following basic differences between management's measurements on a routine basis and measurements carried out for the purpose of verification will be considered.

(i) Confidence in the general nature of the item to be measured.
(ii) Background radiation effects.
(iii) Acceptable time scale for measurement.
(iv) Accuracy required.
(v) Number of items to be measured.

Many of these considerations will be found to be interrelated.

Difference in Confidence in the General Nature of the item to be Measured

A plant operator can justifiably make a number of assumptions regarding the nature of material and measurement conditions which may be unacceptable to an independent inspector. A plant operator may assume, for example, that an effluent from a plutonium separation plant contains plutonium and a certain proportion of other gamma or neutron emitters, but does not have to assume that the response of his instrument may be falsified by having an additional source placed near the detector to give artificially high counts (which is removed if a background count is to be measured), or by having a neutron poison added to depress results. In general an inspector knows much less about a material and the circumstances under which he attempts to make a verification than the operator who handles the material and this must decrease his confidence in his results. Even discounting deliberate falsification, the process of confirming the nature of stored material even when sound analytical data may exist, is not easy. The hazardous nature of the material requires that prior to long term storage it is subjected to multiple containment - in extreme cases this could result in double polythene containment within double steel cans. The geometric arrangement of powder and containment, and the density of the powder are difficult to confirm and hence lower the confidence in a result when the gross item is examined by non-destructive techniques involving γ or neutron emission.

Differences in Background Radiation

On-line techniques such as effluent monitors, instruments to measure profiles in mixer settlers, or even the NaI technique to measure the uranium content of U/Al fuel plates, assume a background within known limits and of certain characteristics. They are also applied at an appropriate location with regard to the production line. Verification measurements frequently require to be made either before or after this stage of plant control, in a storage location with high interfering background, or with instruments which are not able to take advantage of the heavy shielding which may have been incorporated in the plant instrument.

Clearly there is not a specific answer to this problem. It may be possible to shield the instrument sufficiently to perform in the changed environment but any intending verifier must face the possibility that arrangements may have to be made either to use a different technique or to remove material to a suitable location. This apparently simple solution can be very difficult to implement satisfactorily due to local restrictions on movement of material.

Differences in Acceptable Timescale of Measurement

Plant control techniques are devised with a known requirement in terms of speed of measurement and those described have varied from very fast semi-quantitative on-line monitoring systems to methods, such as the measurement of residual fuel in hulls, where the time taken from the decision to make the measurement to receipt of the result could be several days for a single item, largely due to the transport and manipulation involved in transferring the item to and from the measurement station. It is also possible that long counting times may be required by plant operators to achieve the accuracy which they regard as necessary. Safeguarding authorities would certainly agree that accountancy must be based on such measurements, but they may not be compatible with the needs of verification when a large number of items may have to be measured in a short space of time. (See table 1)

Differences in Accuracy Required

At present the major applications of non-destructive techniques to accountancy are limited to the assay of U/Al alloy and the measurement of waste. In each case the accuracy achieved in normal operation is as good or better than could be achieved by any available alternative technique, but if used for verification it might be necessary to sacrifice accuracy to achieve adequate throughput.

The techniques which in the U.K. are aimed primarily at verification, such as gamma enrichment measurements and neutron coincidence counting, can not match the accuracies obtained by present analytical techniques, but when applied to a range of similar items of well controlled isotopic composition, chemical composition and geometry can achieve the present target of providing a result which is no worse than $\pm$ 5% (1σ) on significant quantities of enriched uranium or plutonium. This is inadequate to give in itself sufficient assurance of the contents of a large store and it is accepted that it will always be necessary to support the non-destructive measurements with a small proportion of random sampling followed by more precise analysis.

Differences in number of items to be measured

There have already been references to the effect on verification of the need to verify accumulations of items. Simple gamma devices, used to confirm existing declared values, generally have a rapid response and increased number of items can normally be assayed on a time scale that is satisfactory to the auditor or inspector. It seems likely that any attempt to obtain meaningful measurements on irradiated material would be seriously inconvenienced by any accumulation of materials. The technique may be suitable but the operator's ability to measure items as they move in a routine manner through a programmed route and, having made the measurement, to accept this value at least until it passes through another routine measurement stage, is not available to the verifier.

The use of thermo luminescent dosimeters (see section 3.4) has the advantages that the dosimeters can be taken to

Table 1. Some typical analysis times for Non Destructive Technique

Technique	Time for a single determination*
Monitoring instruments (either gamma, neutron or ultrasonic)	Instantaneous output
Determination of U^{235} or Pu, using NaI	1 - 5 mins.
Segmented gamma scanner measurement of Pu	20 - 30 mins.
Neutron Coincidence Measurements	1 - 10 mins.
Active Interrogation of Soft Waste (Cf^{252} source)	10 mins.
Active Interrogation Using 14 MeV Neutrons	30 - 60 mins.
Gamma Spectrometric Determination of Plutonium Isotopic Composition	Overnight
Calorimetric measurement of Plutonium in Slags	3 days

*This does not include the time taken to present the sample to the measurement equipment, which is frequently more limiting on throughput than the actual analysis time.

the fuel and that large numbers can be used, but the technique requires development before its viability can be proved.

Further Development for Verification Purposes

On the above arguments one is lead to question whether the conventional lines of development of non-destructive analysis are of great utility for verification purposes. Nuclear radiation is attractive for NDA because it is there and because it is penetrating,but when quantitative determination depends upon orientation of the detector relative to the source, assumption that the infinite depth revealed by γ emission is representative of the whole, correction of backgrounds comparable with the indicative count, or uncertain attenuations, the suitability of the techniques for verification rather than initial measurement becomes increasingly questionable.

There is a clear need to develop other, possibly less immediately obvious, lines of approach to the non-destructive analysis of nuclear material that will provide the verification required under practical conditions.

CONCLUSIONS

The applications of NDA techniques have developed in areas where they offer a clear managerial advantage. Frequently this is achieved because it is legitimate to assume certain background parameters which allow a simplified application of a technique to be applied.

Many of these originally developed for plant control have graduated to serve as additional accountancy methods - this particularly applies to the measurement of nuclear material in waste.

It must, however, be recognised that even the improved accuracy and precision achieved by many non-destructive methods which are applied to material which is otherwise difficult to measure does not so improve accountancy that the form of the material examined becomes relatively unimportant. The best accountancy is still achieved on well characterised, easily analysed material.

The pressures on nuclear materials accountancy and control are increasing and increased use will have to be made of NDA methods. The demand will be for more data obtained more quickly and these trends will be intensified if statistical analysis has to be used to compensate for inadequate precision and accuracy. Few of the currently developed techniques appear likely to meet these demands and substantial new thinking may be required.

REFERENCES

(1) R. S. T. Shaw and P. Allenby, "An Assessment of the Eberline Stabilised Assay Meter (SAM 2) Used to Check the U^{235} Concentration of UO_2 Powder in the Oxide Powder Plant" B.N.F.L. Memo 468(S), 1976.

(2) R.S.T. Shaw and P. Allenby, "Further Tests with the Eberline Stabilised Assay Meter (SAM 2) Used to Check the U^{235} Concentration of UO_2 Powder in the Oxide Powder Plant" B.N.F.L. Memo 472(S) 1976.

(3) V.M. Sinclair and W.B. Adam, "Procedures used for the Accountancy of Uranium-235 in the Fabrication of Highly Enriched Uranium Fuels" IAEA-SM-201/62, 1976.

(4) M.F. Banham, "The Determination of the Isotopic Composition of Plutonium by Gamma-Ray Spectrometry" AERE-R 8737, 1977.

(5) H. A. Cole, "An Automatic Drum Scanning System for the Measurement of Plutonium in Waste" Nuclear Instruments and Methods 65, p.45, 1968.

(6) R.A. Cunningham, "The measurement of aged plutonium in waste by 380 KeV gamma counting. A theoretical appraisal" UKAEA PG Report 860(W), 1968.

(7) G.H. Fox and B.J. McDonald, "Measurement of the Plutonium Content of Soft Waste" BNFL Report 55(W) 1974.

(8) G.H. Fox and B.J. McDonald, "A Leached Hull Monitor for Use in the Reprocessing of Oxide Fuels" BNFL Report 213(W) 1976.

(9) G. Birkhoff, L. Bondar and N. Coppo, "Variable Dead Time Neutron Counter for Tamper Resistant Measurements of Spontaneous Fission Neutrons" EUR 4801 1972.

(10) G. Birkhoff, "On the Determination of Plutonium in Solid

Waste Containers by Spontaneous Fission Neutron Measurements" Seminar on the Management of Plutonium Contaminated Solid Wastes, Marcoule 1974, Proceedings 1974.

(11) R. Berg, R. Swennen, G. Birkhoff, L. Bondar, J. Ley and G. Busca, "On the Determination of the Pu^{240} in Solid Waste Containers by Spontaneous Fission Neutron Measurements. Application to Reprocessing Plant Waste" EUR 5158e 1974.

(12) K. P. Lambert, "A Comparison of the V.D.C. and Shift Register Neutron Coincidence Systems for Pu^{240} Assay" AERE 8303 1977.

(13) K. Böhnel, "Die Plutoniumbestimmung in Kernbrennstoffen mit der Neutronenkoinzenmethode" KFK 2203 1975. "Determination of Plutonium in Nuclear Fuels using the Neutron Coincidence Method" (KFK 2203 translated by H. Maneleys) AWRE Translation No. 70(54/4252).

(14) M.M. Stevens, J.E. Swansen and L.V. East, "Shift Register Coincidence Module" LA-6121-MS 1975.

(15) B.J. McDonald, G.H. Fox, W.B. Bremner "Non-Destructive Measurement of Plutonium and Uranium in Process Wastes and Residues" IAEA-SM-201/61, 1976.

(16) W.B. Bremner, A. MacDonald and A.R. Yates, Internal Document UKAEA TRG Memo 6803(D), 1975.

(17) R.C. Asher et al. "Ultrasonic technique for On-Line Surveillance and Monitoring of Process Plant; some Applications in Nuclear Fuel Reprocessing" Society of Chemical Industry Conference, London, September 1977 Proceedings in course of publication.

(18) P.F. Peck "Control Analysis of U^{235} in nuclear fuel by a neutron absorption method" AERE R 8928, 1977.

SAFEGUARDS SYSTEMS FOR NUCLEAR FUEL CYCLE FACILITIES

D. B. Smith
Los Alamos Scientific Laboratory
Los Alamos, New Mexico 87545

ABSTRACT

World-wide demand for nuclear power will require high-throughput facilities to support any of several alternative fuel cycles. Effective safeguards of nuclear materials is a prerequisite for accepted operation of such facilities. This paper describes the functions, elements, and structure of modern safeguards systems that combine advanced materials measurement and accounting techniques with the latest physical protection methods to provide maximum assurance against diversion of nuclear materials.

INTRODUCTION

Nuclear fuels and facilities were recognized as potential targets for theft, diversion, or sabotage right from the beginning of the United States' broad nuclear power program. Accordingly, a substantial national program of nuclear safeguards and security was instituted. This program has served the nation well; however, as much of the world moves toward large-scale utilization of nuclear energy during the last quarter of this century, demands are rising for increasingly stringent controls over the materials that fuel and are produced by the nuclear power industry.

The nuclear power demands of the near future will require high-throughput facilities to support any of several alternative fuel cycles. Spent-fuel reprocessing facilities having the capability to process over 100 kg of plutonium per day have been built, and even larger ones are being designed. The scale of these operations has forced a reassessment not only of facility design and construction and process operation, but also of the safeguards methods employed to prevent unauthorized use of the nuclear materials contained therein.

In 1976, ERDA's Division of Safeguards and Security commissioned the Los Alamos Scientific Laboratory (LASL) to develop and evaluate design concepts for generic safeguards systems intended to provide effective nuclear materials control and accountability for the nuclear fuel-cycle facilities of the 1980s (1-3). These systems are to be integrated into state-of-the-art physical protection and process control systems being designed for these plants by Sandia Laboratories (4,5). The resulting safeguards systems will provide an optimum mix of tight physical security for protection against <u>overt</u> attack on nuclear installations and materials in transit, and stringent in-plant materials control for protection against <u>covert</u> diversion by trusted insiders or blackmail threats by terrorist groups.

SAFEGUARDS SYSTEM FUNCTIONS

The primary objective of safeguards is to prevent the deliberate misuse of nuclear materials. A comprehensive safeguards strategy includes three functions:

- exclusion of all unauthorized persons from the facility, with further exclusion from sensitive areas within the plant;
- confirmation that all materials are present in the facility and in their proper locations; and
- control of all activities involving materials to preclude any that are not specifically authorized.

The system for implementing this strategy must operate without unnecessary disruption of plant operations, compromise of safety requirements, or infringements on employee working conditions.

Traditionally, facility safeguards systems have evolved in two separate areas: physical protection and materials control and accounting. The process-monitoring function is not addressed directly by either of these traditional systems but combines elements of both.

Physical protection has followed the conventional industrial security strategy of exclusion, that is, protecting a secure operating area against unauthorized entry. The physical protection system includes automated equipment and sufficient guard forces to provide effective response in the event of an emergency. It is responsible for access control:

- prevention of forcible entry,
- identification of those authorized to enter, and
- enforcement of security regulations governing those who are admitted.

However, the modern physical protection system expands the conventional security functions to provide, for example, direct protection of materials at those points in the process where material is in the form of discrete and countable items. An important objective in the design of the physical-protection system is to automate its functions whenever possible and to harden the system against subversion.

Process-monitoring may be regarded as an extension of physical protection monitoring and surveillance functions into the process line. The process-monitoring system collects information of the type intended to detect a theft in process from a limited set of on-line measurement equipment, plant-grade instrumentation, and other simple, reliable process-monitoring devices. Its primary function is to detect any movement or presence of material not expected in normal operation of the process. An array of sensors provides information on the status of process valves; presence or absence of material in process, sampler, and decontamination lines, status of valves supplying sample or transfer jets, and pressures in instrument lines. These sensors are all simple, rugged, and relatively inexpensive.

The accountability of materials within a facility and the detection of unauthorized removals have relied, almost exclusively, on material-balance accounting following periodic shutdown, cleanout, and physical inventory. The classical material balance associated with this system is drawn around the entire facility or a major portion of the process, and is formed by adding all measured receipts to the initial measured inventory and subtracting all measured removals from the final measured inventory.

Although conventional material-balance accounting is essential to safeguards control of nuclear material, it has inherent limitations in sensitivity and timeliness. The sensitivity limitation results from measurement uncertainties, so that the system may not detect the loss of relatively large quantities of nuclear material for large plant throughput. The timeliness of traditional materials accounting is limited by the frequency at which the physical inventory is taken. There are practical limits on how often a facility can shut down its process and still be productive. Conventional materials accounting can be augmented by dynamic materials control to achieve a more effective safeguards system.

MATERIALS MEASUREMENT AND ACCOUNTING

The materials measurement and accounting function in a modern safeguards system is an implementation of the DYMAC concept (6,7). It combines conventional chemical analysis, weighing, and volume measurements with the near-real-time measurement and surveillance capabilities provided by NDA instrumentation, and provides rapid and accurate assessment of the location and amount of nuclear material in the facility.

To implement dynamic materials accounting, the facility is partitioned into discrete accounting envelopes, called unit-process accounting areas. A unit process can be one or more chemical or physical processes, and is chosen on the basis of process logic and the ability to draw a material balance, rather than on geography, custodianship, or regulatory requirements. By dividing a facility into unit processes and measuring all material flows, quantities of material much smaller than the total plant inventory can be controlled on a timely basis. Also, any discrepancies are localized to that portion of the process contained in the unit process accounting area.

Computer-generated control charts derived from measurements and process operating characteristics can be used to indicate thefts, losses, or excessive holdup (8). This detailed control forces a potential diverter to steal material in sufficiently small quantities that his individual removals will be masked by measurement uncertainties. Thus, to obtain a usable quantity of material, the diverter must commit many thefts with the concomitant high risk of detection by the accounting system, surveillance instruments, doorway monitors, and the physical protection system.

Material balances drawn around such unit processes are called dynamic material balances to distinguish them from balances drawn after a cleanout and physical inventory. Dynamic material balances are drawn as often as practicable and are based on measurements of significant material transfers into and out of a unit process during its material balance period. Often the balance may be closed only approximately because some in-process holdup and minor materials sidestreams may be measured less frequently than major materials transfers (9). The data are up-dated when holdup and sidestream measurements are made. During the interim, historical data can be used to interpret trends in the material balance data.

Whether in-process material and holdup are measured depends primarily on the <u>variability</u> in the level of these materials during the balance period. If the variability is sufficiently small, the effect on individual balances may be negligible. Furthermore, the added control obtained by measuring small sidestreams of material may not justify the difficulty and expense of making the measurements.

Implementation of near-real-time materials measurement and accounting depends on quantitative measurement of the nuclear material as it crosses each unit-process boundary: it has required development of new measurement technology that permits rapid determination of the nuclear content of entire containers of material. The Atomic Energy Commission recognized this need for new measurement techniques and instrumentation in 1966, and appropriate research and development programs were instituted at that time. The technology is known as nondestructive assay (NDA) and takes advantage of the radiometric signatures available from most nuclear materials. These new NDA techniques complement, and sometimes supplant, the traditional destructive assay methods of analytical chemistry. The time scale for development of NDA has been consistent with the projected growth of the US nuclear industry.

While intensive development efforts have successfully demonstrated the measurement capability of NDA methods in many important nuclear fuel-cycle applications, as is typical of any new technology, some difficulty has been encountered in comparing and evaluating the measurement results. This difficulty is a result of the diversity of applications, equipment, and data reduction methods, and the wide variety of materials used within the nuclear industry. To address the lack of uniform practice in the application of NDA technology, the American National Standards Institute (ANSI) asked the Institute of Nuclear Materials Management to form Standards Committee N15, entitled "Methods for Nuclear Materials Control." Two of the subcommittees under INMM N15 deal with problems that directly affect NDA measurement quality. Subcommittee INMM 8, "Calibration Techniques," has already generated four approved ANSI standards:

- ANSI N15.18-1975, "Mass Calibration Techniques for Nuclear Material Control,"
- ANSI N15.19, "Volume Calibration Techniques for Nuclear Materials Control,"
- ANSI N15.20-1975, "Guide to Calibrating Nondestructive Assay Systems," and
- ANSI N15.22-1975, "Calibration Techniques for the Calorimetric Assay of Plutonium-Bearing Solids Applied to Nuclear Materials Control."

Currently under development by INMM 9 are proposed standards on:

- material categorization,
- container standarization,
- physical (calibration) standards,
- measurement control and assurance,
- NDA techniques and application, and
- automation of NDA measurements.

The writing groups working on these standards comprise several dozen highly qualified individuals with expert knowledge in various aspects of nondestructive assay methods used for nuclear materials control. A total of 29 different organizations are represented: 12 licensees, 2

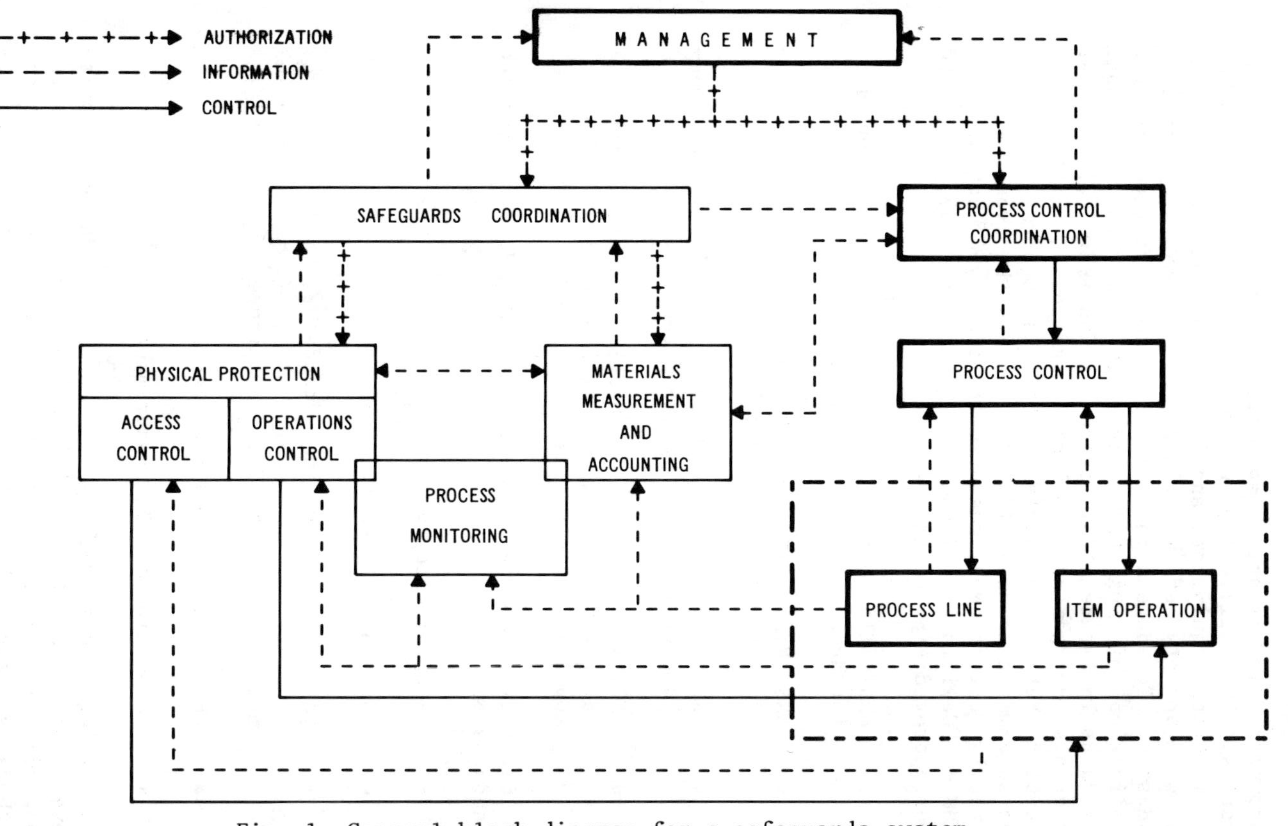

Fig. 1. General block diagram for a safeguards system.

instrument vendors, 11 license-exempt (contractor), and 4 government agencies (10). In addition, each proposed standard must undergo extensive review by the nuclear industry before it is approved as an ANSI standard.

The mechanism within the facility safeguards system that ensures the quality of the measurement data is an organized measurement-control program designed to measure and monitor the accuracy and precision of each instrument in the system and to verify application of NDA techniques in compliance with accepted practices. This measurement-control program is an integral part of the safeguards system in that the computer system records and monitors both measurement and calibration data. A performance history thus is maintained for each instrument.

SAFEGUARDS SYSTEM STRUCTURE

The detailed specification of a coordinated safeguards system must be based on a thorough evaluation of the facility design; however, the fundamental structure of the system should be applicable to any type of facility. A general block diagram for a safeguards system is shown in Fig. 1 (2). At the core of the plant is the process itself (those functions directly related to the process are enclosed in the heavily outlined boxes). The operating condition of each unit process within the process line is specified by the process-control coordination (PCC) unit on the basis of operational authorizations, optimal process-operation considerations, and safety.

Process-control coordination is also responsible for implementing safeguards-related recommendations that impact the flow of bulk materials in the plant. This is necessary to ensure most effective compliance from both the safeguards and process-control viewpoints. Only the handling and storage of discrete material items are directly controlled by the safeguards system. The materials measurement and accounting system (MMAS) and the process-control coordination unit also exchange process-related data information to help improve both process operation and safeguards effectiveness.

The safeguards coordination unit supervises the safeguarding of nuclear material in the facility. As the focal point for safeguards decisions, it interacts with plant management and process-control coordination to ensure effective safeguards. Safeguards coordination recommends modifications to the work authorizations from management according to its assessment of the safeguards status of the plant. For example, if the safeguards status is satisfactory, the management work authorization is unmodified. However, if overt diversion has been detected, or if an attack is imminent, safeguards coordination recommends that PCC control the process to a relatively secure state to further restrict accessibility to materials. Safeguards coordination also transmits continuously the safeguards status to the MMAS for information only, and to the access control unit so that the security force's status and deployment are appropriate.

As the central component of the safeguards system, the safeguards coordination unit has three primary functions: (1) data collection and processing, which is required for (2) safeguards condition assessment, which in turn is the basis for (3) the response determination decision. Safeguards coordination normally relies on automated decision and control, augmented by human supervision in abnormal situations. It must be as simple and reliable as possible, and its decision-making function must be balanced to avoid frequent false alarms that cause unnecessary process disruptions, while maintaining a high probability of effective response to any credible safeguards violation.

CONCLUSION

The threat of diversion of nuclear material by plant personnel or by terrorist groups is an on-going problem facing both domestic and international safeguards. This threat can be addressed effectively by a combination of stringent physical protection and dynamic materials accounting. An integrated safeguards system incorporating the latest in physical protection technology, and advanced materials measurement capability coupled with computer-based accounting, control, and data analysis can ensure the security of materials in nuclear fuel-cycle facilities at reasonable cost, without disrupting production or requiring major process modifications.

REFERENCES

(1) J. P. Shipley, D. D. Cobb, R. J. Dietz, M. L. Evans, E. P. Schelonka, D. B. Smith, and R. B. Walton, "Coordinated Safeguards for Materials Management in a Mixed-Oxide Fuel Facility," Los Alamos Scientific Laboratory report LA-6536 (February 1977).

(2) E. A. Hakkila, D. D. Cobb, H. A. Dayem, R. J. Dietz, E. A. Kern, E. P. Schelonka, J. P. Shipley, D. B. Smith, R. H. Augustson, and J. W. Barnes, "Coordinated Safeguards for Materials Management in a Fuel Reprocessing Plant," Los Alamos Scientific Laboratory report LA-6881 (October 1977).

(3) H. A. Dayem, D. D. Cobb, R. J. Dietz, E. A. Hakkila, E. A. Kern, J. P. Shipley, D. B. Smith, and D. F. Bowersox, "Coordinated Safeguards for Materials Management in a Nitrate to Oxide Conversion Facility," Los Alamos Scientific Laboratory report (to be published).

(4) L. D. Chapman, J. M. de Montmollin, J. E. Deveney, W. C. Fienning, J. W. Hickman, L. D. Watkins, and A. E. Winblad, "Development of an Engineered Safeguards System Concept for a Mixed-Oxide Fuel Fabrication Facility," Sandia Laboratories report SAND76-0180 (August 1976).

(5) W. D. Chadwick, W. C. Fienning, B. G. Self, and G. E. Rochau, "A Concept Definition of an Engineered Safeguards System for a Spent Fuel Reprocessing Facility," Sandia Laboratories report (to be published).

(6) G. R. Keepin and W. J. Maraman, "Nondestructive Assay Technology and In-Plant Dynamic Materials Control--DYMAC," in Safeguarding Nuclear Materials, Proc. Symp. Vienna, 1975 (International Atomic Energy Agency, Vienna, 1976) paper IAEA-SM-201/32, pp. 305-320.

(7) R. H. Augustson, "Development of In-Plant Real-Time Materials Control: The DYMAC Program," Nucl. Mater. Manage. V(III), 302-316 (1976).

(8) D. D. Cobb, D. B. Smith, and J. P. Shipley, "Cumulative Sum Charts in Safeguarding Special Nuclear Materials," Los Alamos Scientific Laboratory internal report LA-UR-76-2749 (submitted to Technometrics).

(9) J. E. Lovett, "In-Plant Dynamic Material Controls: An International Perspective," Nucl. Mater. Manage. V(III), pp. 342-350 (1976).

(10) D. M. Bishop, "New Scope and Goals for N15 Subcommittee INMM-9 (Nondestructive Assay)," Nucl. Mater. Manage. IV(III), pp. 285-298 (1975).

A MORE ACCURATE THERMAL NEUTRON COINCIDENCE COUNTING TECHNIQUE*

Norton Baron
University of California
Los Alamos Scientific Laboratory

INTRODUCTION

At the Los Alamos Scientific Laboratory, passive thermal-neutron coincidence counting is one of several techniques under investigation for the purpose of nondestructively assaying spontaneously fissioning material. Several thermalizing neutron detectors were constructed and their properties investigated. Such detectors have a cylindrical cavity, i.e., well, of about 6-in. diam x 20-in. long into which the sample is placed for counting. ^{3}He filled proportional counting tubes are placed at a fixed radius symmetrically about the well and are embedded in polyethylene (CH_2) moderating material. Designs studied include both one- and two-concentric rings of tubes. The well is generally lined with cadmium to 1) inhibit multiplication caused by thermalized neutrons being backscattered into the well, and 2) provide a system die-away time which is relatively independent of the sample. A typical two-ring detector is pictured in Fig. 1.

The amount of spontaneously fissioning material in the well is determined by comparing its coincidence count rate to that of a known mass, i.e., a standard. Since a sample's multiplication and moderation significantly affect the measured coincidence count rate per unit mass, an accurate comparison measurement demands that these effects be similar for the standard and unknown samples. This traditional approach presupposes a certain knowledge of the sample's matrix material, chemical composition, and geometry, which in fact are often unknown and furthermore necessitates the need for many standards. In such a context, this nondestructive assay (NDA) technique is somewhat expensive and cumbersome to use

*Work performed under the auspices of the U.S. Department of Energy.

effectively. However, thermal-neutron coincidence counting becomes one of the more attractive NDA techniques from considerations of expense, accuracy, and versatility if a procedure can be found to account readily for multiplication and moderation effects within reasonable limits.

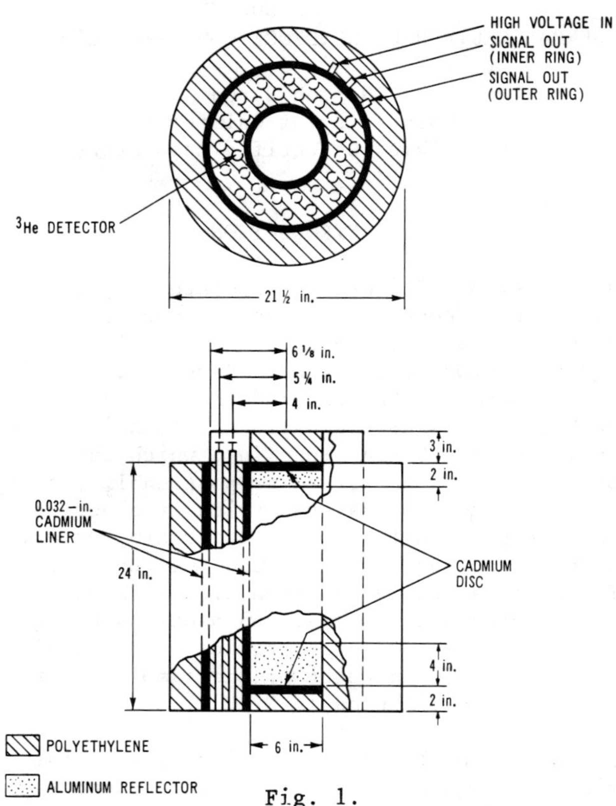

Fig. 1.
A two-ring thermal-neutron well-counter.

Very often NDA measurements must be made on samples of PuO_2 powder that are frequently contaminated with oxides or salts of light weight elements. Such contaminants have large (α,n) reaction cross sections, and the reaction product neutrons can cause significant multiplication. The object here is to investigate 1) the multiplication phenomena caused by both (α,n)- and spontaneous fission (SF)-neutrons associated with such powder samples, and 2) changes in neutron detection efficiency due to variations in neutron moderation among the samples. It is hoped that such work will lead to NDA neutron counting techniques that eliminate the measurement error introduced by failing to account properly for such phenomena.

This technique measures the amount of fertile material in the sample, which is assumed to consist of only isotopes of ^{239}Pu and ^{240}Pu. If there are other fertile isotopes present in addition to ^{240}Pu, then we measure the ^{240}Pu effective mass, M_{240}, which is defined to be the ^{240}Pu mass which will produce the same number of neutrons per second from spontaneous fissions as actually produced in all fertile isotopes in the mixed sample.

THEORY

General

We shall consider four sources of neutrons that contribute to the passive thermal-neutron singles count rate $\dot{T}$ defined as:

$$\dot{T} \equiv \dot{T}_{SF} + \dot{T}_{SM} + \dot{T}_{\alpha n} + \dot{T}_{IM} \quad . \qquad (1)$$

The four contributors to $\dot{T}$ represent neutrons born by

a) spontaneous-fissions ($\dot{T}_{SF}$)
b) fissions induced by SF-neutrons and their progeny; these will be referred to as self-multiplication neutrons ($\dot{T}_{SM}$)
c) (α,n) reactions ($\dot{T}_{\alpha,n}$)
d) fissions induced by the (α,n)-neutrons and their progeny; these will be referred to as induced-multiplication neutrons ($\dot{T}_{IM}$)

The true coincidence count rate for neutrons created by fission process X must be related to the associated singles count rate. This can be expressed functionally as

$$\dot{\eta}_X = f_X(\dot{T}_X) \quad . \qquad (2)$$

Since the measured true coincidence count rate $\dot{\eta}$ is the sum of the count rates of coincident pairs resulting from spontaneous-fissions ($\dot{\eta}_{SF}$), self-multiplications ($\dot{\eta}_{SM}$), and induced-multiplications ($\dot{\eta}_{IM}$), we can write

$$\dot{\eta} = \dot{\eta}_{SF} + \dot{\eta}_{SM} + \dot{\eta}_{IM} \qquad (3)$$

However, the neutron detection efficiency is a function of its energy. Consequently, to make an accurate comparison measurement, it is imperative that our model modify $\dot{\eta}$ to account for variations of neutron energy moderation among different samples. We note that neutrons contributing to true coincidence events are fission products and are therefore born with very nearly the same energy distribution (1). Therefore, we can reason that any variation among different samples of their energy distributions measured by the detector must be due to different moderating properties of the samples. We now define a modified coincidence count rate $\dot{N}$ whose magnitude is independent of the sample's moderation where

$$\dot{N} \equiv \frac{\dot{\eta}}{f(\text{sample's moderation})} \quad . \tag{4}$$

f(sample's moderation) is a function of the sample's moderation, i.e., an energy signature, which will be defined subsequently. Using Eq. 4, an expression for the modified total coincidence count rate analogous to Eq. 3 is defined as

$$\dot{N} \equiv \dot{N}_{SF} + \dot{N}_{SM} + \dot{N}_{IM} \quad . \tag{5}$$

The specific response rate SRR defined as

$$SRR \equiv \frac{\dot{N}_{SF}}{M_{240}} \tag{6}$$

is expected to be constant for a given detector. Using Eq. 5, the quantity $\dot{N}_{SF}$ that must be extracted from the coincidence count rate $\dot{N}$ in order to calculate SRR is given by

$$\dot{N}_{SF} = \dot{N} \cdot \left[\frac{\dot{N}_{SF}}{\dot{N}_{SF} + \dot{N}_{SM} + \dot{N}_{IM}} \right] \quad . \tag{7}$$

Defining the bracketed term in Eq. 7 to be the inverse of the correction ratio CR where

$$CR \equiv \frac{\dot{N}_{SF} + \dot{N}_{SM} + \dot{N}_{IM}}{\dot{N}_{SF}} \quad . \tag{8}$$

allows Eq. 7 to be rewritten as

$$\dot{N}_{SF} = \frac{\dot{N}}{CR} \quad . \tag{9}$$

Substituting Eq. 9 into Eq. 6 gives

$$SRR \equiv \frac{\dot{N}}{CR} \cdot \frac{1}{M_{240}} \quad . \tag{10}$$

Unless $\dot{N}$ and CR can be determined so that the coincidence count rate can be corrected for moderation and multiplication phenomena respectively, one cannot infer accurately the value of M_{240} in an unknown sample by comparing its response to that of a standard.

Multiplication Correction CR

Unfortunately a direct determination of CR is not possible since the quantities $\dot{N}_{SF}$, $\dot{N}_{SM}$, and $\dot{N}_{IM}$ are not measurable separately. As a result, there must be defined a function of some measurable quantity which is equivalent to CR in the sense that it will account adequately for modifications in the measured coincidence count rate due to multiplication phenomena. To do this it is necessary to identify the physical quantities which control these phenomena. This is done by writing more explicit definitions of the singles count rates summed in Eq. 1. The results indicate that these count rates and hence CR can differ among samples because of variations in 1) the probability P for generating random neutrons via the (α,n) reaction (where P is a function of the (α,n) reaction cross section $\sigma_{\alpha n}$ and the intimacy of the mixture μ of the matrix material and plutonium), 2) the energy E of the (α,n) reaction product neutrons which affects the multiplicities (2) associated with the induced multiplication and consequently the coincidence count rate $\dot{N}_{IM}$ (where E is controlled by μ, the (α,n) reaction Q value, and the neutron moderating properties of the sample), 3) the ^{240}Pu effective mass M_{240} which determines the number of SF-neutrons and the partial number of α-particles available to generate (α,n)- neutrons which can induce fissions, 4) the ^{239}Pu mass M_{239} which determines the partial number of α-particles available to generate (α,n)-neutrons which can induce fissions, 5) atom densities ρ_{239} and ρ_{240} of ^{239}Pu and ^{240}Pu respectively which determine the probability that a neutron will induce a fission, 6) path length L of a neutron within a sample.

An attempt will be made to construct an empirical function of these variables that will describe relative values of CR by accounting for relative changes in $\dot{N}_{SM}/\dot{N}_{SF}$ and $\dot{N}_{IM}/\dot{N}_{SF}$ among different samples. To be useful, such a mathematical model of CR must result in a value of SRR that is constant over a reasonable range of values of these variables. Formalistically, CR can be written as a function of these variables where this function is defined as

$$CR \equiv f_{CR}(E, P, M_{239}, M_{240}, \rho_{239}, \rho_{240}, L) \tag{11}$$

Experimental evidence that will be presented subsequently indicates that f_{CR} defined by Eq. 11 is separable, to a good degree of approximation, into the product of two independent functions, one of which depends only on the properties of the matrix material and the other on the properties of the plutonium. Therefore, since E and P are independent of M_{240}, M_{239}, ρ_{240}, ρ_{239}, and L, it is reasonable to assume that f_{CR} can be expressed as

$$CR \equiv f_{\alpha n}(E,P) \cdot f_{Pu}(M_{240}, M_{239}, \rho_{240}, \rho_{239}, L) \ . \tag{12}$$

Physically, $f_{\alpha n}(E,P)$ is that part of CR which accounts for variations in the induced fission rate due to different matrix materials. Its value is proportional to the probability that a random neutron of energy E will be generated by the (α,n) reaction on the matrix material. This quantity will be referred to as the matrix multiplication correction. We note that even among samples having the same neutron moderation, the matrix correction is not expected to be constant for a particular matrix material. This is due to the dependence of P on μ, the intimacy of the mixture of PuO_2 and the matrix material. The less intimate the mixture, the smaller will be the α-particle's energy prior to inducing the (α,n) reaction. This results in a smaller reaction product neutron energy E. Furthermore, in the presence of such a mixture the α-particle must travel farther to collide with a nucleus of the matrix material. This leads to an increased probability that the α-particle will be absorbed by a nucleus other than one of the matrix material thereby reducing the value of P. Thus we can expect perturbations about some average matrix correction $f_{\alpha n}$ due to variations in μ. The greater the intimacy of mixture, the larger will be P and E.

The factor f_{Pu} is proportional to the multiplication caused by SF-neutrons (which originate with ^{240}Pu) in the presence of plutonium densities ρ_{240} and ρ_{239}. In addition both M_{240} and M_{239} are sources of radioactive decay α-particles which are converted by the probability term $f_{\alpha n}$ to (α,n)-neutrons. Consequently f_{Pu} is also proportional to the multiplication caused by (α,n)-neutrons in the presence of ρ_{240} and ρ_{239}. The quantity f_{Pu} will be referred to as the plutonium multiplication correction.

Procedure to Calculate M_{240}

We will now sketch the manner by which the value of M_{240} can be determined for an unknown sample. Using Eq. 12 in Eq. 10 gives

$$SRR \equiv \frac{\dot{N}}{f_{\alpha n}(E,P) \cdot f_{Pu}(M_{240},M_{239},\rho_{240},\rho_{239},L)} \cdot \frac{1}{M_{240}} \quad . \tag{13}$$

where SRR is invariant and $f_{\alpha n}(E,P)$ is a function of the matrix material, its intimacy of mixture, and the sample's moderating property. Both SRR and $f_{\alpha n}$ are assumed to be measurable. Two simplifying assumptions are made which enable one to write $f_{Pu} \simeq f_{Pu}(M_{240})$. Defining the apparent specific response rate ASRR as

$$ASRR \equiv \frac{\dot{N}}{f_{Pu}(M_{240})} \cdot \frac{1}{M_{240}} \quad . \tag{14}$$

and using this definition in Eq. 13 allows us to write

$$ASRR = SRR \cdot f_{\alpha n}(E,P) \quad . \tag{15}$$

Once $f_{\alpha n}(E,P)$ and $f_{Pu}(M_{240})$ are defined and calculated using the results of measurements of standards, a calibration curve of ASRR vs $f_{\alpha n}(E,P)$ can be generated. The amount of ^{240}Pu in a sample for which $f_{\alpha n}(E,P)$ has been measured then can be determined using either the calibration curve or Eq. 15 to obtain ASRR and subsequently solving Eq. 14 for M_{240}.

Plutonium Correction f_{Pu} $(M_{240}, M_{239}, \rho_{240}, \rho_{239}, L)$

The function $f_{Pu}(M_{240}, M_{239}, \rho_{240}, \rho_{239}, L)$ cannot be modeled until we determine the sensitivity of the specific response rate to variations in these parameters. We now make the simplifying assumptions that 1) the PuO_2 powder samples are always in a constant geometry container and the fill heights are constant; therefore the density is proportional to the mass, and 2) all samples have the same isotopic abundances. Invoking these simplifying assumptions allows us to express f_{Pu} as a function only of M_{240} and L where the path length L is dependent on M_{240}. Furthermore, the empirical observation that the coincidence count rate varies non-linearly with the mass of plutonium suggests that we can write

$$f_{Pu}(M_{240}, M_{239}, \rho_{240}, \rho_{239}, L) = \frac{M_{240}^{X}}{M_{240}} \quad . \tag{16}$$

where the exponent X is an empirically determined constant. Substituting Eq. 16 into Eq. 14 gives

$$ASRR = \frac{\dot{N}}{M_{240}^{X}} \quad . \tag{17}$$

It is interesting to examine how restrictive are the simplifying assumptions of constant fill heights and isotopic abundances. Using the empirically determined value of X, one can estimate that variations in the fill height by ±20% affect the value of f_{Pu} by only $\sim$±1%. A similar result is anticipated for variations in the isotopic abundance of ^{240}Pu.

Matrix Correction $f_{\alpha n}(E,P)$

The principal problem in this scheme to assay M_{240} is the identification and measurement of the function $f_{\alpha n}(E,P)$. This quantity must account correctly for changes in CR due to variations in E and P. We note that CR varies monotonically with E since the multiplicity and hence the coincidence count rate $\dot{N}_{IM}$ associated with the induced fission process varies monotonically with the incident neutron's energy (2). Furthermore, the value of CR also varies monotonically with P since the magnitude of $\dot{N}_{IM}$ varies monotonically with the number of random neutrons generated by the (α,n) reaction.

Fortunately a quantity that also varies monotonically with E and P can be defined and measured using a detector like that pictured in Fig. 1. A two-ring detector can provide information concerning the relative energy distribution of neutrons which enter the detector. This information is obtained by measuring the ratio of the neutron singles count rates in the outer and inner rings, $\dot{T}^O/\dot{T}^I$, where superscripts O and I refer to the outer and inner rings respectively. Using Eq. 1, this ratio can be written as

$$\frac{\dot{T}^O}{\dot{T}^I} = \frac{\dot{T}^O_{SF} + \dot{T}^O_{SM} + \dot{T}^O_{IM} + \dot{T}^O_{\alpha n}}{\dot{T}^I_{SF} + \dot{T}^I_{SM} + \dot{T}^I_{IM} + \dot{T}^I_{\alpha n}} \quad . \tag{18}$$

As previously mentioned, the SF-, SM-, and IM-neutrons are all fission products and to a good approximation their energy distributions at birth do not vary among different samples. However, the larger the energy E of the (α,n)-neutron reaction product the harder will be the singles' spectrum and the larger will be the ratio $\dot{T}^O/\dot{T}^I$. Furthermore, inspection of Eq. 18 shows that if the average (α,n)-neutron energy $\overline{E}$ is greater than the average fission neutron energy $\overline{E}_f$,

$$\overline{E} > \overline{E}_f \quad , \tag{19}$$

then an increased value of P causes equal percentage increases in both $\dot{T}^O{}_{\alpha n}$ and $\dot{T}^I{}_{\alpha n}$ thereby increasing the ratio $\dot{T}^O/\dot{T}^I$. This effect is somewhat moderated by a concurrent increase in $\dot{T}^{O(I)}_{IM}$ where

$$\dot{T}^{O(I)}_{\alpha n} > \dot{T}^{O(I)}_{IM} \quad . \tag{20}$$

Therefore, both CR and $\dot{T}^O/\dot{T}^I$ have the same qualitative dependence on E and P.

Prompted by this qualitative similarity we assume that a function $f(\dot{T}^O/\dot{T}^I)$ can be defined so that

$$f^{O(I)}_{\alpha n}(E,P) \equiv f^{O(I)}(\dot{T}^O/\dot{T}^I) \tag{21}$$

for samples containing matrix materials that meet the criterion of Eq. 19 and whose (α,n) reaction Q values vary monotonically with expected values of P. We use the specific relation

$$f^{O(I)}(\dot{T}^O/\dot{T}^I) \equiv (\dot{T}^O/\dot{T}^I)^{\delta^{O(I)}} \tag{22}$$

to define $f_{\alpha n}^{O(I)}(E,P)$ in Eq. 15 where δ^O and δ^I are constants associated with the outer and inner rings respectively.

Detector Constants

Upon substituting Eq. 16 into Eq. 13 we can write the specific response rate as

$$SRR^{O(I)} = \frac{\dot{N}^{O(I)}}{f_{\alpha n}^{O(I)} \; M_{240}^X} \tag{23}$$

Using Eqs. 21 and 22, the above relation can be rewritten as

$$SRR^{O(I)} = \frac{\dot{N}^{O(I)}}{(\dot{T}^O/\dot{T}^I)^{\delta^{O(I)}} \cdot M_{240}^X} \tag{24}$$

and equating the subsequent expressions of SRR for samples A and B yields

$$\left[\frac{(\dot{T}^O/\dot{T}^I)_A}{(\dot{T}^O/\dot{T}^I)_B}\right]^{\delta^{O(I)}} = \left[\frac{\dot{N}_A^{O(I)}}{\dot{N}_B^{O(I)}}\right] \cdot \left[\frac{(M_{240})_B}{(M_{240})_A}\right]^X \tag{25}$$

The exponents in Eq. 25 are obtained empirically in the following manner. A choice of δ^O determines values of X_i for several standard samples A_i comparisoned to standard B. A value of δ^O is found which minimizes the standard deviation calculated for the several values of X_i. Using this value of δ^O and the counting data of several standards in Eq. 24 enable a least squares determination of SRR^O and X^O. This procedure is repeated using the counting data of the inner ring to determine δ^I, SRR^I, and X^I (which is very nearly equal to X^O). The final value of X is taken to be the average of X^O and X^I and the final values of SRR^O and SRR^I are determined using Eq. 24.

Using standards of PuO_2 powder mixed with different matrix materials, the singles and coincidence count rates were measured using a two-ring detector like that pictured in Fig. 1. The results of these measurements are listed in Table I. Since all of these samples were fired at the same temperature during their preparation, they are expected to have equivalent neutron moderation properties. In such a situation, there is no need to modify the measured coincidence count rate by a factor that accounts for varying moderation among these samples. We therefore define f(sample's moderation) $\equiv$ 1 in Eq. 4 and calculate the detector constants to be

$$\delta^I = 3.00 \qquad (26\text{-a})$$

$$\delta^O = 4.00 \qquad (26\text{-b})$$

$$X = 1.058 \qquad (26\text{-c})$$

$$(SRR^I)_{Av} = 24.94 \pm 2.24\% \qquad (26\text{-d})$$

$$(SRR^O)_{Av} = 19.62 \pm 1.07\% \qquad (26\text{-e})$$

Also listed in Table I are values of $ASRR^{O(I)}$ calculated using Eqs. 17 and 26-c. Pictured in Fig. 2 are plots of these values as a function of $f_{\alpha n}^{O(I)}(E,P)$ defined by Eqs. 21, 22, 26-a, and 26-b. The linear relation between $ASRR^{O(I)}$ and $(\dot{T}^O/\dot{T}^I)^{4(3)}$ indicates that the matrix correction defined by Eqs. 21 and 22 can be determined in the presence of an arbitrary mixture of these matrices.

Response of "Outer + Inner" Rings

We can improve the counting statistics and thereby reduce the measurement uncertainty by calculating a single response

TABLE I

COINCIDENCE COUNTING MEASUREMENT RESULTS[1]

SAMPLE ID	DOMINANT MATRIX MATERIAL	MASS OF Pu IN PuO_2 SAMPLE (grams)	EFFECTIVE MASS OF ^{240}Pu (grams)	TRUE COINCIDENCE COUNT RATES CORRECTED FOR BKGD & DEAD-TIME		"SINGLES" COUNT RATE RATIOS $\dot{T}^O / \dot{T}^I$	APPARENT SPECIFIC RESPONSE RATES		SPECIFIC RESPONSE RATES			SRR^B CORRECTED FOR MODERATION[d] κ
				OUTER RING $\dot{N}^O$	INNER RING $\dot{N}^I$		OUTER[a] RING $ASRR^O$	INNER[a] RING $ASSR^I$	OUTER[b] RING SRR^O	INNER[b] RING SRR^I	BOTH[c] RINGS SRR^B	
STD 4	SiO_2	60.02	3.837	34.61	54.08	.8084	8.34	13.04	19.53	24.68	88.13	30.20
STD 5	MgO	60.03	3.838	39.93	60.83	.8407	9.62	14.66	19.27	24.67	87.55	31.88
STD 6	SiO_2	120.0	7.672	73.96	113.09	.8117	8.57	13.10	19.73	24.49	88.19	31.83
STD 7	MgO	120.0	7.672	85.25	132.19	.8424	9.87	15.31	19.61	25.61	90.03	31.42
STD 8	MgO	240.1	15.35	175.4	276.47	.8367	9.76	15.38	19.90	26.25	91.88	30.82
STD 9	(none)	480.3	37.39	426.32	645.96	.8281	9.24	14.00	19.65	24.66	88.33	32.58
385	(none)	459.0	43.39	480.07	743.05	.8222	8.89	13.76	19.45	24.76	88.10	30.88
382	(none)	556.0	54.93	619.18	949.04	.8234	8.93	13.69	19.44	24.53	87.63	31.44
381	(none)	615.0	64.97	747.04	1141.92	.8228	9.03	13.80	19.69	24.77	88.63	32.01
447	(none)	779.0	81.36	981.29	1486.17	.8273	9.34	14.15	19.95	25.00	89.61	33.09

[1] No CH_2 placed about samples.

[a] Calculated using Eqs. 17, 26-c.

[b] Calculated using Eqs. 24, 26a-c.

[c] Calculated using Eq. 32.

[d] Calculated using Eqs. 32, 34, 35.

due to the counting data of both rings of a two-ring detector. To do this we define

$p^{O(I)}$ ≡ the average probability that one member of a pair of coincident neutrons will be detected by a counter in the outer (inner) ring.

If the average coincidence detection efficiency of SF-neutrons in the outer (inner) ring is defined as

$$\overline{\varepsilon^{O(I)}} \equiv \frac{SRR^{O(I)}}{a} \tag{27}$$

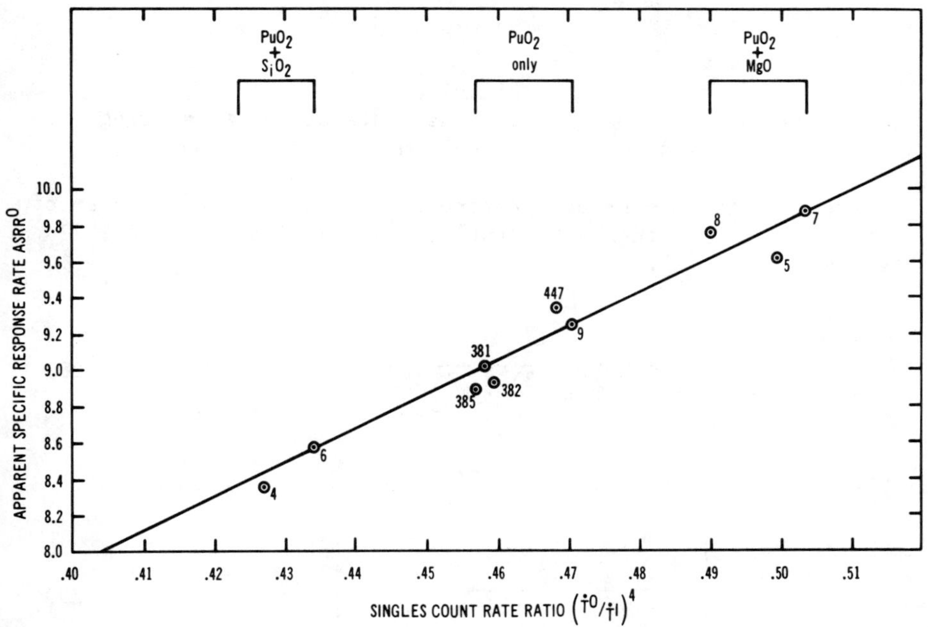

Fig. 2-a.
The apparent specific response rate of the outer ring versus the outer ring's matrix correction.

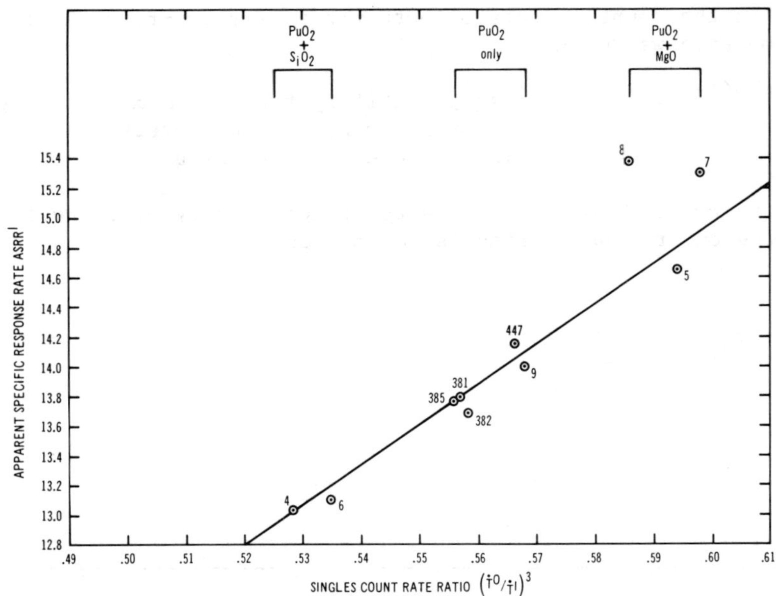

Fig. 2-b.
The apparent specific response rate of the inner ring versus the inner ring's matrix correction.

where a is the number of spontaneous fissions per s per gram of effective ^{240}Pu, then substituting Eqs. 24, 26a-c into Eq. 27 gives

$$\overline{\varepsilon^{O(I)}} = \frac{\dot{N}^{O(I)}}{a \cdot (\dot{T}^O/\dot{T}^I)^{4(3)} \cdot M_{240}^{1.058}} \tag{28}$$

Therefore, the probabilities $p^{O(I)}$ are

$$p^{O(I)} = \left[\frac{\dot{N}_{SF}^{O(I)}}{a \cdot (\dot{T}^O/\dot{T}^I)^{4(3)} \cdot M_{240}^{1.058}} \right]^{1/2} \tag{29}$$

The average efficiency of counting a pair of coincident neutrons by both rings of a two-ring detector can be written

$$\overline{\varepsilon^B} = (p^I)^2 + (p^O)^2 + 2 \cdot p^I \cdot p^O \tag{30}$$

The specific response rate SRR^B which results from combining the responses of both rings is defined as

$$SRR^B \equiv \overline{\varepsilon^B} \cdot a \tag{31}$$

Substituting Eq. 29 into Eq. 30 and using the result in Eq. 31 allows us to write

$$SRR^B = \frac{\left[\dfrac{\dot{N}^I}{(\dot{T}^O/\dot{T}^I)^3} + \dfrac{\dot{N}^O}{(\dot{T}^O/\dot{T}^I)^4} + 2\left[\dfrac{\dot{N}^I \cdot \dot{N}^O}{(\dot{T}^O/\dot{T}^I)^7}\right]^{1/2}\right]}{M_{240}^{1.058}} \tag{32}$$

where the constant SRR^B is denoted as the calibration factor. Using the data of Table I, values of SRR^B are calculated and listed there.

Correction for Moderation

As was mentioned earlier in the discussion concerning Eq. 4, the neutrons contributing to true coincidence events are expected to be born with nearly the same energy distribution. Consequently, the ratio $\dot{\eta}^I/\dot{\eta}^O$ of the true coincidence count rates measured by the inner and outer rings of a two-ring detector should be a function of the moderating property of the sample's material. Referring to Eq. 4 we now define

$$f(\text{sample's moderation}) \equiv f_{MOD}(\dot{\eta}^I/\dot{\eta}^O) \tag{33}$$

Let us now examine the validity of this assumption.

In addition to the data already referred to in Table I, there is another data set listed in Table II for the same samples about which were placed the equivalent of 559 grams of CH_2. Values of SRR^B are calculated for this data using Eq. 24 and the detector constants in Eqs. 26a-c. The results are listed in Table II. As was done for the analogous calculation in Table I, we assume f(sample's moderation) ≡ 1.

Pictured in Fig. 3 is a plot of these values of SRR^B vs $(\dot{\eta}^I/\dot{\eta}^O)^{2.4}$. We observe a grouping of the data. Thus it is reasonable to expect that a function of the ratio $\dot{\eta}^I/\dot{\eta}^O$ can be determined that will account for moderation effects in

TABLE II

COINCIDENCE COUNTING MEASUREMENT RESULTS[1]

SAMPLE ID	DOMINANT MATRIX MATERIAL	MASS OF Pu IN PuO_2 SAMPLE (grams)	EFFECTIVE MASS OF ^{240}Pu (grams)	TRUE COINCIDENCE COUNT RATES CORRECTED FOR BKGD & DEAD-TIME		"SINGLES" COUNT RATE RATIOS $\dot{T}^O / \dot{T}^I$	APPARENT SPECIFIC RESPONSE RATES		SPECIFIC RESPONSE RATES			SRR^B CORRECTED FOR MODERATION [d]
				OUTER RING $\dot{N}^O$	INNER RING $\dot{N}^I$		OUTER[a] RING $ASRR^O$	INNER[a] RING $ASSR^I$	OUTER[b] RING SRR^O	INNER[b] RING SRR^I	BOTH[c] RINGS SRR^B	κ
STD 4	SiO_2	60.02	3.837	36.09	64.54	.7488	8.70	15.56	27.67	37.05	128.76	31.91
STD 5	MgO	60.03	3.838	42.53	76.23	.7846	10.25	18.37	27.05	38.04	129.25	31.85
STD 6	SiO_2	120.0	7.672	75.49	135.18	.7531	8.74	15.66	27.18	36.66	126.97	31.36
STD 7	MgO	120.0	7.672	89.12	159.19	.7824	10.32	18.44	27.55	38.50	131.19	32.60
STD 8	MgO	240.1	15.35	186.75	331.56	.7799	10.39	18.44	28.07	38.87	133.00	33.53
STD 9	(none)	480.3	37.39	440.72	781.08	.7686	9.55	16.93	27.38	37.30	128.59	32.56
385	(none)	459.0	43.39	497.22	894.15	.7666	9.21	16.56	26.66	36.76	126.03	30.82
382	(none)	556.0	54.93	651.46	1170.96	.7674	9.40	16.90	27.11	37.39	128.18	31.37
381	(none)	615.0	64.97	772.66	1380.41	.7690	9.34	16.68	26.69	36.67	125.93	31.28
447	(none)	779.0	81.36	1038.01	1830.26	.7731	9.88	17.43	27.67	37.72	130.00	33.32

[1] 559 grams CH_2 placed about samples

[a] Calculated using Eqs. 17, 26-c.

[b] Calculated using Eqs. 24, 26a-c.

[c] Calculated using Eq. 32.

[d] Calculated using Eqs. 32, 34, 35.

each sample. Consequently we define the modified coincidence count rates $\dot{N}^{O(I)}$ that appear in Eq. 5 to be

$$\dot{N}^{O(I)} \equiv \frac{\dot{\eta}^{O(I)}}{f_{MOD}(\dot{\eta}^{I}/\dot{\eta}^{O})} \tag{34}$$

where

$$f_{MOD}(\dot{\eta}^{I}/\dot{\eta}^{O}) \equiv (\dot{\eta}^{I}/\dot{\eta}^{O})^{2.4} \tag{35}$$

for the detector pictured in Fig. 1. Therefore, values of SRR^B calculated using Eqs. 32, 34, and 35, are independent of sample moderation. We denote these moderation-independent values of SRR^B as κ. Values of κ are calculated for the data presented in Tables I and II and also listed there.

The mean specific response rates with and without 559 g of CH_2 placed about the samples are calculated and listed in

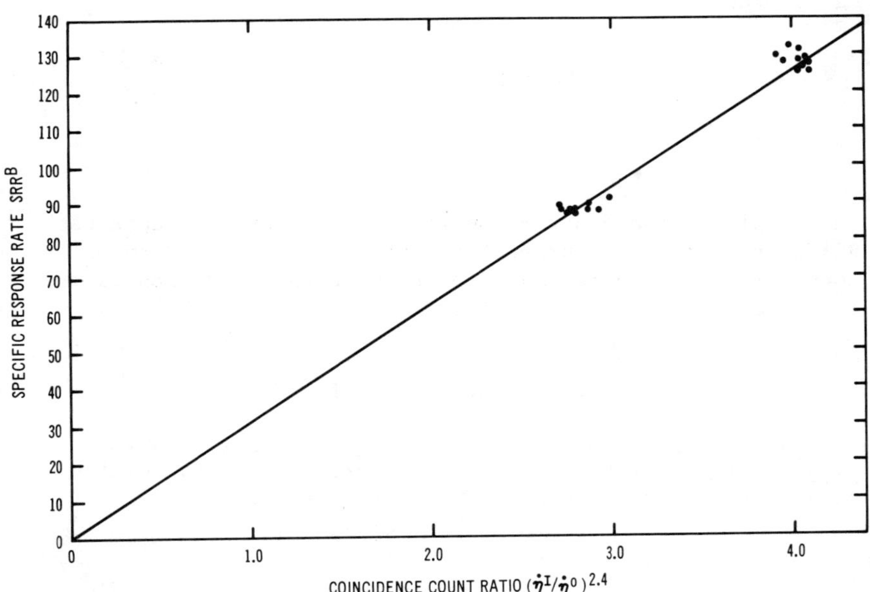

Fig. 3
The specific response rate SRR^B as a function of the coincidence count rate ring ratio.

Table III. The ratio of these mean specific response rates, for which no moderation correction has been made, is calculated as

$$\left[\frac{SRR^{B}_{Av}\ (559 \text{ grams of } CH_2)}{SRR^{B}_{Av}\ (0 \text{ grams of } CH_2)}\right] = 1.45 \pm 3.33\%$$

Similarly the ratio of the mean specific response rates corrected for moderation, which are also listed in Table III, is calculated as

$$\left[\frac{\kappa_{Av}\ (559 \text{ grams of } CH_2)}{\kappa_{Av}\ (0 \text{ grams of } CH_2)}\right] = 1.01 \pm 3.96\%$$

Thus we eliminate the 45% increase in specific response rates caused by the 559 grams of CH_2 moderating material. Furthermore, the mean value and associated standard deviation of the 20 specific response rates κ corrected for sample moderation, listed in the last columns of Tables I and II are calculated to be

$$\kappa_{Av} \pm \sigma_{\kappa_{Av}} = 31.84 \pm 2.73\%$$

This small standard deviation for samples having such grossly different moderating properties indicates the validity of correcting for moderation effects by using a function of the coincidence count rates $\dot{\eta}^{I}/\dot{\eta}^{O}$.

The Effective Mass of ^{240}Pu

The effective mass of ^{240}Pu in an unknown sample for which $\dot{\eta}^{O(I)}$ and $\dot{T}^{O(I)}$ have been measured can be calculated using Eqs. 32, 34, and 35. The final expression is

$$M_{240} = \exp\left\{\frac{\ell n\left(\dfrac{\dfrac{\dot{N}^{I}}{(\dot{T}^{O}/\dot{T}^{I})^{3}} + \dfrac{\dot{N}^{O}}{(\dot{T}^{O}/\dot{T}^{I})^{4}} + 2\cdot\left[\dfrac{\dot{N}^{I}\cdot\dot{N}^{O}}{(\dot{T}^{O}/\dot{T}^{I})^{7}}\right]^{1/2}}{SRR^{B}_{Av}}\right)}{1.058}\right\} \quad (36)$$

TABLE III

MEAN VALUES OF SPECIFIC RESPONSE RATES

	NO POLYETHYLENE ABOUT SAMPLE		559 GRAMS POLYETHYLENE ABOUT SAMPLE	
	SRR^B (a)	κ (b)	SRR^B (a)	κ (b)
$\left(\text{SPECIFIC RESPONSE RATE}\right)_{Av}$	88.8	31.6	128.8	32.1
$\sigma\left(\text{SPECIFIC RESPONSE RATE}\right)_{Av}$	±1.5%	±2.7%	±1.7%	±2.8%

a) Calculated using Eq. 32.

b) Calculated using Eqs. 32, 34, 35.

where

$$\dot{N}^{O(I)} = \frac{\dot{\eta}^{O(I)}}{(\dot{\eta}^{I}/\dot{\eta}^{O})^{2.4}} \tag{37}$$

$$SRR^B_{Av} = \frac{\sum_{i=1}^{i_{max}} (SRR^B)_{Std\ i}}{i_{max}} \tag{38}$$

and $(SRR^B)_{Std\ i}$ is calculated using Eq. 32.

EXPERIMENTAL RESULTS

Using a two-ring well counter like that pictured in Fig. 1 and the theory described above, a standard deviation of less than 3% was obtained for the system's specific response rate. This figure reflects the expected accuracy for an assay of effective mass of ^{240}Pu. These results are based on the analyses of 20 samples of PuO_2 powders combined with different matrix materials among which 1) the probability for (α,n)-neutron induced fissions varied by 15%, 2) the multiplication varied by 28%, 3) a variable moderation effect changed the specific response rate by 45%, 4) plutonium densities varied by a factor of 6.5, and 5) masses of total plutonium varied from 60- to 779-grams.

CONCLUSIONS

Using passive thermal neutron coincidence counting techniques, the accuracy of nondestructive assays of fertile material can be improved significantly using a two-ring detector similar to that pictured in Fig. 1. It was shown how the use of a function of the coincidence count rate ring-ratio can provide a detector response rate that is independent of variations in neutron detection efficiency caused by varying sample moderation. Furthermore, the correction for multiplication caused by SF- and (α,n)-neutrons is shown to be separable into the product of a function of the effective mass of ^{240}Pu (i.e., plutonium correction) and a function of the (α,n) reaction probability (i.e., matrix correction). The matrix correction is described by a function of the singles count rate ring-ratio. This correction factor is empirically observed to be identical for any combination of PuO_2 powder and matrix materials SiO_2 and MgO because of the similar relation of the (α,n)-Q value and (α,n)-reaction cross section among these matrix nuclei. However the matrix correction expression is expected to be different for matrix materials such as Na, Al, and/or Li. Nevertheless, it should be recognized that for comparison measurements among samples of similar matrix content, it is expected that some function of the singles count rate ring-ratio can be defined to account for variations in the matrix correction due to differences in the intimacy of mixture among the samples. Furthermore the magnitude of this singles count rate ring-ratio serves to identify the contaminant generating the (α,n)-neutrons. Such information is useful in process control.

REFERENCES

1. Fast Neutron Physics, Part II, by Marion & Fowler, Interscience Publishers, 1963, pg. 2036.

2. Physics of Nuclear Kinetics, by G. R. Keepin, 1965, pg. 54.

HAND-HELD SPECIAL NUCLEAR MATERIAL MONITOR FOR PERSONNEL AND VEHICLE SEARCHES

Walter E. Kunz
Carl N. Henry
David R. Millegan
Los Alamos Scientific Laboratory

ABSTRACT

A hand-held special nuclear materials monitor is being developed for personnel and vehicle searches in areas having widely varying gamma-ray background intensities. The monitor produces an alarm when the counts accumulated during any counting interval exceed the trip level, which is the sum of the stored background count and a fixed incremental count. Tests were conducted by rotating a standard 10g ^{235}U test source past the monitor and recording the detection efficiency for a series of trip level settings at each of nine gamma background radiation intensities. The results show that it is possible to exceed the required 50% detection efficiency at background levels that range from 0 to 400 counts per 0.3-s counting interval.

INTRODUCTION

The Department of Energy (DOE) has become increasingly concerned with preventing diversion of special nuclear materials (SNM). Regulations now in effect require the search of all personnel and vehicles leaving areas containing significant amounts of SNM.

One personnel-search technique involves the use of a radiation-detecting doorway monitor (approximately $15,000 cost) that can quickly scan all outgoing pedestrians. For vehicle searches, a gateway monitor requires much larger detector array than does the doorway monitor and is, consequently, more expensive. For locations where the traffic is low, a hand-held monitor for personnel and vehicle searches would cost less (approximately $1000 and, in the event of a positive indication, could more quickly narrow the search for any SNM. This report describes such a hand-held radiation monitor developed at the Los Alamos Scientific Laboratory (LASL).

DESIGN CRITERIA

Sensitivity

Hand-held monitors used to detect SNM require high gamma-ray sensitivity. Proposed monitor standards stipulate the ability to detect, with 50% or greater probability, a 10g ^{235}U sphere transported at speeds of about 0.5 m/s at distances of at least 0.25 meters. Such detection requires enough gamma-ray sensitivity to accumulate a significant number of counts from the uranium source in a time interval of 1 second or less. In general, this requirement rules out the use of Geiger counters and other detectors of comparable sensitivity. Medium-sized (38 mm by 38 mm) NaI(Tℓ) scintillator/photomultiplier combinations meet this sensitivity requirement. The normal background-radiation count rate from such a detector is about 100 counts/s. The count rate produced by a 10g sphere of ^{235}U at 0.25 m from such detectors is approximately equal to normal background.

Signal Readout

A hand-held monitor with a detector sensitivity equal to that noted above must offer a useful system for signal readout. Visual readout from a count rate meter and audible monitoring of the direct or scaled count rate are the most common modes. For a count rate meter to be useful in detecting small amounts of SNM, the response time must be less than 1 second if the detector is moving past the source at approximately 0.5 m/s with the distance of closest approach about 0.2 meters. With use of a short response time, the meter reading will fluctuate rather widely because of statistical variations in the background count rate. Detection of small amounts of SNM thus becomes difficult, even when the operator gives complete attention to the meter, but during searches, the operator not only must watch the meter but also must hold the monitor about 0.2 m from the subject, an exercise that requires full attention. Consequently, the visual method of signal readout is not suitable for this application. Making the count rate or a scaled count rate continuously audible by use of loudspeaker or earphone is a common method of readout for handheld monitors. However, because the ear is not very sensitive to minor count rate variations, an untrained operator finds this approach difficult to use at marginal detection levels or in a high-noise-level environment, but if the count rate meter is fitted with an alarm that sounds when a preset count rate is exceeded, the system becomes more practical. The operator can then devote full attention to the search.

DELTA RATE MONITOR

Description

Monitor design at LASL has been directed toward producing instruments for personnel and vehicle searches, which can be used effectively by operators with little special scientific training. Detection of a small amount of SNM, which produces a small increase in count rate above that from the background radiation, is difficult for even a highly-trained operator when using the conventional methods of signal readout. Therefore, we have designed monitors with digital logic systems that free the operators from having to decide whether the count rate has increased by a significant amount. These monitors sound an alarm each time the background count per interval is exceeded by a certain amount. The monitors are silent except when the alarm is operating, which adds to the ease of operation in a noisy or distracting environment.

Our earlier monitor, which has been described previously in LASL Report LA-6359, is now available commercially. It operates by storing a trip level that is some preselected multiple (normally 1.4) of the background count rate per time interval. Each time the trip level is exceeded during consecutive time intervals, an alarm sounds. This monitor performs quite well when used in areas where background radiation intensities vary by less than factors of two. Because the trip level is a multiple of the background count rate per interval, each doubling of the background intensity requires a similar doubling of source strength to maintain the same detection probability. Thus, the monitor's sensitivity falls to unacceptable levels at high background radiation intensities. (Background radiation as used in this report means all gamma radiation present except that produced by the source being searched for.)

We are testing a laboratory prototype of an improved SNM monitor, called the Delta Rate Monitor (DRM), which is capable of maintaining adequate sensitivity during background intensity variations as great as a factor of ten. The DRM differs from our previously described monitor only in the method of obtaining the trip level. The trip level in the earlier version is a preset multiple of the stored background count; the trip level in the current version (DRM) is the stored background count plus a fixed increment.

The DRM contains a conventional 38-mm long by 38-mm diameter NaI(Tℓ) scintillator/photomultiplier, a pulse amplifier, and a lower level pulse-height discriminator (LLD). Gamma rays interacting with the NaI crystal cause scintillations that are converted to electrical pulses and amplified by the photomultiplier, further amplified by the amplifier, and fed to the LLD. Amplifier gain is adjusted so that the 60-keV gamma-ray pulses from ^{241}Am just exceed the LLD setting. All pulses exceeding the LLD setting produce 5-V logic pulses.

The digital logic system stores a background count for the preselected counting interval and adds it to the fixed increment to produce the trip level. When the background is stored, the system begins to compare the number of counts accumulated during each counting interval with the trip level. If the trip level is equalled by the accumulated counts during the counting intervals, an audible alarm is produced. The alarm continues until the end of the counting interval, at which time it stops. The monitor immediately begins to accumulate new counts, and if the trip level is equalled again, the alarm sequence is repeated.

During the background update operation, logic pulses are stored digitally in accordance with the setting of the counting interval, the trip level update mode, and the background store time multiplier. The lowest counting interval setting is 0.15 seconds. The settings can be increased by factors of two up to the maximum setting of 9.6 seconds.

The trip level update mode can be set for manual or automatic operation. The lowest background store time multiplier is 16. Multipliers can be increased by factors of two up to the maximum multiplier of 1024. The fixed increment number is selected by appropriate switches.

Mode of Operation

Assume that a 0.3-s counting interval, a background store time multiplier of 16, a fixed increment of 14, and a manual trip level update mode are selected. First, the operator presses the background update button causing the trip level update process to begin. Background counts are collected during an interval that is the product of the 0.3-s counting interval and the background store time multiplier of 16. If the background count rate is 100 counts per second, then 480 counts are collected and scaled by 16 (divided by 16 and truncated to an integer). The result, 30, is added to the fixed increment of 14 to become a trip level of 44. After this process has been completed, 16 x 0.3 s = 4.8 s is the time required for update with these control settings, the monitor is ready for use.

The monitor now counts continuously for 0.3-s intervals, with no dead time between intervals, and on the average accumulates 30 counts per interval. If during any counting interval the accumulated count equals 44, the trip level, an audible tone is switched on and sounds until the next counting interval begins. This process is repeated during every 0.3-s interval. If a gamma-emitting radioactive source is placed close enough to the monitor to raise the average number of counts per interval to twice the trip level, the operator will hear a series of tones that begin at about 0.3-s intervals. These tones will begin about half way through the counting intervals and will last for about 0.15 seconds until the end of the interval. If the source is moved closer, the time that the tone remains on during the interval becomes longer; moving the source farther away decreases the tone length. This effect assists the operator in locating a radioactive source. With the source removed completely, an occasional tone is heard because of statistical fluctuations in the background count rate. With this monitor, an occasional tone does not detract from the ease of operation and assures the operator that the system is working.

Operation in the automatic trip level update mode is the same as for the manual mode, except the trip level is automatically redetermined at time intervals that are the product of the background store time multiplier and the count interval.

PERFORMANCE TEST

We have used the previously described sensitivity standard for hand-held monitors in testing the operation of the DRM. Testing is conducted with an automated arrangement that determines the probability of source detection. As the source is rotated past the monitor by means of a 0.796-m radius arm attached to a 0.1-rps motor, timer-scalers record: the elapsed time, the number of times the source passes the monitor, the number of gamma counts, and the number of detections. A detection is defined as the first alarm occurring during a 1.75-s period centered 0.3 seconds after the time when the source is closest to the detector. Thus, no more than one detection can occur per pass even though several counting intervals may produce alarms. Background radiation levels above normal are produced by positioning a ^{137}Cs source at appropriate distances from the monitor.

The prototype DRM incorporates a visual readout for the stored background count. An appropriate incremental count, switch selected, is added to this background count to produce the trip level. "Delta Count" is the trip level less the mean background per interval. We assume that the average background rate, obtained during the 40,000-s or longer counting period required for each run, is indistinguishable from the mean. The average background count rate is determined by subtracting the previously determined count rate produced by the ^{235}U source from the count rate obtained from the entire run. At each background radiation level, various delta count values were used to produce detection probabilities from approximately 50% to the highest probability compatible with an acceptable false alarm rate. Each run consisted of more than 4000 passes of the source past the monitor and determined one detection probability.

Figure 1 illustrates the effect of different delta counts on detection probability for our local background rate of 38 counts per interval and for a rate of 78, produced by the addition of radiation from a ^{137}Cs source. Similar plots were obtained for each of six higher background count rates. The data obtained from all runs with detection probabilities less than 99% are shown in Figure 2, where detection probability as a function of the background count rate is plotted for delta counts ranging from 20 to 42. The plotted curve for each delta count is extrapolated upward to the background count rate which produces a 3.0% false alarm rate, which we arbitrarily define as an excessive rate. These calculated false alarm rates agree well with the experimentally determined ones.

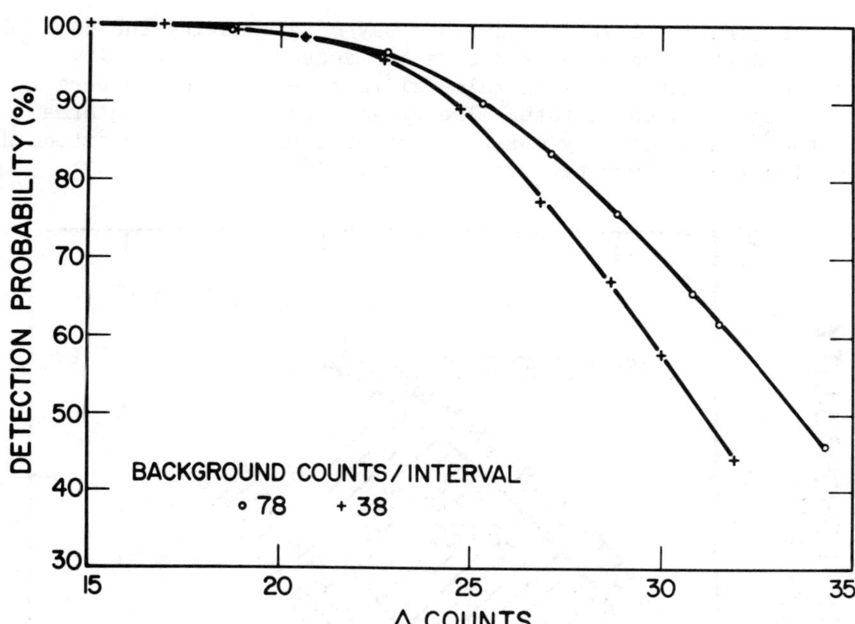

Fig. 1 Detection probability versus delta counts for background count rates of 38 and 78 per interval.

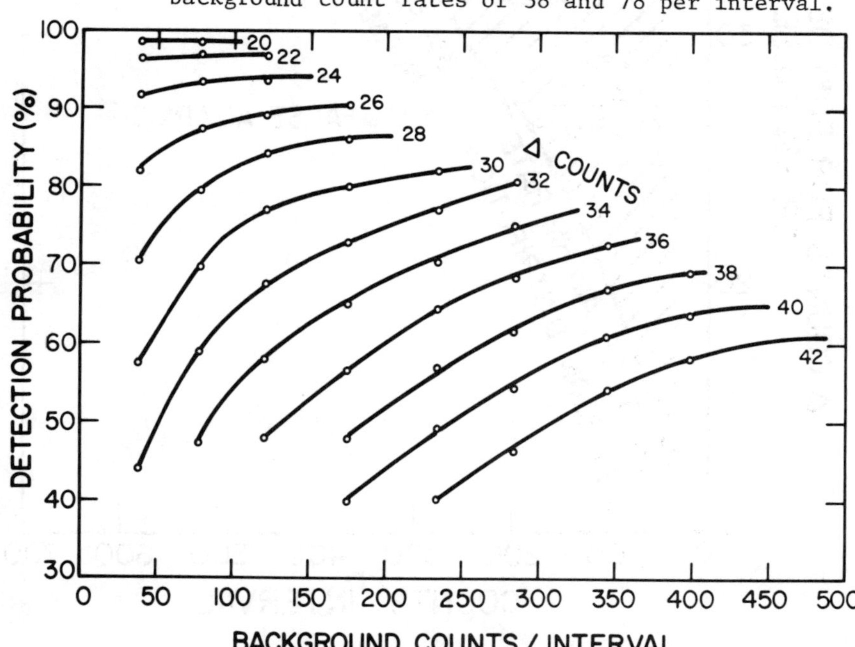

Fig. 2 Detection probability versus background counts per interval for delta counts ranging from 20 to 42.

Figure 3 illustrates another way of displaying the same data. Delta counts to produce a 50% detection probability and delta counts for a 3% false alarm rate are plotted versus the background count rate. The cross hatched operating area is the area covered by our data and obviously could be extended to higher and lower background rates.

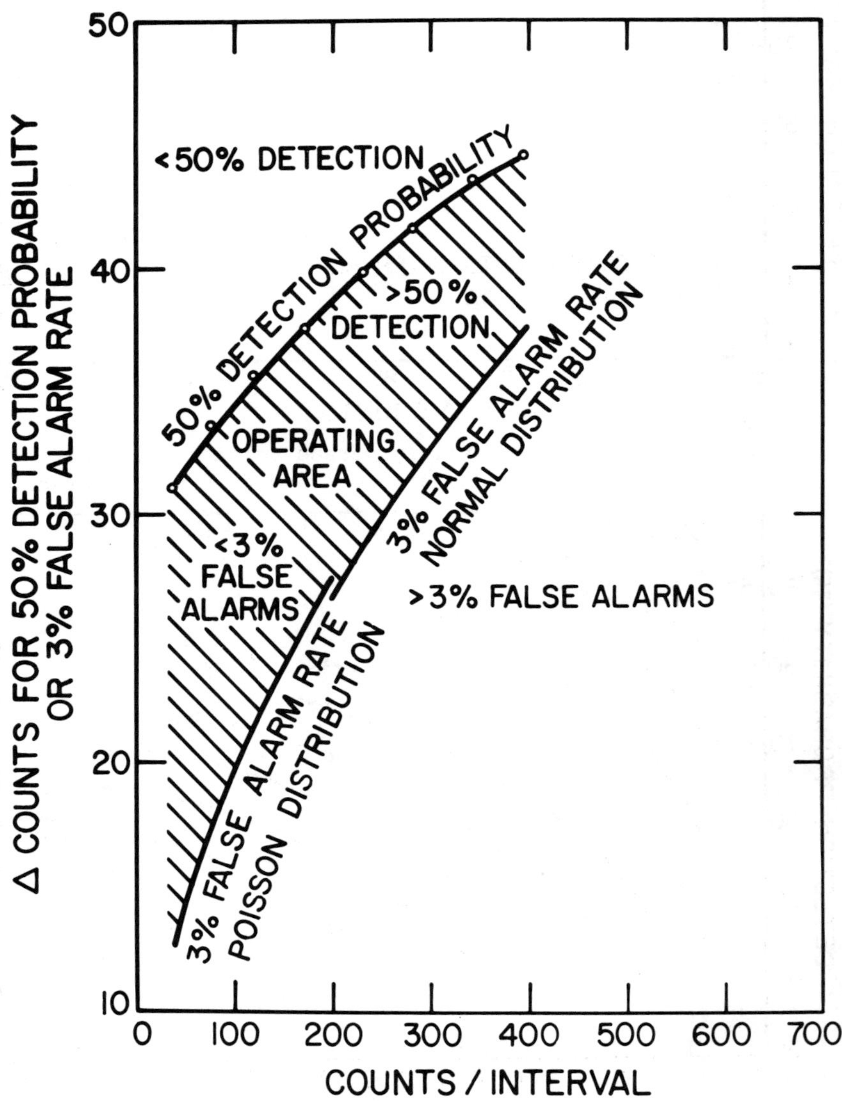

Fig. 3 Plot of delta counts for a 50% detection probability and also for a 3% false alarm rate.

We must now address the problem of setting trip levels for our operational DRM that will remain within the operational envelope shown in Figure 3. In the following discussion we neglect the rather small errors caused by the digital logic system of the monitor, which operates with integers only. With this system, the background count rate determined during the update operation is rounded down to the next lower integer, the incremental count is always integral, and thus, the trip level, the sum of the two, is an integer. The use of non-integral values for these quantities simplifies our discussion. The optimum trip level is the sum of the delta count for the maximum permissible false alarm rate of 3% and the particular mean background. The trip level so obtained is the lowest possible if the false alarm rate is not to exceed 3%, and it is optimum because it produces the highest detection probability. In like manner, the maximum permissible trip level is that which produces a 50% detection probability at this same background rate. From Figure 3 we find for a background rate of 38 counts per interval that the maximum permissible delta count is 30.5, and that the optimum delta count is 12.3. The range of allowable delta counts at this background rate is bounded by these two values; thus, the total range is 12.2. The range of allowable delta counts, and hence the range of allowable trip levels, decreases with increasing background rates. At a background rate of 400 counts per interval, the range is only 6.5 counts.

One method for obtaining the trip level is to determine the background count rate and add to it a multiple of the square root of the background count rate; the sum is the trip level. This method is adequate if long enough background accumulation times are used. To obtain a confidence level of 95% that the trip level will lie within the operating range at background rates of 350-400 counts per interval requires a background accumulation time of approximately 2-1/2 minutes. This long an update time requires operation in the automatic update mode. Update times of 5- to 10-s are adequate for backgrounds up to about 150 counts per interval. We will not use this method of trip level determination because the addition of the digital logics to determine the square root doubles the complexity of the DRM's logic system. When low-current microprocessors become available, we may reconsider this decision. (We require approximately 100 hours of operations between recharges of our battery operated monitors, hence the need for low-current microprocessors.)

Our operational DRM's trip level will be set by storing a background and then adding an appropriate incremental count with a digital switch. The operator can rapidly set this switch to any number from 0 to 99 because separate thumb wheels are used for setting the units and the tens. The operator sets a number for the incremental count, which produces a fairly high false alarm rate (3 to 6 or more per minute, which is 1-1/2- to 3-% or more). The trip level set in this manner for any particular background rate up to 400 counts per interval will be in the operating range. We extend the range below the 3% false alarm rate curve because a false alarm rate somewhat above 3% increases detection probability with little decrease in the ease of operation. Background update time can be set for the minimum of 16 x 0.3s = 4.8s because rather large variation of the stored background from the mean can be accommodated with this method of trip level update. The trip level is not reset during operations unless a significant change in the false alarm rate indicates the need for a new trip level update operation.

We intend to test the operational DRM rather extensively for detection sensitivity under field conditions with different operators before specifying the range of background radiation levels for feasible use.

APPLICABILITY OF TEST RESULTS TO OTHER MONITORS

The results of the testing of the DRM shown in Figures 1-3 are applicable to our earlier hand-held monitors and to any monitor using a 38-mm diameter by 38-mm long NaI(Tℓ) crystal/photomultiplier detector, provided gamma-ray pulses which exceed the LLD setting of approximately 50 keV are accumulated into 0.3-s counting intervals with alarm production occurring when the count equals the mean background plus the delta count.

CONCLUSIONS

The laboratory prototype DRM will pass the hand-held monitor sensitivity test when operated in background radiation fields producing counts per 0.3-s intervals from 0 to 400. Field evaluation awaits the availability of an operational model.

A SYSTEM FOR THE NONDESTRUCTIVE EVALUATION AND ASSAY OF BOILING WATER REACTOR FUEL RODS

John P. Stewart

Nondestructive evaluation and verification of Boiling Water Reactor (BWR) fuel rods at General Electric's Nuclear Fuel Fabrication Facility was initiated in 1969 with the installation of a passive rod scanning operation. This operation eventually employed 42 passive rod scanning systems performing ^{235}U distribution or enrichment verification measurements on the entire fuel rod output of the Wilmington, North Carolina, facility. These systems were manually operated, utilizing manual loading and unloading and visual evaluation of chart-recorder data.

In early 1970, the Los Alamos Scientific Laboratory (LASL) Fuel Rod Scanner was introduced into the Wilmington Facility as a joint effort of the then USAEC and the General Electric Company. The LASL Fuel Rod Scanner was designed to assay the fissile or ^{235}U content of BWR fuel rods; in contrast, the passive rod scanning system measured the fissile content per unit length of fuel column. This joint effort demonstrated the feasibility of accountability fuel rod scanning within the uncertainties required by the site accountability system. Thus, concurrent with the joint USAEC-GE accountability scanning effort, the General Electric Company developed and built at the Vallecitos Nuclear Center, and installed at the Wilmington Facility, a neutron-activation based fuel rod scanning system.

Initial operation of the system, called the Cf-252 Active Scanner, began in 1972 with the scanner being operated manually. That is, chart recordings of the fuel rod ^{235}U enrichment and UO_2 fuel column density traces were evaluated manually, similar to the passive fuel rod scanner, by scanning operation personnel using visual templates.

This operation differed from the passive operation in two respects; an added density evaluation and higher throughput.

Subsequently, an on-line dedicated mini-computer was installed and software designed to provide data acquisition and processing and system automation. This system configuration has evolved through modification of the operational software system to accommodate changing BWR fuel rod designs and further automation of scanning functions. A fully automated fuel rod transport system was installed in 1975 completing handling system development. The system processor was replaced in late 1977 with a processor capable of faster execution times to accommodate higher system scanning throughput. The system configuration, as of December 1977, is discussed in the following paragraphs. The system verifies BWR fuel rod ^{235}U enrichment values and distribution; UO_2 fuel column density and integrity; fissile (^{235}U) content, and internal assembly integrity.

II. Cf-252 FUEL ROD SCANNING SYSTEM

Experimental data obtained in 1971-72 were used to develop the Cf-252 fuel rod scanning system[1] at the Vallecitos Nuclear Center (VNC). It was demonstrated that sufficient counting rates could be obtained from the thermal flux irradiation of a BWR-type fuel rod and subsequent counting of the U-235 fission product activity to perform measurements of enrichment and U-235 content for quality assurance and accountability purposes. With the feasibility of the technique demonstrated, a scanning system was devised, with Cf-252 as a source of neutrons, which fulfilled the following requirements: (a) thermal neutron flux in the vicinity of $5x10^6$ $n\text{-}cm^{-2}\text{-}sec^{-1}$ with a U-235: U-238 fission ratio of at least 10^4; (b) minimum source-to-detector spacing was required to optimize the signal, but yet sufficient to give an acceptably low background count rate from the Cf-252 source; (c) provision for adequate shielding of personnel; and (d) multiple irradiation channels for production fuel scanning flexibility and optimum source utilization, but with spacing adequate to prevent excessive adjacent channel interaction and to accommodate a fuel rod handling system.

I. System Design and Operation

Based on the above studies, the irradiator was designed and built at VNC (Figure 1). The 9-ton irradiator cask is cylindrical (4-1/2 long by 5 feet in diameter) and can accommodate six fuel rods simultaneously in holes or channels located symmetrically about the long axis of the cylinder on a 1-foot diameter circle. Approximately 3 milligrams of Cf-252, contained in a standard SR-CF-100 capsule, is located in the radial center of the irradiator, but off set axially to minimize source-to-detector effects. Source transfer is made in a completely shielded system from a specially designed shipping cask made to be compatible with the irradiator cask. Neutron moderation and shielding are provided by water-extended-polyester (WEP-65). Lead and boron were added to the outer region of the shielding to minimize the capture gamma ray component. Polyethylene was placed in the region between the source and detector to enhance neutron shielding. Since the count rates from irradiated fuel rods are too high for standard NaI detector assemblies, NE102 plastic hole-through scintillators and fast phototubes were used as detectors for the measurement of the induced fission product gamma activity. The detectors are embedded in a shield of lead and tungsten alloy (Figure 2). Since the gamma ray activity measured by the enrichment detectors is actually a function of the U-235 content per unit length, rather than enrichment, it is necessary to monitor the density of the fuel rod. Therefore, a Cs-137 source is located at the radial center within a shielding assembly at the entry of the irradiator. The 662-kev gamma ray beam from the Cs-137 is collimated to the center line position of each fuel rod, and the attenuation of the beam is used to monitor fuel column density and uniformity. A set of NE102 plastic detectors is used to monitor the collimated beam intensity (Figure 3). Proceding these detectors is a set of hole-through Na(Tl) detectors embedded in the Cs-137 source shielding assembly to monitor the U-238 daughter activity and any residual fission product activity which may be present when rescanning fuel rods (Figure 3).

In designing the counting system, consideration was given to minimizing the counting loss especially at the high enrichment and high count rate end of the scale. Since count rates for irradiated fuel are in the range of 50 to 500 MHz, a 20-nsec paired pulse resolution is required to obtain a counting loss of less than one percent. For this reason, the system uses fast (100 MHz) discriminators to process these signals. Events are counted with 100 MHz CAMAC scalers which were chosen as the most straightforward means of interfacing the detectors to an on-line computer (Figure 4). These scalers make the interfacing quite simple and inexpensive to pre-CAMAC alternatives; which would have required that a special interface be designed and built. Strip chart recorders permitted the utilization of the system for production scanning while the computer software was being developed and debugged.

2. Data Retrieval and Processing

The computer utilized in the Cf-252 fuel rod scanning system is a Data General Corporation Eclipse S230 (Figure 5), which is 1/0 Bus compatible with the CAMAC controller. The Eclipse S230 is a fast (200nsec memory cycle time) mini-computer and is enhanced greatly on this system by a cartridge disk system, a 9-track magnetic tape, and a high speed electro-static printer/plotter. This combination of software and high speed peripheral equipment has made the system ideal from a software development standpoint. The rod scanning software used FORTRAN language for all data processing routines and operates using Data General's Real Time Operating System (RTOS). RTOS is an all-core, multi-task priority scheduling monitor that allows simultaneous but independent data gathering and processing from each of the eighteen scanning detectors. A real time clock controls the reading and resetting of the CAMAC scaler each 1/8-inch along the fuel rod. Rod data (e.g., rod number and design enrichment) are input by either the CRT console or the computer teletype. Raw data are stored on the disk until the rod traversal (approximately three minutes) is completed. Then, the raw data as well as identification, station number, date, and time are transferred to magnetic tape for permanent storage. The data are corrected for background, density, and channel cross-talk effects, and then digitally filtered to optimize the signal-to-noise ratio. The enrichment, density, and U-235 data are then checked for compliance to quality assurance and accountability limits. If these limits are exceeded, a blinking message is issued to the scanner operators on one of two CRT display consoles (Figure 6), and a plot (Figure 7) of enrichment and density is produced on the electrostatic printer. A summary of these measurements is also printed on the plot and on the CRT console screen. Appropriate disposition of the fuel rod is then taken by scanner operation personnel. The plot plus measurement summary becomes the permanent record for each fuel rod.

3. Calibration

The Cf-252 fuel rod scanning system is calibrated with an automatic calibration software routine, code-named "Autocal." This routine enables the computer to self-calibrate or "auto-calibrate" the scanning system. When recalibration is required, the scanning system operator selects the appropriate fuel rod standard nearest to the production rod enrichment being scanned, places the standard on the load rack, and inputs the standard number to the computer by the CRT console. The operating system software routinely scans each rod serial number and compares these rod numbers to those listed in the standard rod library maintained in the computer memory. When a match is made between the rod serial number input and a standard rod number, the computer calls the Autocal software routine. The Standard Rod is then scanned and the measured values for enrichment, density, and U-235 compared via the Autocal software with the assigned values for enrichment, density, and U-235 in the standard rod library. The Autocal software then adjusts the constants in the enrichment, density, and U-235 models to calibrate the system. The measured values prior to calibration adjustment for enrichment, density, and U-235 are also printed on the electrostatic printer/plotter summary as a measure of system drift since the last calibration. The scanning system is calibrated every twenty-four hours of production operation and, additionally, between either enrichment or fuel design changes. The fuel rod standard nearest to the design enrichment of the rods to be scanned is used to calibrate the system. The Autocal summaries are collected, input to a time-share computer file, and system performance statistics derived for each channel on a weekly basis.

4. System Performance

The fuel rod scanning measurement of performance of the Cf-252 Active Scanner is maintained as a rigid system of measurement controls; routinely overinspected to quantify actual measurement or assay uncertainties; and adversary-tested using appropriately designed and fabricated fuel rod standards to verify scanner effectiveness to detect off-specification conditions. The system of measurement controls consists of routine, periodic scanning of fuel standards and verification of the measurement of ^{235}U content, UO_2 column density, and ^{235}U enrichment against independently established values for these quantities. Over-inspection activities include an intensive equipment preventative maintenance program to verify proper equipment operation, including detectors, analog counting electronics, CAMAC modules, and data processing system. Also included are statistical evaluation of calibration and verification standards data to detect changes in system measurement uncertainties, including drift, measurement bias, and precision.

To facilitate operation of the Cf-252 fuel rod scanning system in the most efficient manner, a committee consisting of representatives from the operations, maintenance, quality, engineering, and development functions meet routinely to review operational and measurement problems.

III. FUEL ROD STANDARDIZATION

Full length fuel rod standards have been fabricated for use in the calibration and qualification of the Cf-252 fuel rod scanning system. The materials used in these standards were selected from production materials and processed on production line equipment under carefully controlled and audited conditions. During the fabrication process, samples were taken for submittal to several independent analytical laboratories. The resultant analytical results were statistically tested and combined to characterize the fissile content and associated limit of error for each standard rod. This standards program has produced standards which are characterized in fissile content, uranium content, density, and enrichment. The fabrication and qualification process is described below.

1. Fabrication Process

The fabrication process for fuel rod standards is initiated by the issuance of and agreement to fabrication guidelines. These guidelines delineate the responsibilities, process guidelines, and sampling plans to be followed during the fabrication of the standards. The guidelines are then published for management/technical review and concurrence. Following this process, UO_2 powder is selected from the in-process inventory, blended for enrichment homogeneity, and sampled for isotopic, %U, and metallic impurities. The analysis of these samples assures that the UO_2 powder selected, regardless of its origin, meets manufacturing and quality specifications. Each blending and sampling process is audited by quality assurance personnel. The powder is quarantined while in storage waiting processing.

Before pressing, each pellet press to be used is cleaned according to standard enrichment cleanout procedures. The cleanout is audited by quality personnel to ensure the integrity of the cleanout. Each powder lot is then pressed into green pellet boats. Each boat is further checked for enrichment to ensure avoidance of enrichment mixup errors. While waiting sintering, each pellet boat is clearly marked as standards material and quarantined. When the sintering process is complete, the pellets are ground to a diameter tolerance and stored waiting loading operations in special quarantine cabinets. Zirconium tubing, for use as standard rod tubeshell, is then selected from the inprocess stock and subjected to ultrasonic gaging to ensure uniformity of wall thickness relative to quality specifications. The tubeshell is then serialized utilizing numbers supplied by the materials control function and the tubing placed in the rod loading area to await loading.

Before the rod loading operation begins, each fuel stack weighing scale is carefully calibrated using a mass standard traceable to a NBS mass standard. During the rod loading process, the scale calibration is checked before, during, and after weighing of the fuel stacks of each enrichment lot to determine scale drift. Sufficient weighings of each fuel stack are performed to characterize the weight of each fuel stack. Also samples for % Uranium, enrichment, and metallic impurities are taken to determine the density and U-235 content of each standard rod. Each rod is then loaded, outgassed according to current manufacturing specifications, and the final end plug welded in place. The finished fuel rod standards are then placed into rod storage cabinets under quarantine until the laboratory and weight data has been processed to assign a U-235 value to each standard. Each of the aforementioned process steps is audited by quality assurance personnel to ensure conformance to the process guidelines and the actual process steps documented.

2. Standard Rod Qualification Process

Samples taken during the standard rod loading process are submitted to several independent analytical laboratories for analysis of percent uranium enrichment, and metallic impurities. The resultant analytical values are statistically tested and combined to characterize the uranium and U-235 values for each standard rod using the following method:

a. Cochran's test for equality of variances. For example, the data from a particular laboratory would be excluded if its sample variance is significantly larger than the others.

b. A fixed-effects analysis of variance (ANOVA) was used to analyze the data not excluded by Cochran's test. If an F test on the resultant among laboratories and within-laboratories variances indicated no significant differences between these two variances, then all data points are regarded as being from the same population. The mean and its associated variance are then assigned to the standards as appropriate.

c. Duncan's Multiple Range Test is used if the F-test in b. is indicated that the among laboratories variance was significantly greater than the within laboratories variances. Duncan's multiple Range Test is then used to see if this group of heterogeneous means could be separated into subgroups of nonheterogeneous means to give the best estimate of the overall mean. If no distinction can be made, values for the assigned mean and its associated variance are determined as if all data points are from the same population. If the mean of one laboratory is significantly different from the means of the other laboratories, then data from the significant laboratory is eliminated. The remaining data points are then treated as one sample to determine the mean and its associated variance to be assigned to the standards.

During the useful life of a fuel rod standard, its fuel stack integrity, tubeshell uniformity, and fuel column design are routinely checked and the standard replaced if it ceases to adequately represent the production fuel rod being assayed and evaluated. Records are kept of the evaluation process during the life of the standard as part of the scanner history.

IV. CONCLUSION

The Cf-252 fuel rod scanning system is used for the routine evaluation and assay of production fuel rods as part of the quality inspection process at the Wilmington Facility. Routine scanning operations, standardization, and measurement control are subject to continual evaluation and are consistent with recent Nuclear Regulatory Commission Guides and relevent ANSI Standards.

V. REFERENCES

[1]Duckard, Ruiz, et al; "Nuclear Fuel Rod Active Gamma Scan System," NEDO-20580 (Class 1), July 1974.

Fig. 1 Cf-252 Irradiator

Fig. 2 Enrichment Detectors

Fig. 3 Density and Background Detectors

Fig. 4 CAMAC Subsystem

Fig. 5 Eclipse Computer

Fig. 6 CRT Terminal

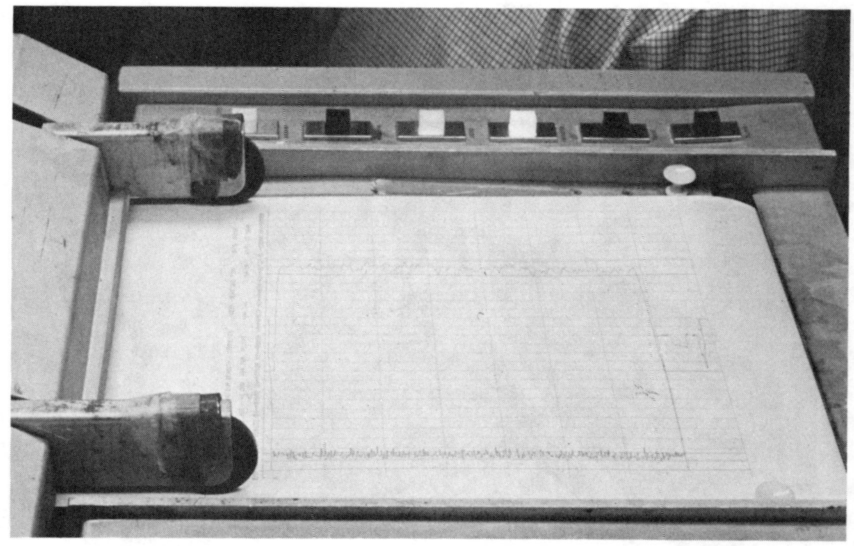

Fig. 7 Output Data Plot

SEARCH UNIT SPECIFICATIONS FOR IMPROVED REPEATABILITY OF ULTRASONIC EXAMINATIONS

A. S. Birks
W. E. Lawrie

ABSTRACT

Results obtained from ultrasonic examination of materials is dependent on the interaction of the search unit sound beam and the resultant beam profile of the reflector. Repeatability of test results can be improved significantly when search unit parameters affecting the radiation pattern, such as frequency, size and operating condition, fall within specified tolerances.

This paper discusses typical ultrasonic examination measurement errors resulting from the selection and use of standard commercial search units. Echo dynamics are considered to provide the technical basis for establishing tolerances on critical parameters that contribute to variations in examination results. A group of randomly selected search units are evaluated using interim specifications and tolerances.

Recommendations are made to develop search unit tolerances which will be accepted throughout the nuclear industry.

INTRODUCTION

Ultrasonic examination techniques are of vital importance to the nuclear industry both during fabrication and throughout the operating life of the nuclear steam supply system, providing a sensitive method for detecting and evaluating potentially hazardous flaws. The critical nature of ultrasonic examinations has resulted in a variety of programs sponsored by various technical groups to determine if UT equipment provides the repeatability demanded by this application.

Variability of UT examination test data has been reported by several technical committees.[1-2] The inconsistency of these examinations and the absence of suitable justification suggests that there may not be sufficient data available to identify the basic factors contributing to these variabilities.

A potential factor presently unavailable to NDE engineers is relevant engineering data on ultrasonic equipment components. Similarly, there are only nominal specifications available to describe the performance of these components for an intended application. At this time there are no established or accepted industry wide tolerances on UT equipment. This situation prohibits the selection of optimized ultrasonic system components, thus setting the stage for wide examination variabilities.

The wide interest in developing appropriate specifications and testing procedures for evaluating ultrasonic system components is demonstrated by numerous technical programs that are now underway under the auspices of the National Bureau of Standards, the U. S. Naval Research Laboratory, ASTM, ASNT, SAE and PVRC committees and the manufacturers and users of ultrasonic equipment. One of the most dynamic programs relevant to UT is now being conducted by the Pressure Vessel Research Committee (PVRC) Sub Committee on Nondestructive Testing. The main thrust of this program is directed toward defining the capabilities of ultrasonic examination and developing more definitive equipment specifications.

Although much effort is being expended determining the parameters to be measured and the optimum method of measurement, little progress has been made in setting the tolerances to be specified. It is understood that in Germany the search units center frequency must be within ten percent of the nominal frequency. This tolerance is essential to permit the accurate use of AVG/DGS diagrams and nomograms for flaw size estimation.

NEED FOR STANDARDIZATION

Is there truly a need for standardization? The results of the first PVRC Round Robin Test[1] tentatively indicates that there was little agreement among participating examination teams. Additional round robin efforts by other technical societies and government sponsored research further confirms the presence of wide data variances.

Table 1 - Results of straight beam examination of weld sample.

STRAIGHT BEAM SEARCH UNIT	NOMINAL FREQUENCY (MHz)/SIZE (IN.)	MEASURED CENTER FREQUENCY (MHz)	FLAW RESPONSE – % DAC AND APPROX. AREA (IN.^{2})				
			1	2	3	4	5
			0.018	0.019	0.019	0.012	–
A	2.25/0.75	1.8	75	70	37	125	95
B	2.0/1.0	2.0	70	100	–	–	100
C	4.0/1.0	3.0	25	25	–	50	40
D	5.0/0.75	3.5	15	10	20	15	15
E	5.0/0.75	3.5	17	12	–	30	30

Tests conducted by Babcock and Wilcox on a typical weld further demonstrates and correlates search unit variables by the spread of resulting examination data. These tests were conducted on a six inch thick weld mock-up simulating a typical pressure vessel weld. The data shown in Table 1 was obtained using a single instrument and five randomly selected off-the-shelf commercial search units.

The results of this test strongly indicate that significant variations in examination results may occur by varying the search unit portion of the ultrasonic equipment. This data was obtained using an ultrasonic instrument with a broadband receiver whereby the search unit spectrum determined the actual examination frequency. If a tuned receiver were used, the search unit would only partially control the effective frequency. The measured operating frequency spectrum of these five search units indicated that their actual peak or center frequencies were far from those specified on their labels. To determine which search unit or frequency provided the most accurate evaluation of those flaws, radiography and destructive sectioning was performed. In all cases, the search units with lower operating frequencies provided the most accurate prediction of actual flaw area.

As a result of this work, a larger group of new and used 2.25 MHz search units typically used for weld examination were collected and evaluated to determine the variance in operational parameters using a number of tentative specifications. These specifications were selected intentionally to evaluate parameters that influence the results of ultrasonic examination of thick welds in steel pressure vessels. The parameters and tolerances specified are shown in Table 2.

Table 2 - Tentative search unit specifications and tolerances.

PARAMETER	*TOLERANCE*
CENTER FREQUENCY f_o	± 10% f_N
FREQUENCY BANDWIDTH (6dB)	± 25% f_o
BEAM EXIT POINT	± 0.025 INCH
REFRACTED ANGLE	± 2°
LATERAL BEAM SYMMETRY	± 0.125 INCH AT Y_o^+
Y_o^+ POSITION	± 10% CALCULATED
SENSITIVITY	30 dB RESERVE

The tolerances on the search unit parameters listed in the above table were selected strictly on a best effort basis.

The Ad Hoc tolerances actually consist of a compromise between the tolerances required to provide repeatability and the tolerances that can be reasonably attained by search unit manufacturers. These tolerances should be used only on an interim basis.

The results of evaluating eighteen straight beam and angle beam 2.25 MHz search units is summarized in Table 3. Five of the eighteen units or 28% were rejected for "off-frequency" conditions as defined by the tolerances listed in Table 2. Five percent (5%) were rejected for asymmetry and low gain respectively resulting in an overall rejection rate of 38.8%. It should be noted that only 28% of the units passed the tentative bandwidth tolerances, suggesting that most manufacturers are offering highly damped, broadband units. This deviation from the established tolerances indicates that the tentative tolerances must be subjected to further study and definition.

The objectives of setting tolerances on search unit parameters is to insure that ultrasonic examinations conducted using nominally similar search units and instruments, will result in comparable amplitudes. Since no scientific basis was used in establishing the tolerances listed in Table 2, there are no assurances that those selected are adequate. The tolerance on each search unit characteristic must therefore be established in terms of the allowable variations in examination response obtained using search units with acceptable tolerances.

Table 3 - Results of evaluating search unit tolerances.

Size	Nominal Angle	Center Freq. MHz	Bandwidth MHz–%fo	Sensitivity	Lateral Beam Symmetry		Beam Exit Point	Actual Angle
1 X 1 in.	45°	2.00	0.7–35%	26 dB	3/16 in.	3/16 in.	OK	45°
1 X ½ in.	45°	2.30	0.85–37%	30	1/4	9/32	OK	45°
1 X ½ in.	45°	2.45	0.05–42%	26	5/16	5/16	OK	45°
1 X ½ in.	60°	2.25	0.9–40%	28	5/16	1/4	OK	62°
1 X ½ in.	60°	2.10	0.93–44%	32	1/4	1/4	OK	60°
1 X 1 in.	60°	2.20	0.62–28%	32	1/4	3/16	OK	60°
1 X 1 in.	60°	2.35	0.62–26%	44	3/16	3/16	OK	60°
1-1/8 in. D	0(LW)	2.05	0.6–29%	16	9/16-9/16	9/16-9/16	OK	0°
1-1/8 in. D	0(LW)	2.0	0.5–25%	16	9/16-9/16	19/32-19/32	OK	0°
1-1/8 in. D	0(LW)	2.05	0.7–34%	16	9/16-9/16	9/16-9/16	OK	0°
1-1/8 in. D	0(LW)	2.05	0.45–22%	18	11/32-11/32	13/32-13/32	OK	0°
1-1/8 in. D	0(LW)	1.95	0.5–26%	REJECTED–OFF FREQUENCY				
1-1/8 in. D	0(LW)	1.7	0.5–29%	REJECTED–OFF FREQUENCY				
1 X ½ in.	45°	1.95	0.5–26%	REJECTED–OFF FREQUENCY				
1 X 1 in.	45°	1.7	0.4–24%	REJECTED–OFF FREQUENCY				
1 X 1 in.	45°	2.6	0.6–23%	REJECTED–Low Gain/Asymetric Beam				
1 X 1 in.	60°	2.5	0.45–18%	REJECTED–OFF FREQUENCY				
1 in. D	45°			REJECTED–Low Gain, Could Not Measure Spectrum				

NOTE: Y_o^+ position not measured.

Caution must be exercised in setting strict tolerances since the interaction of various parameters may influence the allowable tolerances on each search unit. The selection of characteristic tolerances to insure repeatability of ultrasonic examination may also introduce difficulties in measurement of the parameters. For example, it is difficult to measure a refracted angle of a small, low frequency search unit used to generate a 60° shearwave.

MEASUREMENT OF PARAMETERS

As was indicated earlier, many organizations are addressing the problem of how to adequately measure the important parameters of an ultrasonic instrumentation system. Consequently, it is not necessary to go into detail on how each of the parameters should be measured. It is necessary, however, to examine the methods for measuring these parameters in terms of the objectives of the measurement. For example, if the frequency characteristics of the search unit are to be measured to determine the uniformity of search unit characteristics, a different and perhaps better approach might be used than if the repeatability of ultrasonic examination results is to be obtained. Pseudo continuous electrical signals applied to a search unit from laboratory instrumentation, slowly swept in frequency, might be used if comparison of search unit frequencies is the primary objective. However, if repeatability of ultrasonic results is to be attained, the only frequencies of interest are those that reach the detector in the ultrasonic instrument, since all other information will be lost as a result of the inherent filtering resulting from the interconnection of components forming the complete ultrasonic examination system. The characteristics of the search unit evaluated independent of the remainder of the system may not be representative of the objectives of this work.

Certainly all of the parameters listed in Table 2 should be measured and tolerances should be set on acceptable results. Perhaps two types of measurements are required, one that provides a relatively simple shop or field evaluation and a more complete and exact measurement using laboratory equipment. The shop or field evaluation procedures must, however, be sufficiently sensitive to determine whether the characteristics are in or out of tolerance.

BASIS FOR SETTING TOLERANCES

Tolerances on search unit characteristics (really ultrasonic system characteristics) should be set to provide repeatability in ultrasonic examination. For example, an acceptable tolerance of ± 2dB on the ultrasonic response from a particular defect might be established. All other parameters should then be selected so as to result in errors less than 2dB because of variability in characteristics.

The field of echo-dynamics is suggested as the basis for selecting tolerances on search unit characteristics. Here the term echo-dynamics refers to the variation in ultrasonic response for defects of various sizes and orientation that might be obtained with particular search units and ultrasonic instruments. In simple terms, the ultrasonic response obtained from a defect is dependent upon the interaction of the radiation patterns or beam profiles of the search unit and the defect. The radiation pattern obviously must be those that contribute to the cathode ray oscilloscope indication used for interpretations of results. It is for this reason that only the frequencies selected from the search unit spectrum reaching the detector are of importance. Similarly, each of the characteristics can be examined to determine how variations in the characteristic will alter the interaction of the radiation patterns and in turn affect the ultrasonic response.

It has long been assumed that the use of a reference hole to set the gain would result in uniform response from different ultrasonic systems. When using either a side drilled hole or a flat bottomed hole, the axis of the radiation pattern from the transducer approaches the reference hole at normal incidence. Consequently, the reflection propagates back along the axis of the search unit radiation pattern. The use of the reference hole thus provides an adjustment of sensitivity only in one direction along the axis of the beam. This is illustrated in Figure 1. However, when the beam is used to evaluate material defects, particularly the more critical planar defects, the axis of the impinging sound beam may not be at normal incidence to the plane of the defect. Maximum response will be obtained as shown in Figure 2, when the line joining the center of the search unit radiating surface to the center of the defect is off-axis on both the search unit and defect radiation patterns. For defect the same size and shape as the reference hole, the response will be reduced because of the off-axis operation. Although Figure 2 is drawn for normal beam longitudinal waves, the concept applies equally to angle beam shear waves. It should be quite clear that any change in characteristics that alters the radiation patterns of either the search unit or the defect acting as a reradiator will alter the ultrasonic response. By quantifying the relationship between characteristics and change in response, the tolerances on the characteristics that will provide uniformity of response within any specified amount can be derived.

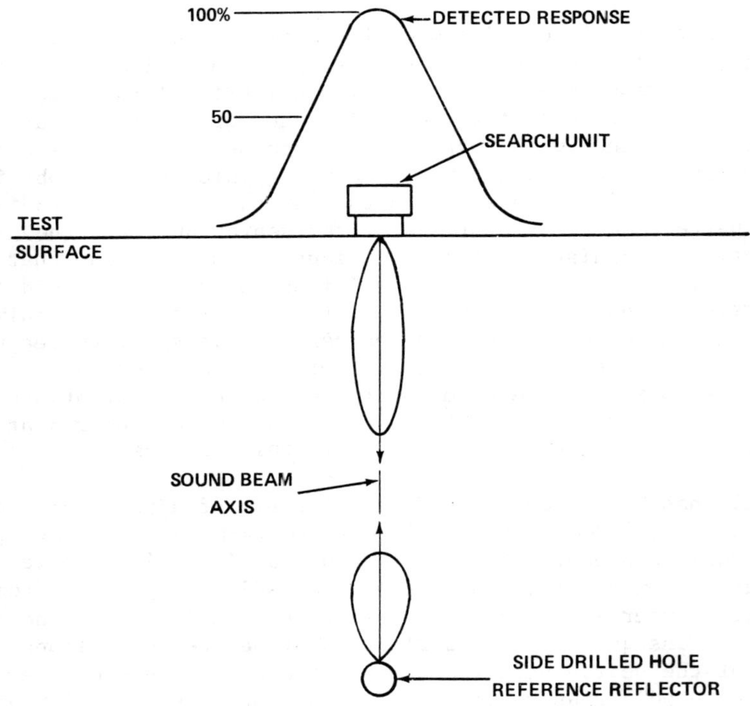

Fig. 1 - Sound beam configurations during calibration.

TOLERANCE ON FREQUENCY AND BANDWIDTH

Farfield radiation patterns for circular radiators with uniform phase and amplitude excitation over the surface can be computed from the equation

$$\text{Sin}\ \theta = K \frac{V}{fd}$$

Where: θ is the off-axis angle at which the response is reduced by a specified amount.

K is a constant which is dependent upon the specified reduction in response and shape of the transducer.

V is the ultrasonic velocity

f is frequency

d is the diameter of the radiator

With a table of values of K corresponding to various reductions in response, this equation can be used to calculate the typical lobe shaped farfield radiation pattern. If the frequency is decreased, the width of the lobe will increase. Thus, if we replace the radiation pattern shown in Figure 2 by a much broader radiation pattern corresponding to a lower frequency calibrated to the same reference, the response from misoriented reflectors will be increased.

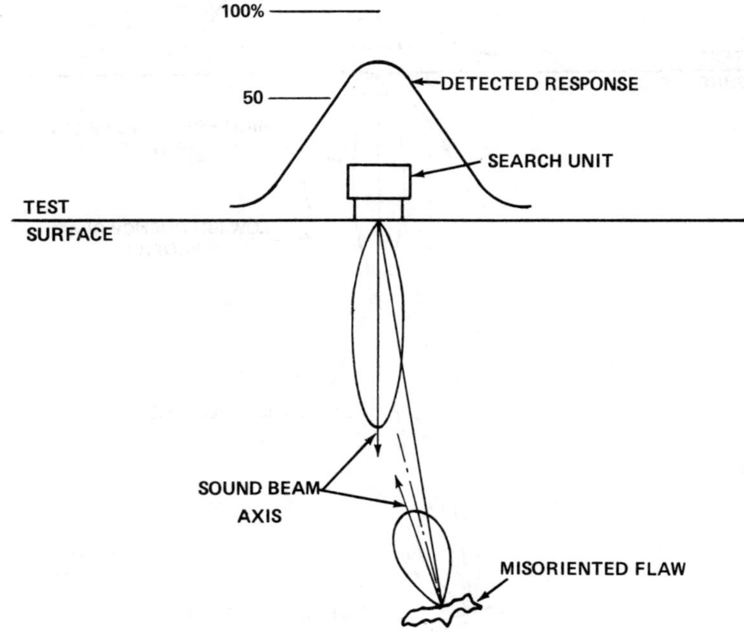

Fig. 2 - Interaction of sound beams for misoriented flaws.

This analysis applies not only to the radiation pattern of the transducer but also that of the reflector.[3] In principle the reflector acts like a radiator with the pressure amplitude and phase distribution produced by the search unit acting as a transmitter. The only control attainable over the defect radiation pattern is through control of the search unit frequency. If the frequency is reduced, then both the search unit and defect radiation patterns become broader and the off-axis response will increase as shown in Figure 3. This results in a higher ultrasonic response at lower frequencies for misoriented defects. The response from misoriented defects will still be below the response expected if the orientation was optimum. For repeatability, however, the tolerance on frequency must be based on the change of response with frequency for badly oriented defects.

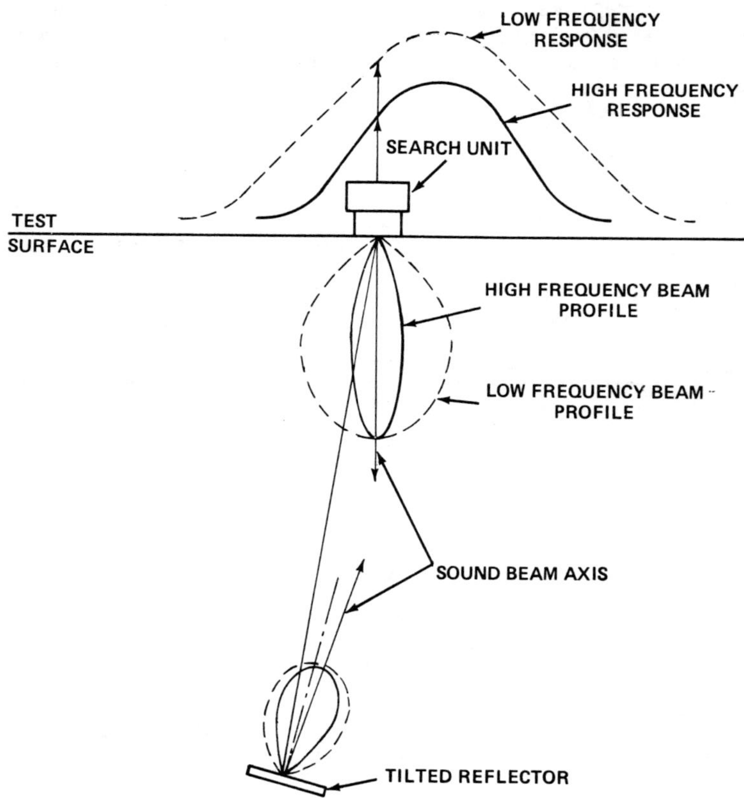

Fig. 3 - Comparison of high and low frequency response.

A tilted defect acts like a low pass filter. Because the radiation patterns are broader for lower frequencies than for higher frequencies, the response at lower frequencies within the band generated by a search unit will be greater than that for the higher frequencies. Thus, for a badly oriented defect the center frequency can be expected to shift downward provided that the band is sufficiently wide to permit such a downward shift. If an examination was being conducted with two systems operating at the same center frequency, but with substantially different bandwidths, the system with the broader bandwidths would give the higher response. This is because the lower frequency components which provide higher response are present in one system and not present in the other. If repeatability is to be attained then bandwidth as well as center frequency must be specified and must be controlled.

TOLERANCE ON REFRACTED ANGLE

The axis of the radiation pattern from either normal beam or angle beam search units should be in the nominal direction within a small tolerance. For example, if the beam axis of the beam from a straight beam search unit is not normal to the surface, then the interaction between the search unit radiation pattern and the defect radiation pattern will be altered. If the beam was down and slightly to the right as shown in Figure 4b, the responses would be reduced. In contrast, if the beam axis pointed down and to the left as shown in Figure 4a, response would be increased, since in effect the misorientation angle has been reduced.

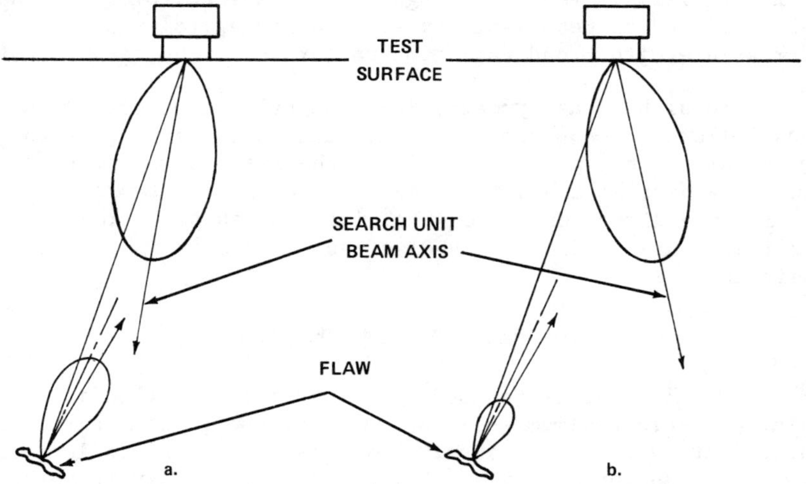

Fig. 4 - Effect of rotating out-of-tolerance sound beam.

TOLERANCE ON BEAM EXIT POINT

The tolerance on the beam exit point is not directly related to radiation pattern considerations. However, if the beam exit point is incorrect, it may indicate that the active element is improperly positioned within the search unit case. As a consequence errors will be introduced in the flaw location measurement. This implies that both the exit point and the beam axis should be carefully checked.

TOLERANCE ON SENSITIVITY

In many ultrasonic tests, sensitivity is not a critical parameter. Frequently the ultrasonic instrumentation will have variable gain controls or attenuators with a range of 80 to 90 decibels. Typically 40 to 50 decibels of the available gain may be used for calibration and scanning, leaving in reserve 30 to 40 dB. If adequate reserve gain is available, then the search unit is satisfactory.

TOLERANCE ON LATERAL BEAM SYMMETRY

The tolerance on lateral beam symmetry is similar to that of the refracted angle, in that if the radiation pattern is not symmetric, then the responses will be dependent upon the actual beam pattern. The response obtained from a defect would then depend upon which side of the radiation pattern was used for detection and evaluation of the defects. Beam symmetry tolerances must be established if repeatible sizing of large flaws is required. Again, an absolute tolerance should not be placed on beam symmetry since a larger tolerance is permissible with broad radiation patterns.

Lateral beam as symmetry is frequently associated with bond failure between the active element in a search unit and a wear plate or a plastic lens. If the beam asymmetry is caused by partial unbonding, then the change in diameter will result in a direct change on the width and shape of the beam profile. This condition may also cause a change in the Y_0^+ position.

TOLERANCE ON Y_0^+ POSITION

The Y_0^+ point is defined as the position of the first axial pressure maximum while approaching the search unit from an infinitely large distance. This position which is located at a point approximately $d^2/4\lambda$ will be correct only if both the effective frequency and the dimensions of the transducer are as specified. If the diameter of the transducer is in fact the nominal dimension, then the tolerances on frequency and Y_0+ position should be the same.

SUMMARY

This paper has identified typical derived or measured operational characteristics of search units and has recommended tolerances on these parameters. Evidence of variances in examination results and the interrelationship of measured operational parameters of off-the-shelf search units having only nominal specifications has been shown. Currently the absence of specified tolerances on search unit characteristics limits the repeatability of ultrasonic measurements. Any characteristic that contributes to the radiation pattern must be controlled, if responses are to be repeatable. The most significant problem resulting from the absence of search unit tolerances appears to be wide differences between the nominal and measured operating frequencies of off-the-shelf search units.

The importance of this parameter is underscored by evaluating the response of a 1/4" square planar reflector at 10° from normal incidence. The calculated response of such a flaw examined by search units operating on-frequency and at both extremes of the center frequencies, 1.7 and 2.6 MHz, as reported in Table 3 are shown in Table 4. Note that the range of response varies 9.2 dB, a ratio of nearly 3 to 1! The deviation from the expected response using the correct 2.25 MHz frequency ranges from +4.9 to -4.3 dB. This is an unacceptable variation that far exceeds the desired 2 dB or 20% tolerance on response tentatively established by current code committees.

The significance of controlling one characteristic, such as frequency, cannot be established without reference to other characteristics. There is also a need for setting specifications and tolerances on each component and the complete ultrasonic system, if repeatable examination results are to be obtained.

Table 4
Response from 1/4″ reflector using frequency extremes from Table 3.

TEST FREQUENCY	0° REFLECTOR CALIBRATION RESPONSE	10° INCLINED FLAW RESPONSE
1.70 MHz	0 dB	– 5.4 dB
2.25	0	–10.3
2.60	0	–14.6

RECOMMENDATIONS

Presently various technical groups are at work developing programs or proposing recommended practices which will lead to consistent examination performance. The tentative tolerances set forth in this paper should be maintained as a milestone beyond which further efforts must be expended to develop adequate and appropriate tolerances for specific examinations. The final tolerances on each characteristic parameter should be set in accordance with the allowable variability in ultrasonic response. Support from various agencies must be provided to conduct the research necessary to determine these tolerances.

The anticipated results will provide the required control of the search unit portion of the ultrasonic system via adequate specifications, tolerances and tests to assure these parameters are maintained.

REFERENCES

1. Buchanan, R. A.

"Analysis of the Nondestructive Examination of PVRC Plate-Weld Specimen 251 J - Part A", Welding Research Council Bulletin #221, November, 1976.

2. ASNT Sonics Committee Meeting Agenda Item #3, Round Robin Report, ASNT Spring Conference March, 1977.

3. Serabian, S. and Lawrie, W. E.

"A Detection Model for Pulse Echo Ultrasonics", ASNT Fall Meeting September, 1977.

ACOUSTIC PROPERTIES OF STAINLESS STEEL WELD METAL*

D. S. Kupperman and K. J. Reimann
Materials Science Division
Argonne National Laboratory
Argonne, Illinois 60439

ABSTRACT

Ultrasonic inspection of stainless steel welds is particularly difficult because of the coarse grain and anisotropic nature of the weld metal. The ultrasonic attenuation in the weld material is much greater than that experienced in the base metal. Spurious ultrasonic signals and refracted beams from the dendritic weld structure can lead to difficulties in data interpretation. Thus, adequate flaw identification and characterization is inhibited. In this paper the acoustic properties of stainless steel weld metal is investigated so that a better understanding of the origin of the ultrasonic inspection problems can be obtained and to provide information which may lead to increased effectiveness of ultrasonic inspections of stainless steel welds.

Variations in the velocity of sound, ultrasonic frequency spectra, and attenuation have been measured for longitudinal and shear waves as a function of propagation direction through the weld metal. Furthermore, for shear waves the effect of varying the angle of polarization on flaw detection has been examined.

A comparison of the effectiveness of longitudinal and shear waves for detecting artificial flaws is presented. For shear waves with mutually orthogonal polarization, differences in frequency spectra and variations of ∿25% in velocity of sound have been noted. Furthermore, the detectability of

*Work supported by the U. S. Department of Energy.

artificial reflectors is dramatically affected by variations in polarization angle of the shear waves used to examine them. In the above cases, normal-incidence shear waves were employed. However, the disadvantage in their use is that a viscous couplant must be applied. In one example, shear waves (2.25 MHz) with a polarization orthogonal to that which is conventional can be successfully used to "detect" a side drilled hole in the weld metal where conventional shear waves cannot. In this case, longitudinal waves (2.25 MHz) were also effective for detection of the reflector, but the shear waves are more advantageous because of better resolution and the absence of mode-conversion problems associated with longitudinal waves. In another example, longitudinal waves were less effective in detecting a side drilled hole than shear waves of the optimum polarization. Micrographs of the weld coupons are presented as well as X-ray diffraction data used to establish the preferred orientation of the weld-metal samples.

INTRODUCTION

Ultrasonic inspection of austenitic stainless steel weld metal is difficult, and the results obtained are not easily interpreted.

The problems encountered in the inspection of stainless steel originate because of the absence of a phase transformation in the stainless steel weld upon cooling, as in, for example, ferritic steels. As a result, slow cooling leads to the formation of large grains (dendrites) in the weld metal. The dendrites tend to grow with a <100> orientation and essentially along the lines of heat dissipation. Since the beam scattering depends upon anisotropy and the ratio of grain size-to-wavelength, the severity of the attenuation in stainless steel weld metal may be reduced by welding techniques that tend to produce a finer dendritic structure. However, for coarse-grain materials, proper selection of the acoustic mode when the weld metal is inspected may enhance the reliability and sensitivity and lead to an improved ultrasonic testing method.

In a typical V-type weld shown schematically in Fig. 1, the dendrites grow perpendicular to the base metal "V" for a distance of a few millimeters and then move upward (Z direction) following the lines of heat dissipation. Assume the

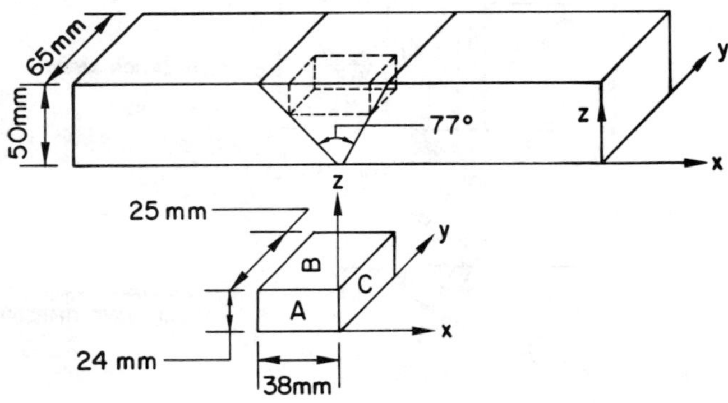

Fig. 1. A Weld-Metal Coupon for Acoustic-Properties Data.

dendrites are long, cylindrical, single crystals with the vertical axis in the <100> (Z) direction, and the dendrites are randomly oriented about this axis. We now have a model that describes the region of the weld where the dendrites are vertical. The effectiveness of the model can be checked by comparing the predicted and measured ultrasonic velocities in various propagation directions.

EXPERIMENTAL PROCEDURE

Experimental results were obtained for a "V" type multi-pass shielded metal arc weld and 1-cm^3 coupons from welded Type 304 stainless steel tensile specimens made available for this study.

The welded plate (300 x 150 x 50 mm with a 77° included angle), from which other specimens were cut, was Type 304 stainless steel with Type 308 stainless steel weld metal. This specimen was cut in half along the long dimension. From one half, a 24 x 25 x 38-mm weld-metal coupon was removed for ultrasonic examination, as shown in Fig. 1. A second sample was prepared (Fig. 2) to demonstrate the effect of shear-wave polarization on defect detection.

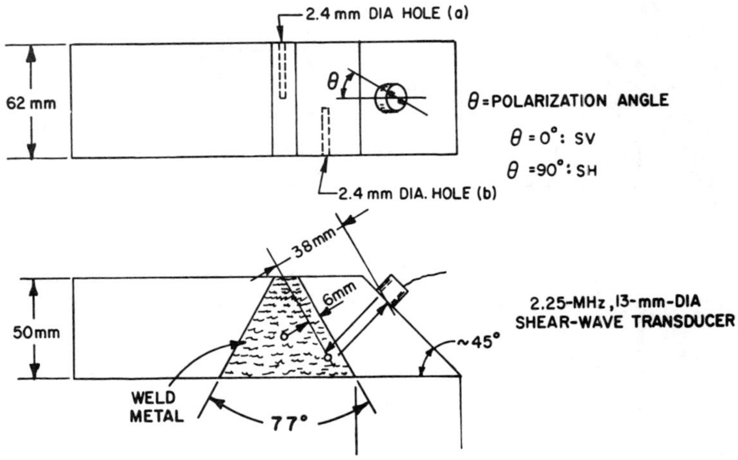

Fig. 2. Stainless Steel Weld Sample (Base Metal 304 SS, Weld Metal 308 SS) Used for Evaluation of Effects of Variation in Shear-Wave Polarization Angle on Detection of Side-Drilled Holes

Velocity of sound data were acquired using a Panametrics 5053 ultrasonic time intervalometer. Normal incidence shear-wave transducers and longitudinal wave probes (2.25 MHz, 13-mm dia) were employed. A Panametrics shear-wave couplant (No. 5070) was used to couple shear waves directly with the weld metal.

The results obtained from metallography studies are typical of stainless steel weld metal. Dendrites up to 10 mm in length and 1 mm across can be observed.

RESULTS AND DISCUSSION

Velocity

Predicted values of velocity can be estimated from the single-crystal elastic constants C_{11}, C_{12}, and C_{44}. These values are taken from Matsumoto and Kumura (1).

$$C_{11} = 21.6 \times 10^{11}\ \text{dynes/cm}^2$$

$$C_{12} = 14.5 \times 10^{11}\ \text{dynes/cm}^2$$

$$C_{44} = 12.9 \times 10^{11}\ \text{dynes/cm}^2.$$

The velocity of sound along the principal axes of the single crystal (face-centered cubic) are found from the relationships shown in Table 1 (2), where V_ℓ is the velocity, V_t is the transverse velocity, and ρ is the material density. The indicated maximum and minimum velocities occur along principal axes. The velocity of sound along principal axes calculated from the elastic constants are shown in Table 2.

Table 1. Velocity of Sound along Principal Face-centered-cubic Single-crystal axes Relative to Elastic Constant.

Direction	Equation	Mode[a]
<100>	$\rho V_\ell^{\ 2} = C_{11}$	L (Min)
	$\rho V_{t_1}^{\ 2} = \rho V_{t_2}^{\ 2} = C_{44}$	T (Max)
<110>	$\rho V_\ell = 1/2\ (C_{11} + C_{12} + 2C_{44})$	L
	$\rho V_{t_1}^{\ 2} = C_{44}$	T (Max)
	$\rho V_{t_2}^{\ 2} = 1/2\ (C_{11} - C_{12})$	T (Min)
<111>	$\rho V_\ell^{\ 2} = 1/3\ (C_{11} + 2C_{12} + 4C_{44})$	L (Max)
	$\rho V_{t_1}^{\ 2} = \rho V_{t_2}^{\ 2} = 1/3\ (C_{11} - C_{12} + C_{44})$	T

[a]L = longitudinal and T = transverse.

Table 2. Single Crystal Velocity of Sound vs Propagation Direction for Longitudinal and Transverse Modes in Type 308 Stainless Steel Weld Metal (from Available Elastic Constant Data).

Propagation Direction	Mode	Velocity (10^5 cm/s)
100	L	5.23
110	L	6.26
111	L	6.57
100	T	4.04
110	T	4.04;2.12
111	T	2.9

In general, two shear modes are possible for the two shear-wave polarizations. However, for the <100> and <111> propagation directions, only one velocity is present. From single-crystal data, we can now estimate the velocity of sound in the weld metal for propagation parallel, perpendicular, and skewed to the dendrites. For example (see Fig. 1), when propagation is parallel to the dendrites (Z axis), the velocity is that of the single-crystal <100> direction. For propagation perpendicular to the dendrites (X and Y directions), the velocity will lie between the <100> and the <110> values (for longitudinal and shear waves polarized <110>). This occurs because the dendrites are assumed to be randomly oriented about the <100> (Z) axis. For shear waves polarized in the <100> direction, all velocities in the <100> plane are the sam e as found in the <100> direction. For propagation at a $\sim$45° angle to the dendrites (45° to Z), the velocity will vary approximately between the <110> and <111> values. We can estimate the weld-metal velocities for any given direction by taking the simple average of the extremes of velocities possible for that direction. The theoretical results and the experimentally determined values are shown in Table 3.

The predicted and measured values in Table 3 are in reasonably good agreement. One important exception is the presence of a weak, slow shear wave for propagation along the dendritic axis. Otherwise, the model prediction and measured velocity values are within 10% of each other.

Table 3. Comparison of Model Prediction and Measured Velocity of Sound for Stainless Steel Weld Metal. Dendrite <100> axis is in Z Direction.

Mode	Propagation Direction	Polarization	Model Prediction (10^5 cm/s)	Measured Value (10^5 cm/s)
Long.	X	-	5.7	5.7
Long.	Y	-	5.7	5.6
Long.	Z	-	5.2	5.4
Long.	45° to Z axis	-	6.4	6.2
Shear	X	Z	4.0	3.7
	X	Y	3.1	3.0
	Y	Z	4.0	3.9
	Y	X	3.1	2.9
	Z	X	4.0	3.9;2.9
	Z	Y	4.0	3.8
	45° to Z axis	Y	3.5	3.5
	45° to Z axis	45° to Z axis	2.5	2.3

Attenuation

Several factors may affect the attenuation of ultrasonic waves. The primary factors that must be considered for the weld metal are (a) grain-boundary scattering (depends on wavelength and polarization), (b) mode conversion (different not only for longitudinal and shear waves but also for shear waves of varying polarizations), and (c) the guiding of beams from their wave normal (characteristic of single-crystal and preferred texture material). The latter phenomenon is discussed in more detail by Tomlinson et al. (3) in another paper at this meeting. The general problem of anisotropy is described by Miller and Musgrave (4) and others (5,6). They point out that the energy in a wave tends to propagate in directions of maximum velocity. For stainless steel, longitudinal waves tend to be guided to the <111> direction and transverse waves to <100> and <110> (<100> polarization) directions.

Experimental observations for longitudinal waves indicate that waves propagating ∿45° to the long axis of the dendrites have the least attenuation. Ultrasonic pulse-echo signals obtained from 1-cm^3 weld-metal coupons shows that ultrasonic waves propagating 45° to the dendrites are 3-6 dB larger in amplitude than waves propagating parallel or perpendicular to the dendrites. These results suggest that the convergence of ultrasonic beams along the propagation direction of maximum velocity may be a major factor in attenuation. Also, the wavelength increases with velocity, and thus the attenuation associated with Rayleigh scattering [$\alpha(1/\lambda^4)$] will vary considerably and could also be a major factor in the net attenuation.

The results for shear waves are not as clear. Maximum shear velocities occur for (a) waves propagating parallel to the dendrites and (b) perpendicular with polarization parallel to the dendrites. The experimental observation is that shear waves propagating in the latter mode (across the dendrites) had transmitted signals (2-cm path) at least 6 dB larger than all other signals, including propagation 45° to the dendrites. For shear waves traveling parallel to the dendrites, the difference in attenuation for various polarizations may be due to variation in mode-conversion characteristics, since the dendrites are not exactly parallel to the wave-propagation direction, or that the dendritic structure is not as symmetric about the Z axis as assumed. For shear waves traveling perpendicular to the dendrites, grain-boundary scattering may be an important factor, since the acoustic-impedance mismatch across the dendrites varies considerably for different polarizations. The mismatch is least for waves polarized along the dendrites and greatest for polarization across the dendrites. If this is the case, it would help explain the large variation in attenuation with polarization for shear waves propagating across the dendrites.

Flaw Detection

The significant variations in acoustic-propagation characteristics, particularly for shear waves, suggest flaw detection in stainless steel weld metal may vary not only with mode (longitudinal or transverse) but also with polarization when shear waves are employed. The polarization can be such that a shear wave is excited (slow or fast) or two shear waves, each with less amplitude, can be generated. The presence of two shear modes will, of course, make signal interpretation even more difficult.

The detectability of artificial reflectors placed in stainless steel weld metal is, in fact, affected by variations in the polarization of the shear waves that are used in the test. Two examples of this effect are presented. A schematic of the sample used is shown in Fig. 2. The orientation of the shear-wave transducer and the notation (7) used to indicate the two shear-wave polarizations are presented. The sample simulates ultrasonic inspection that employs "full-V" acoustic paths.

A 2.25-MHz, 13-mm-dia Panametrics broadband shear-wave transducer was used to detect a 2.4-mm-dia side-drilled hole, as shown in Fig. 2 (hole "b"). With horizontally polarized (SH) shear waves (all particle motion parallel to the reflecting plane), the reflection from the hole could be readily distinguished from the noise (Fig. 3). However, with vertically polarized (SV) shear waves (particle motion perpendicular to the reflecting plane), the reflection from the hole could not be readily observed (Fig. 4). The SV shear waves are used for conventional shear-wave inspection in which a longitudinal-mode crystal and wedge are used to generate shear waves via mode conversion at the wedge/metal surface interface. With a 2.25-MHz, 13-mm-dia, gamma-series longitudinal-wave transducer manufactured by Aerotech, the hole could also be readily distinguished from the background noise (Fig. 5), although the longitudinal wave has less resolution because of the longer wavelength and ringing. These results are consistent with those presented earlier. The ultrasonic wave propagates parallel to dendrites for a distance of ∿3 mm and then at ∿45° to the dendrites for another 3 mm. For either case, the horizontal polarization has the least attenuation. The longitudinal attenuation is greatest for the parallel and least for the skewed dendrites.

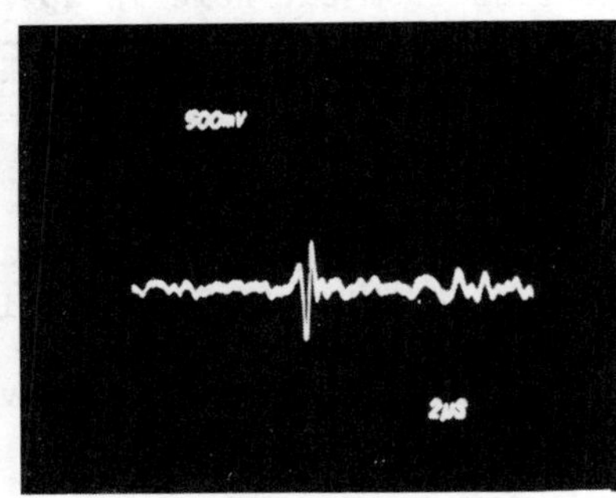

Fig. 3. Ultrasonic Signal from 2.4-mm-dia Hole Shown in Fig. 2 (Hole "b") Using SH Shear Waves (2.25 MHz).

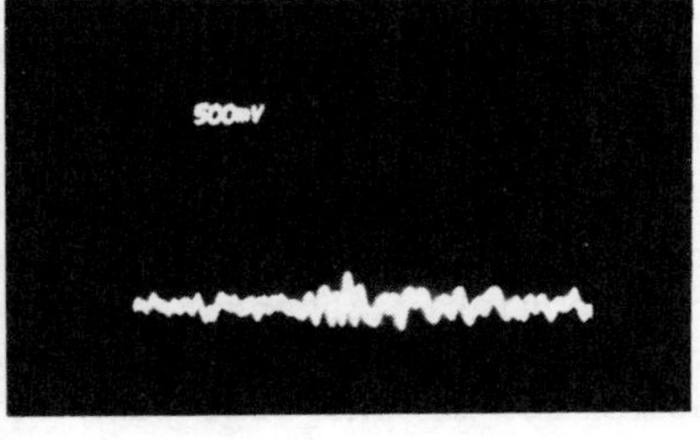

Fig. 4. Ultrasonic Signal from 2.4-mm-dia Hole Shown in Fig. 2 (Hole "b") Using SV Shear Waves (2.25 MHz).

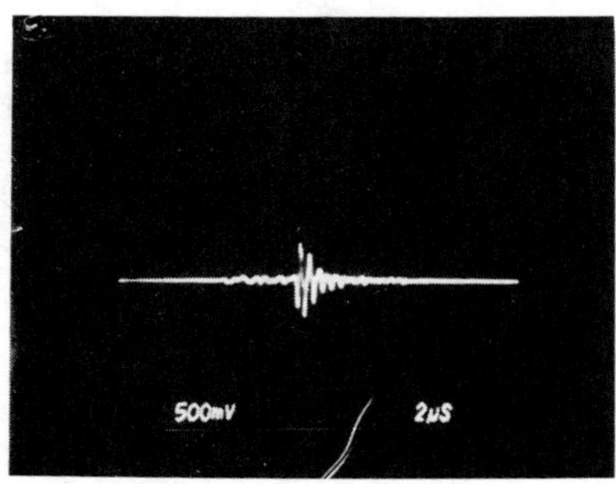

Fig. 5. Ultrasonic Signal from 2.4-mm-dia Hole Shown in Fig. 2 (Hole "b") Using Longitudinal Waves (2.25 MHz).

A comparison of longitudinal and conventional shear-wave testing (SV shear waves) leads to the conclusion that longitudinal-wave testing may be the better method for defect detection in stainless steel weld metal. However, with SH shear waves, the difference between shear and longitudinal waves is not as apparent. To determine whether these relationships are dependent upon the direction of acoustic-wave propagation and microstructure of the weld metal, as suspected, another hole was drilled in the same weld. In this case, the hole was placed in the center of the weld metal (Fig. 2 hole "a") and normal incidence waves were used for detection. The radio-frequency (rf) signal obtained from the hole using SH shear waves is clearly distinguishable from the noise, similar to that shown in Fig. 3. No reflection from the hole is apparent when SV shear waves were used. The results for longitudinal waves are shown in Fig. 6. The signal is distinguishable from the noise but not as clearly as in the case of the SH shear waves.

Again, the results are consistent with previous observations. Longitudinal waves propagating parallel to the dendrites can be highly attenuated (or deviate from the wave normal), whereas this propagation direction is favorable for shear waves polarized in the weld-pass direction (Y).

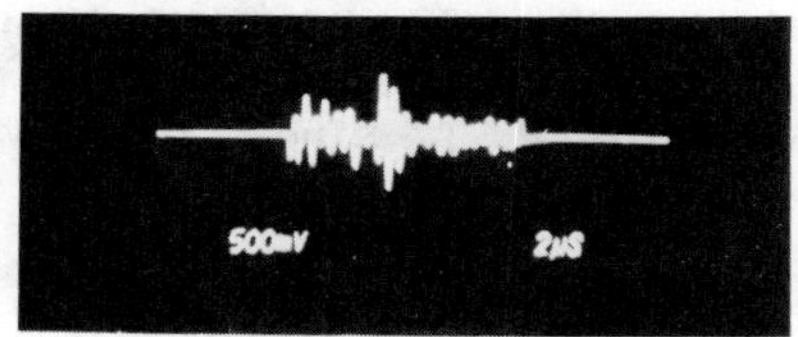

Fig. 6. Ultrasonic Signal from 2.4-mm-dia Hole Shown in Fig. 2 (Hole "a) Using Longitudinal Waves.

SUMMARY

A simple model describing the propagation of ultrasonic waves in Type 308 stainless steel weld metal is, with one important exception, consistent with velocity of sound data for both longitudinal and transverse waves. The model assumes the weld consists of a collection of long single-crystal dendrites with a <100> orientation for the long axis. These dendrites are arranged parallel to each other with a random orientation about the long axis. Our attenuation data are consistent for longitudinal waves with the hypothesis (4) that the least attenuation occurs in the direction of maximum velocity. It has been demonstrated that inspections of welds with ultrasonic waves 45° to the axes of the dendrites are carried out best with horizontally polarized shear waves or longitudinal waves, whereas inspections parallel or perpendicular to the dendrites is most effective with shear waves of the proper polarization.

REFERENCES

(1) S. Matsumoto and K. Kumura, "The Relationship Between Grain Size and Ultrasonic Attenuation Coefficient in Austenitic Stainless Steel and Iron," Trans. Natl. Res. Inst. for Metals, Vol. 14, No. 4, 1972, pp. 155-164.

(2) M. F. Markham, "Measurement of Elastic Constants by the Ultrasonic Pulse Method," Brit. J. App. Physics Supp., Vol. 6, 1967, pp. S56-S63.

(3) J. R. Tomlinson, A. R. Wagg, and M. J. Whittle, Second International Conference on Non-Destructive Evaluation in the Nuclear Industry, Salt Lake City, February 13-15, 1978.

(4) G. F. Miller and M. J. P. Musgrave, Royal Society of London, Proceedings, Vol. 236, Sec. A, 1956, p. 352.

(5) B. R. Dewey, L. Adler, R. T. King, and K. V. Cook, Exp. Mech., Vol. 17, No. 11, 1977, p. 420.

(6) Nondestructive Testing Development Program Quarterly Progress Report for Period Ending December 31, 1976, Oak Ridge National Laboratory Report, ORNL-5253, p. 50.

(7) L. C. Lynnworth, "Ultrasonic Probes Using Shear Wave Crystals - Principles," Mater. Eval., Vol. 25, 1967, p. 12.

ON-LINE FLAW- AND DIM-INSPECTION DATA ACQUISITION AND EQUIPMENT CONTROL FOR HEAT EXCHANGER TUBES

K. Abend
Nukem GmbH
West Germany

Increasing quality requirements, and, at the same time higher production speed for quality tubes require new or remarkable improved inspection techniques.

For the flaw inspection at quality tubes new test flaws were developed, i.e. the so called "saw tooth" defect for heat exchanger tubes etc. High accuracy-specification concerning the tube geometry require 100% measurement of outer diameter, wall thickness, inner diameter, ovality, eccentricity, wall variation etc. with an accuracy in the μ-range (1 μ = 1/1000 mm = 0.00004 inches). The use of on-line signal processing of the combined flaw inspection and dimension measurement allows automated decisions about further manufacturing of the tube like "accepted," remanufacturing, additional polishing or re-inspection. In some quality tube plants one has installed today in Europe, even in Eastern-Bloc-Countries, inspection systems with automated saw cutting machines to provide optimized tube length after inspection. The saw cutting machines are controlled on-line by the inspection equipment.

The following paper describes rotating transducer systems for high resolution flaw inspection, dim-measurement and gives an example for an on-line data processing.

FLAW INSPECTION AT TUBES

In fig. 1 are shown typical test flaws for the UT-tube inspection, rectangular notch, V-notch and saw-tooth. Depending on material quality and surface, inside and outside flaws are to be detected with the same sensitivity, 5% of the wall thickness. The minimum flaw depth detected in industrial applications is about 10 μ. With the introduction of the saw tooth flaw and the required 100% inspection, inside and outside of the tube, the four-direction inspection technique was adapted.

In fig. 2 the operation of the four-direction inspection technique is demonstrated. Two line-focused transducers, 180° opposite to each other are inspecting the tube 100%. Every transducer covering 100% tube surface. For longitudinal flaws and for transversal flaws one needs four line-focused transducers. For the so called 45° flaws one can either change transducer directions or add additional transducer channels. Beside of the excellent reproducibility of the inspection, even at high inspection speed $\pm$ 2 dB, one gets with this technique possible identifications of typical natural flaw sizes *(reference 1). With the use of this inspection technique in rotating systems, transducers rotating around the tube, one gets a test point distance at the tube surface.

$$D = \frac{U \times L}{f}$$

D = test point distance
L = tube circumference
U = rotations per time
f = UT pulse repetition rate

In fig. 3 are shown test point distances for various diameters and rotation speeds, one can see an increasing rotation speed requires an increased pulse rep.-rate. For rotational speeds of 6000 and more pulse rep.-rates of 20 kHz per channel are standard.

DIMENSION MEASUREMENT

The outer diameter of tubes is measured today with the pulse travel time measurement (pulse echo technique) with a pulse rep.-rate of 5 kHz and an accuracy of $\pm$ 4 μ (fig. 4). To compensate the velocity change in water because of temperature change, the reference transducer has become most common.

For wall thickness measurement the same principle is used for the wall range down to 0,8 mm. For wall thicknesses below this value the use of the resonance method with lower repetition rate but accuracy around $\pm$ 1,5 μ is standard. The pulse echo method allows the wall thickness and OD-measurement simultaneously with the same transducer. For lower wall thicknesses one can use a second set of transducers. The inner diameter of the tube can be calculated with the same

* (Materialprufung, Bd. 19, 1977, Nr. 5, May, S. 178-180)

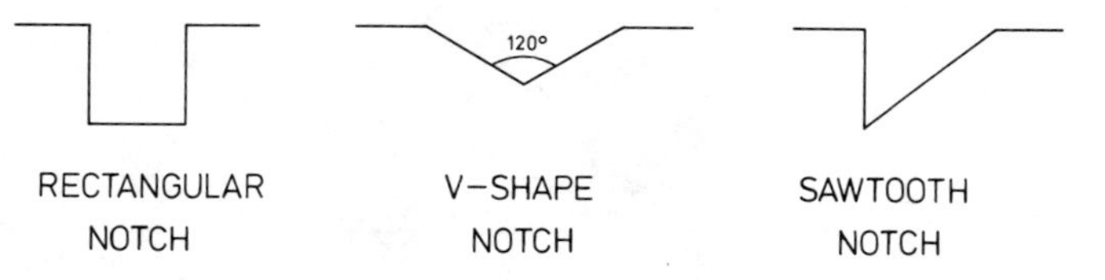

Fig. 1. Typical flaw notches, tube inspection

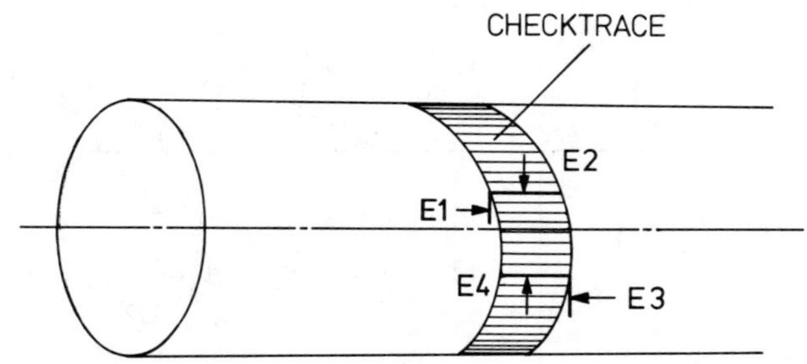

Fig. 2. 4-direction inspection

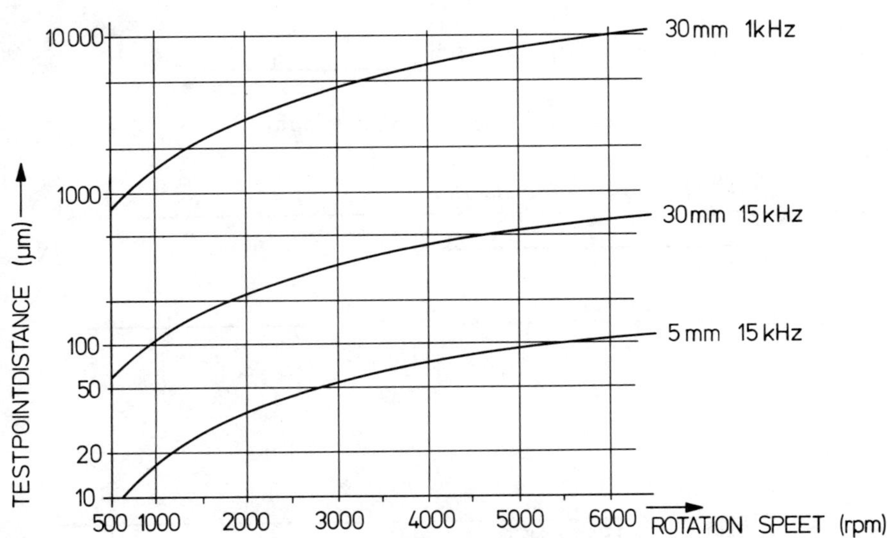

Fig. 3. Test point distances at the tube surface

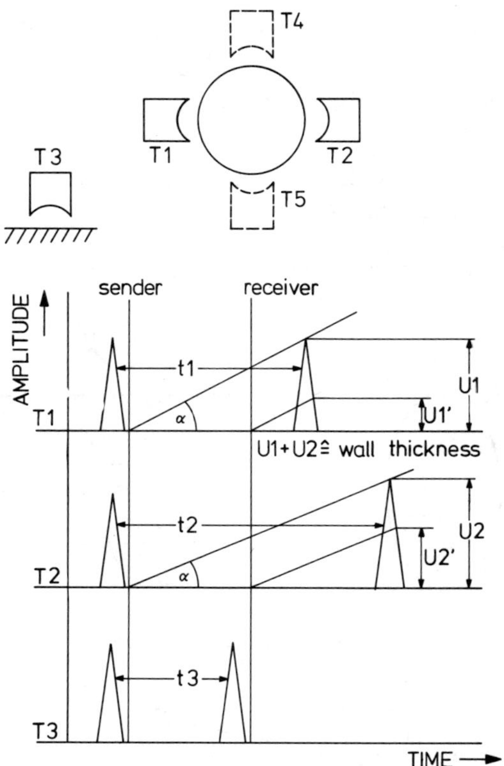

Fig. 4. DIM - measurement

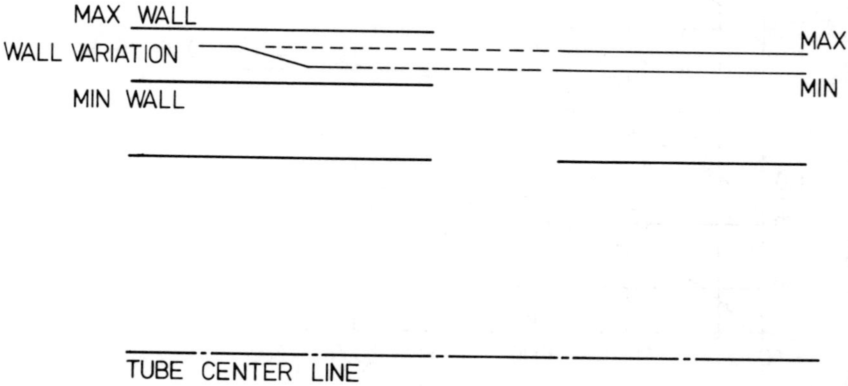

MAX WALL

WALL VARIATION

MIN WALL

MAX

MIN

TUBE CENTER LINE

Fig. 5. Def. wall variation

frequency as the outer diameter and wall thickness measurement, using equation

$$ID = OD - (WT1 + WT2)$$

With storage and processing of the measurement values per rotation one gets the ovality, eccentricity and wall variation. The wall variation is the change in wall thickness per tube circumference and/or per tube length unit within a given, larger max/min-tolerance range of the tube. In Europe one often has as specifications a smaller max/min-tolerance range and no wall variation (fig. 5).

DESCRIPTION OF ROTATING TRANSDUCERS FOR HEAT EXCHANGER AND FUEL CLADDING

In fig. 6 one can see two rotating transducer systems. One system for the four-direction flaw inspection and one system for the dim-measurement as described under sub-point 2. Depending on the test flaw size and tube dimension, the system is able to inspect up to 50 m/minute as standard speed. With different rotation body sizes one can overlap the other diameter range from 6 to 200 mm.

ON-LINE DATA PROCESSING, FLAW INSPECTION

A typical rep.-rate for flaw inspection is 15 to 20 kHz. With four channels one gets a relative high data rate to be transferred to the computer. If one tries to produce a similar output as the analog recording, one has to handle this data rate and is wasting computer memory space for storage of every single measurement. To prevent the situation one can use two alternatives:

1. Together with the length measurement of the tube, to indicate where the flaw occurs, one has a hardware data reduction and can transfer only the maximum flaw indication. The computer takes the digitized maximum flaw indication and the corresponding length information for storage and later printout (fig. 7).

2. Together with the length measurement one can use discriminators, monitoring the analog output. As soon as the discriminator is activated, the descrimination level is exceeded by the analog indication, the actual signal height is transferred to the computer together with the tube length information.

Fig. 6. Two rotating transducer systems

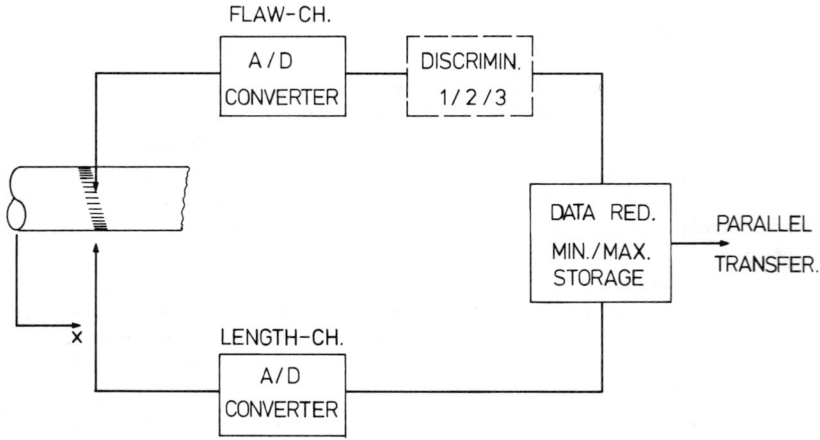

Fig. 7. Hardware data reduction, flaw channel

The data processing is done in the same way as in the previous example. Using this technique one has two advantages. The first advantage is given by the possibility of installing several discriminators at every channel to get informations more similar to the analog recorder (fig. 8). With reject level 1 in

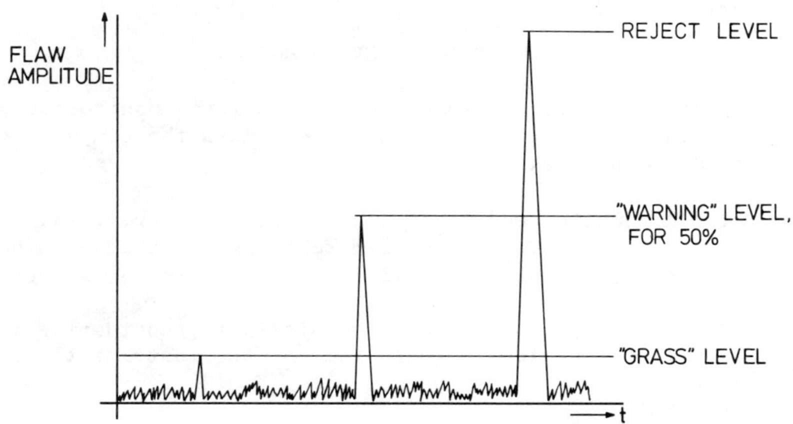

Fig. 8. Flaw discrimination with 3 discriminators

fig. 8 one can recognize indications close to the grass, level 3 is the actual reject level. The second advantage is the possibility to have a computer program controlled check of the test tube flaw indication height. A drift of this indication within 24 hours or any required time period can be corrected by the computer program. This is also possible for flaw indications.

ON-LINE DATA PROCESSING, DIM MEASUREMENT

The repetition rate for the dim-measurement is 5 kHz per OD, the two wall thicknesses, each. If the ID is calculated by the analog computer at the instrument one gets additional 5 kHz. For every installation we made until today we used a hardware data reduction. We store the maximum and minimum indication per dimension value and per tube circumference. Only the max and min-values per circumference are transferred to the computer. The same transfer is used for ovality, eccentricity, wall variation etc. In this way one has a data reduction to be transferred to and stored at the computer of at least 50:1. An additional data reduction is possible by hardware. Instead of transfer the max/min values per circumference one can store these values and transfer max/min per given tube length.

Using the advantage of the on-line computer one can control the test piece indications and compensate by program any possible drift within the dynamic range of the measurement instrument.

ON-LINE COMPUTER OPERATION

Beside the data storage, the reject level monitoring and the digital printout of the measurement data the computer can take additional work.

1. Automatic reading of tube numbers and providing the printout with the proper tube number, charge number, date, operator, identification, instrument number etc.

2. Tube counting, production statistic considering flaw free tube length, dead time of the instrument etc.

3. Optimizing

   The production length for heat exchanger tubes is 40 to 50 m. The required tube length within the heat exchanger may vary between 1 - 20 m. The computer can optimize between the different tube lengths and the flaw free inspected tube length. A control program fed by these data, controls the tube length sawing machine. In fig. 9 one can see a block diagram of a flaw and dim-inspection machine for heat exchanger tubes with an on-line PDP11 system. The computer controls up to five saw machines on-line.

CONCLUSIONS

With a reasonable data reduction one can use today small computers or microprocessors for economical data storage and printout of the test documents. Together with connected processing machines one can control these machines on-line, i.e. with optimizing programs. The cost of hardware systems, including flaw marking and sorting is about equal to the cost of microprocessors for the same purpose. Any kind of optimizing is possible with microprocessors or mini computers.

EXPERIENCE WITH FLAW SIZE ESTIMATION BY ULTRASONIC HOLOGRAPHY WITH NUMERICAL RECONSTRUCTION

H. WÜSTENBERG
E. MUNDRY
J. KUTZNER
BUNDESANSTALT FÜR MATERIALPRÜFUNG (BAM), BERLIN

INTRODUCTION

One of the many possibilities for producing and reconstructing acoustical holograms consists in the digital storage of an acoustical hologram and its numerical reconstruction by use of computers. Due to the progress and the cheap availability of computer hard and soft ware for a quick execution of the fast fourier transformation, the numerical reconstruction of a hologram must now be considered as a severe competitor for other methods reconstructing acoustical holograms.

At our Institute we started with numerical reconstructions from another point of view. We tried to reconstruct defect sizes and orientations from an echodynamic pattern, that is the record of the echo-amplitude from a defect versus the probe displacement, as demonstrated in fig. 1 (1). Such echodynamic patterns are ambiguous. A pattern can correspond to a manyfold of different defect arrangements. In the case of fig. 1, the crack produces nearly the same interference structure - indicated by maxima and minima in the echodynamic pattern - as two slag-inclusions with a distance in depth corresponding to the crack depth extension. The ambiguity of echodynamic patterns forced us to look for better solutions. One possibility to establish a better reciprocity between a reflecting defect and his acoustic response is to produce an acoustic image. This can be done by a system with acoustic lenses or by holography for instance with scanning probes.

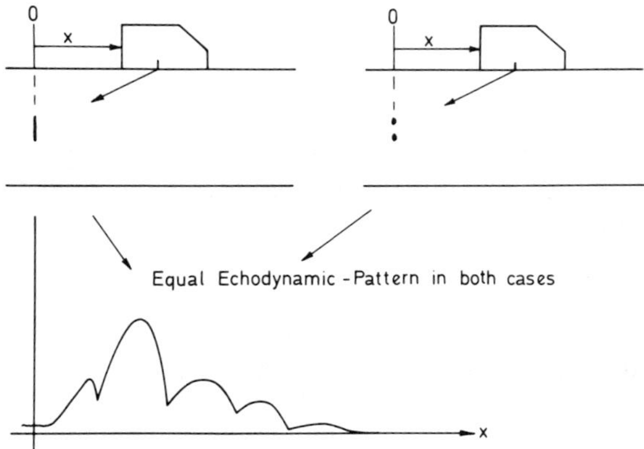

Fig. 1. Echodynamic Pattern of Two Defect Situations.

A scanning with focusing probes can be regarded as equivalent to the production of an acoustic image with the beam diameter as resolution.

There is some hope that also pattern recognition techniques, using not only one physical relationship as for instance the interference behaviour of defects, may help to derive from echodynamic records valuable results for the interpretation of ultrasonic indications. (2, 3)

Instead of using pattern recognition techniques we decided to replace the recording of only the amplitude by the use of amplitude and phase records of the echo. In this case we can apply the mathematically unique relationshipsbetween the scanned pattern at a surface or a surface line and the acoustical intensity distribution in the area of the defects which are the basis of the holography (4, 5). By the use of the acoustical holography we can strongly reduce the ambiguity of the simple amplitude patterns. Such an imaging can be executed also by the use of focusing probes with strongly concentrated sound beams in the defect zone. The advantages of focusing probes - uncomplitated system and theory, therefore easy to introduce in practical applications especially if probes with a direct coupling are used-are accompagnied with some drawbacks. Typical drawbacks of such focusing probes are the restriction of the usable area, the relatively small aperture in connection with a limited resolution power and sometimes difficulties in the handling of large probes. The aperture of a holographic scanning can be chosen as large as necessary for the needed resolution, the defect area can be defined by adjusting the equipment. Therefore we decided to apply not only focusing probes, but also to develop a numerical reconstruction of amplitude and phase data , collected along the line of the probe movement during the scanning of the defect. Doing so we replaced the evaluation of an echo-dynamic pattern by the linear holography. Meanwhile we extended the numerical reconstruction from a line to a surface scan, but experiences with surface holograms and their numerical reconstruction are in our institute just at the beginning, so I will restrict this report to linear holograms for determining the extension of defects in length and in depth.

2. EQUIPMENT FOR NUMERICAL STORAGE AND RECONSTRUCTION OF HOLOGRAMS

Fig. 2 shows a scheme of a numerical reconstructed holography. At the surface of the test object we

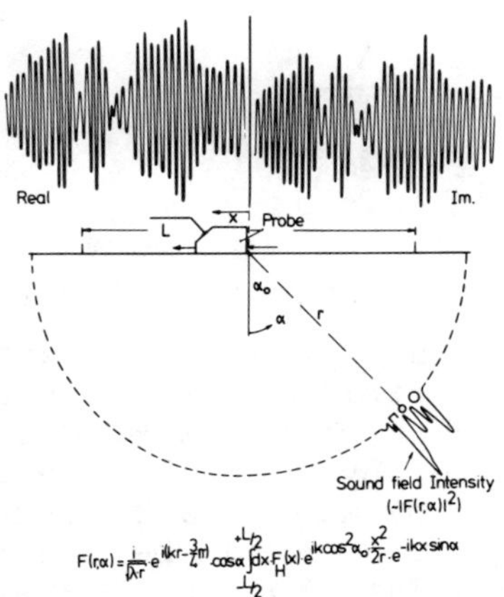

Fig. 2. Numerical Reconstructed Holography.

scan the echoes received from the defective area and measure their complex echo signal. The best way to do this for a numerical reconstruction is to measure the real and the imaginary part of an echo. The distribution of a real and imaginary part on the surface is demonstrated at the upper part of fig. 2. From this sound field distribution at the surface of our object $E_H(x)$ we can derive the sound intensity distribution on a circle of a radius r by the reconstruction formula indicated in the lower part of fig. 2. For the reconstruction formula we must regard the radius r and the angle α_o, under which we observe our defective area from the center of the aperture L. One reconstruction is only valid for angles in the range of the observation angle α_o. But it should be mentioned that a numerical reconstruction is not restricted to a more or less paraxial situation of defects.

Fig. 2a summarizes the calculation procedure. The

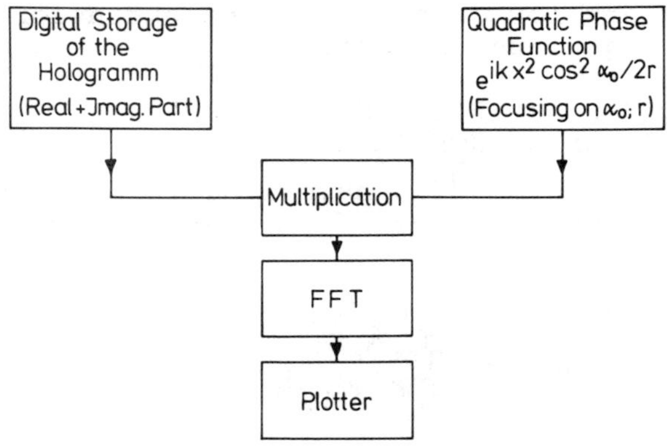

Fig. 2a. Numerical Reconstruction Procedure.

digitally stored datas - real and imaginary part of the echo - are multiplied with a quadratic phase function, which provides a kind of focusing on angle and distance. By a little change of this phase function, just by adding a curvature correction, curved surfaces can be regarded. After the multiplication, a Fast Fourier Transformation is executed. The result, the amplitude or intensity distribution on a circle of radius r around the center of the aperture, can be plotted in an arbitrary form. The simpliest method has been represented in figure 2.

The most important parameters, which determine the accuracy and the performance of the numerical reconstruction are listed in fig. 3. We must not only observe the resolution, determined by the optical diffraction theory, but also the divergency angle of the scanning probe sound field. It is useless to apply a large aperture L if not sufficiently divergent beams of the probes are used.

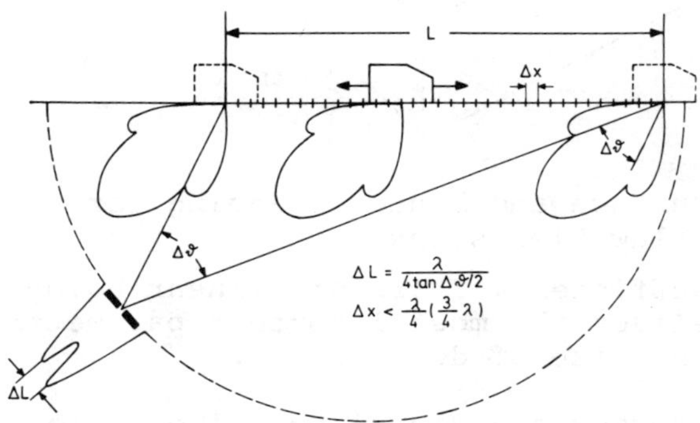

Fig. 3. Resolution and Sampling Rate for a Linear Holography in a Single Probe Technique.

The distance between two samples should be smaller than half of the minimal period length of the real or imaginary part at the surface. For inclined incidence we can assume that a quarter of the wave length represents a good estimate for this distance. For normal incidence we can apply larger distances between the samples in the range of $3/4\lambda$.

This fact is also used during the production of surface holograms as shown in fig. 4. The distance between the lines can be three times larger than the sampling point distance along a single line. That means, that e.g. for a 80 x 50 mm^2 surface, 1,6 mm wave length and 45° about 8200 sampling points are needed.

In order to determine the defect extension in depth only one scan along a line is needed. The same is necessary for the determination of the defect length. Here one scan parallel to the weld

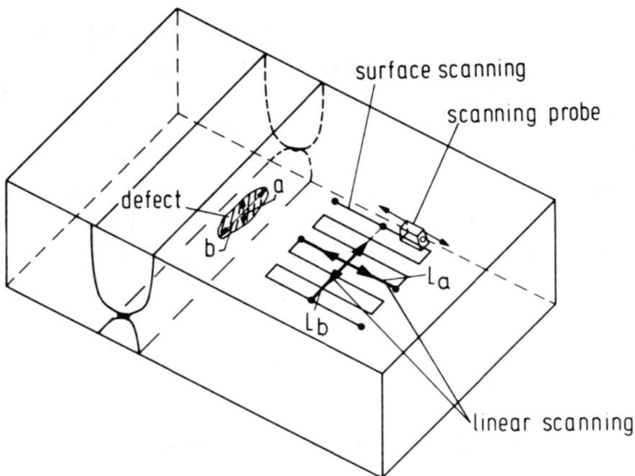

Fig. 4. Surface and Linear Scanning for Critical Flaw Dimensions.

seam may be sufficient. Therefore linear holography can deliver the most important parameters to describe the size of defects.

In order to improve the signal to noise ratio we are using for the linear holography a probe with two transducers, one for the determination of the defect extension in depth and the other for the determination of the length. This is demonstrated in fig. 5. In the lower part of fig. 5 the directivity pattern in the azimuthal plane is demonstrated for the depth scanning transducer.

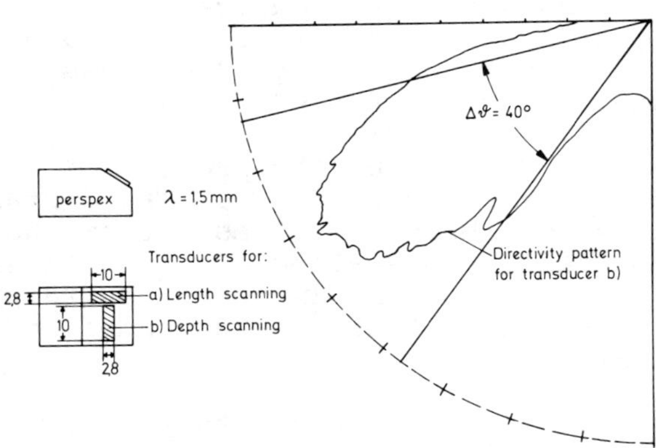

Fig. 5. Probe for Linear Holography with Inclined Incidence.

The directivity pattern has been measured with a contactless electrodynamic pick-up.

Fig. 6 is demonstrating a scheme of our holographic equipment, with which we have measured the holograms shown in the following slides. A probe in direct contact is moved by a manipulator and excited by a phase-synchronized transmitter.

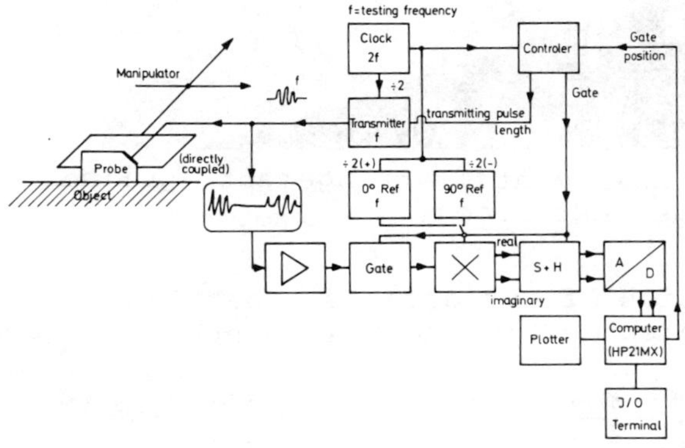

Fig. 6. Equipment for Numerical Reconstructed Holography.

A phase sensitive receiver is measuring real and imaginary part of echo indications within the gate. This gate is positioned by a computer in order to guarantee, that only indications, which are in the defect area of interest, are analysed. The real and imaginary parts are converted into digital signals and fed into the memory of the computer. By choosing the distance r and the observation angle α_0 a reconstruction as shown in fig. 2 is executed for a certain area.

Fig. 7 (photograph) shows the experimental arrangement applied on a large test-block. The probes are directly coupled.

In the following, some examples on natural defects are demonstrated. Fig. 8 shows the reconstructed holograms of the depth and length scans on a slag inclusion, which is represented on the radiographic film as indicated in the lower part of fig. 8. The wall thickness of the object was 255 mm and the

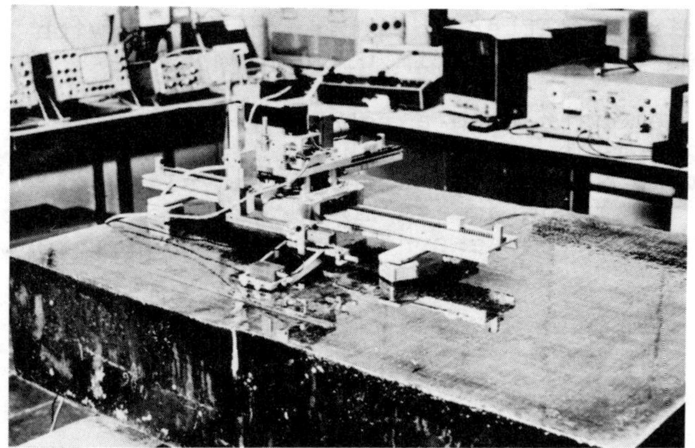

Fig. 7. Experimental Arrangement Applied on a Large Test Block.

depth situation of our defects about 120 mm. The weld will be cutted in the next month.

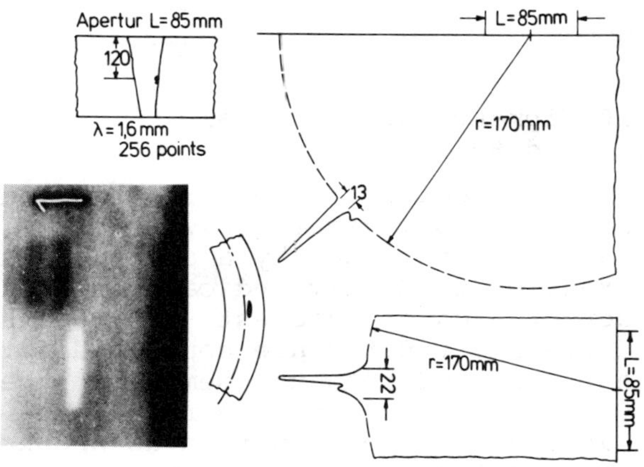

Fig. 8. Holographic Defect Reconstruction of a Slag Inclusion.

Fig. 9 shows two reconstructions of holograms measured on relatively thin-walled welds. Left-hand side a tube with a diameter of 920 mm and a wall-thickness of 16 mm and a longitudinal seam has been the test object. A crack on the transition area between base material and weld seam at the outer side has been scanned with a 4 MHz transverse wave probe with about 45° angle of incidence. From the reconstruction we can derive a crack depth of about 3 mm. The evaluation must take into account, that the crack is represented in the reconstruction with double real size due to the reflection at the outer surface. Right-hand side (fig. 9) is shown a 40 mm thick weld which we scanned on the same way as shown on the left-hand side. A lack of fusion of about 7 mm is represented in the reconstruction. We used the probe demonstrated in fig. 5 (2 MHz, 50°).

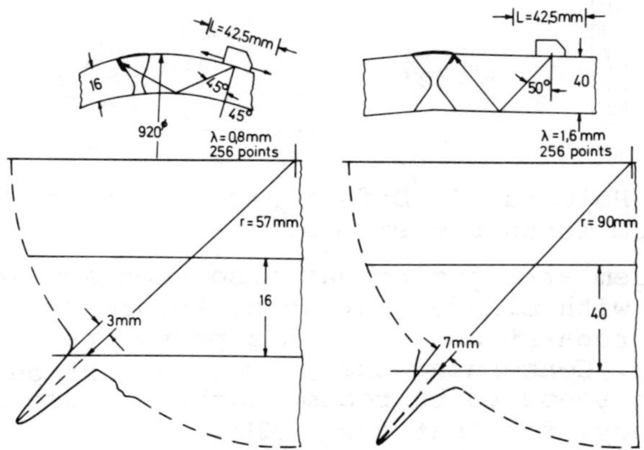

Fig. 9. Holographic Defect Reconstruction of Near Surface Cracks.

It must be mentioned that the use of the corner effect has some inherent limitations, e.g. the maximum depth should not be greater than 5 according to our first experience.

For both cases in fig. 9 the real crack depth has been examined by cutting and has been proven in good agreement with the values derived from our holographic reconstruction. (Fig. 9 left-hand side: measured: 3 mm, real: 2,6 mm, right-hand side: measured: 7 mm, real: 7 mm)

In fig. 10 a tandem arrangement has been used to measure the extension in depth and length of a defect more or less perpendicular to the surface. This defect is positioned in one of the hsst-blocks, which are examined in Europe in the frame of a Round robbin measurement program. For the verification of the defect data we must wait for the cutting of the object.

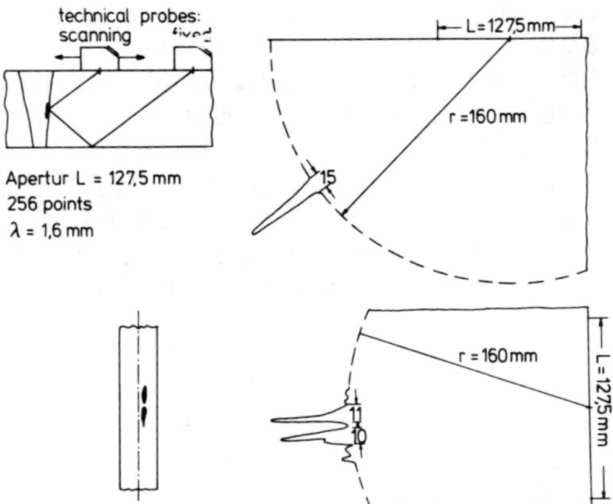

Fig. 10. Holographic Defect Reconstruction for Tandem Technique Probes.

The same tandem arrangement has also been applied on specimens with cladded surfaces. According to the surface preparation 1 or 2 MHz probe frequency has been used. Concerning the **as welded surface** conditions it seems to be recommendable to use a lower frequency, for instance 1 MHz.

According to fig. 11 a numerical holographic reconstruction can also be used for the determination of crack depths of small cracks.

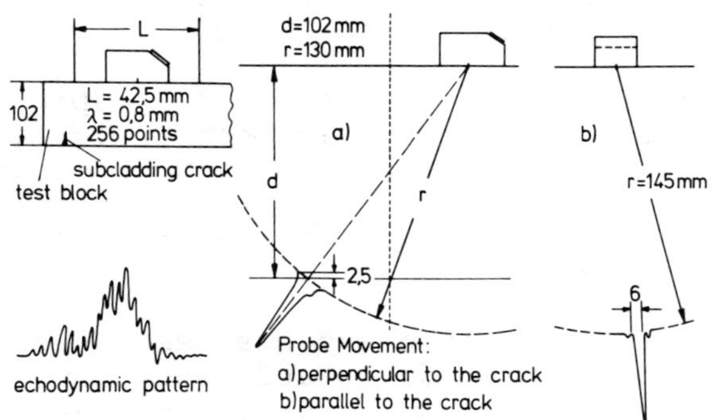

Fig. 11. Holographic Defect Reconstruction of a Small Crack

We have applied the holographic scanning procedure with a 4 MHz 45° probe on an undercladding crack of about 2,5 mm in depth in a 102 mm thick wall. The cladding has been eliminated by machining. The reconstructions are shown in fig. 11. The defect length has also been evaluated by magnetic particle inspection to about 6 mm.

The restrictions inherent to all acoustic imaging methods are defining unaccuracies directly connected with physical datas: 1.5 times the lateral resolution defines the smallest flaw size which can be determined within an accuracy commonly higher than ten percent. A separation better than about 1.5 times the resolution cannot be expected.

For focusing probes the accuracy of defect sizing depends from the real flaw size as indicated in fig. 12. Here the reflector size is plotted versus the half value extension. For sizes higher than 1.5 times the beam diameter accuracies higher than ten percent are achieved. A similar accuracy must be expected for the holographic reconstruction.

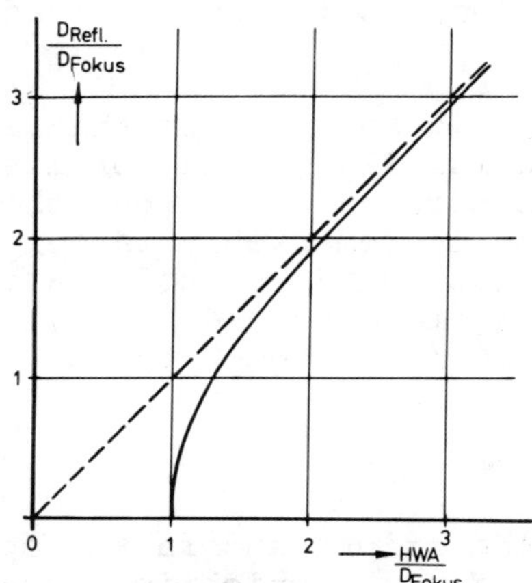

Fig. 12. 6 dB-Drop Extension for Focusing Probes.

4. CONCLUSIONS

In the examples of the last figures we have demonstrated some practical applications of acoustical holography with numerical reconstruction. Our first impression is that due to the remarkable lateral resolution of the holographic procedure and the direct evaluation of numerical reconstructions this technique will be one of the future instruments to interpret ultrasonic indications. Compared with focusing probes a holographic procedure has the advantage of better adaptability to different practical conditions. One disadvantage of the holographic procedure is, that it guarantees only a poor longitudinal resolution.

A numerical reconstruction has, compared with an optical reconstruction, the advantage to be nearly on line in the case of a linear holography and to avoid misinterpretations for instance due to not correctly determined magnification factors.

The presentation of the sound field intensity as a curve may help to better distinguish multiple and different reflectors in the regarded defectuous area.

For the time being it seems to us that a numerical reconstruction is ideally suited for linear holography. For a surface holography the time for the reconstruction is remarkably larger than for a simple linear scan. Hard and soft ware improvements are necessary to obtain a nearly on line reconstruction also for a numerical surface holography.

4. ACKNOWLEDGEMENTS

Many of the considerations and facts reported here are results of work, which is sponsored within the reactor safety research program by the ministry of research and technology of the Federal Republic of Germany.

REFERENCES

(1) H. Wüstenberg, J. Kutzner and G. Engl, "Dependence of Echo Amplitude on Defect Orientation in Ultrasonic Examinations" 8. WCNDT, Cannes, France 6.-11.9.1976

(2) J. Rose, P. Mast and L. Niklas, "The Potential of "Simulearning" in Flaw Characterization" Brit. Journ. of NDT, XVII, 6 (1975), 176-181

(3) A.N. Mucciardi, R. Shankar and M.J. Buckley, "Application of Adaptive Learning Networks to Nondestructive Evaluation Technology" NAECON '75 Record, 460-469

(4) J. Kutzner and H. Wüstenberg, "Akustische Linienholographie, ein Hilfsmittel zur Fehleranzeigeninterpretation in der Ultraschallprüfung" Materialprüfung 18 (1976) Nr. 6 Juni, 189-194

(5) J. Kutzner and H.Wüstenberg, "Akustische Holographie in Tandemanordnung, ein Hilfsmittel zur Fehleranzeigeninterpretation in der Ultraschallprüfung" Materialprüfung 18 (1976) Nr. 12 Dez., 462-465

AN ACOUSTIC TRANSPONDER FOR CALIBRATING ULTRASONIC EQUIPMENT

Henry H. Chaskelis
Naval Research Laboratory
Washington, D. C. 20375

INTRODUCTION

The initial ultrasonic inspection of materials and structures must be carried out with thoroughness and precision commensurate to the critical damage results which would occur due to various failure modes. For extended life or reusable items; subsequent maintenance inspections must either be equivalent to the initial test or follow the growth pattern of the various anomalies previously found but which are not yet critical. For the nuclear industry the potential damage due to failure is most serious. Therefore it is not only crucial to implement a very thorough nondestructive evaluation (NDE) program, but also it is most prudent to be thoroughly cognizant of the possible tolerance degradation of the inspection system itself. This report addresses such a topic with respect to the use of ultrasonic inspection systems. Furthermore as a partial solution to the complex problem of tolerance degradation a recently developed acoustic transponder device is discussed. This device is an outgrowth of a previously reported work (Ref. 1). Currently there are six units which are being evaluated in the field.

Tolerance Degradation of an Ultrasonic Inspection System

The term tolerance degradation refers to the situation whereby, during the examination for one unknown parameter (the flaw), the existence of other unknowns (instrument or test peculiarities) prevents the unique or accurate determination of the desired parameter. For example, interpolating the echo strength from a flaw relative to that of several known flaw requires that the amplifier be linear. Failure to verify the linearity of the amplifier may be the cause of inaccurate readings. In practice there are many degradation modes that may act interdependently, which makes an analysis rather difficult. To alleviate this problem it is desirable to group the various tolerance degradation modes and relate them to the actual measurement which is undertaken.

For ultrasonic testing, the basic quantities measured are the pressure amplitude, the frequency variations in pressure amplitude (spectrum), the relative phase, and a time-dependent parameter (velocity). These measurements are made and recorded with respect to some coordinate axes which serve as reference points at the surface of the inspected object. Of these five measurable quantities* the most commonly used today for industrial NDE are pressure amplitude, time, and location (coordinate axes). The degradation modes therefore refer to the possible inaccuracies which may occur, due to either test technique or instrumentation causes, when measuring these quantities. To further aid in systematically dealing with degradation modes it is worthwhile to separate the domains under which these occur. These are the acoustic field, electromechanical interaction, and the electronic components.

THE ACOUSTIC DOMAIN

An acoustic source coupled to some medium will radiate a field in accordance with certain physical laws. In standard ultrasonic practice the reflected (and some scattered) portion of this field is used to detect and evaluate the homogeneity of the medium. This reflected field is dependent on wavelength, materials constraints (impedance) and geometric factors. Therefore the accuracy and consistency of the measured intensity, time, and spatial coordinate will be proportionally affected with respect to the a-priori knowledge of, or variation in, these three acoustic domain factors. For example, suppose a fairly accurate thickness measurement is desired by use of the pulse echo technique. Time (T) is the usual quantity measured and, the most general relationship to part thickness is:

$$T = \frac{Nt \sec}{V_2}\left[\sin^{-1} \frac{V_2}{V_1} \sin \alpha\right]$$

Here N refers to the number of traverses though the part thickness t, α is a directional angle and V_2 and V_1 are velocities of the specimen and coupling medium respectively. If one of the geometric factors is not precisely known, say the direction in which the source emits, then the thickness calculation will be in error by a multiplicative amount related to the secant of the angle in error (Fig. 1). Assuming the piezoelement is misaligned in the transducer housing by 1 degree of arc, the accuracy can be no better than the third decimal place if $V_2 = V_1$. The situation is more grave if one attempts to "average out" the time measurement over a multiple bounce path (N) or uses the Nth and $N + 1$ multiples as a measurement base.

ELECTROMECHANICAL INTERACTION

The conversion between electrical and mechanical energy occurs at the transducer. By virtue of certain material and orientation properties the transducer both affects and is affected by contact with electrical and mechanical forces.

*Fundamentally this could be reduced to just three; amplitude, space and time.

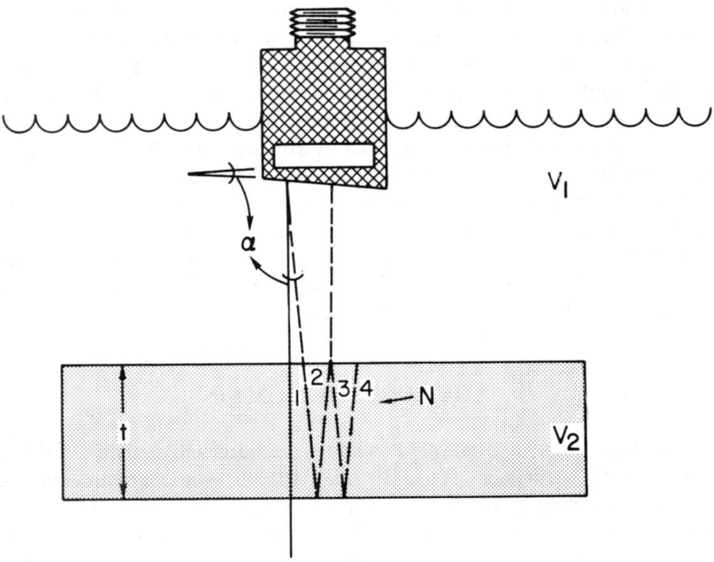

Figure 1 — Accuracy of measured thickness t depends not only on knowledge of instrumentation precision and material velocity V_2, but also on acoustic path. A normal ray would follow the solid line but angulation errors α will cause the ray to follow the dashed line. This error will be dependent also on the velocities V_1 and V_2 and the number of half-bounces N.

As a source of acoustic energy, its dimensions, homogeneity, and coupling coefficients (materials constants) determine the radiation field generated. However, the impedance difference due to mechanical contact also causes a change in the electrical characteristics. The converse also occurs. Therefore both the electrical and mechanical impedances influence the acoustic field generated. Likewise, when used as a receiver, the charge created by the presence of an acoustic pressure field is affected by the impedances.

Because of these interactions it is possible to generate errors in the accuracy of two of the measured quantities; amplitude and time. For example, a change in the mechanical impedance (as in immersion versus contact testing) can change the transducer center frequency and bandwidth (Fig. 2). This effectively will change the directivity pattern of the acoustic wave in the medium. In turn the return echo from an otherwise identical situation may be different.

ELECTRONIC COMPONENTS

At this juncture one enters into the more familiar yet least resolved problem area, the electronics package. Basically the standard ultrasonic electronic package consists of a pulser, receiver stages, display section (space and time base), and the associated power supplies. Both the pulser and first receiver stage

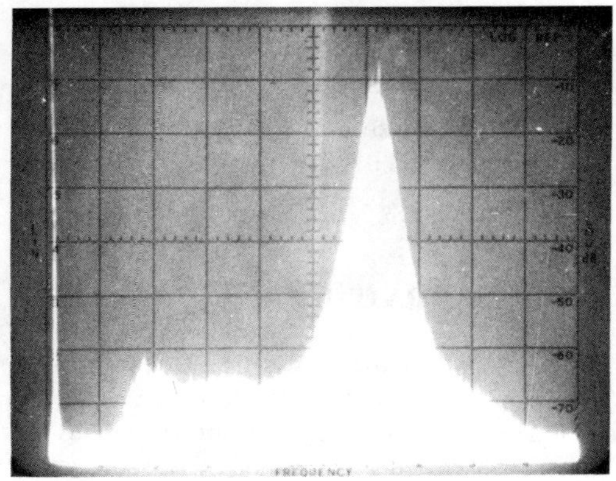

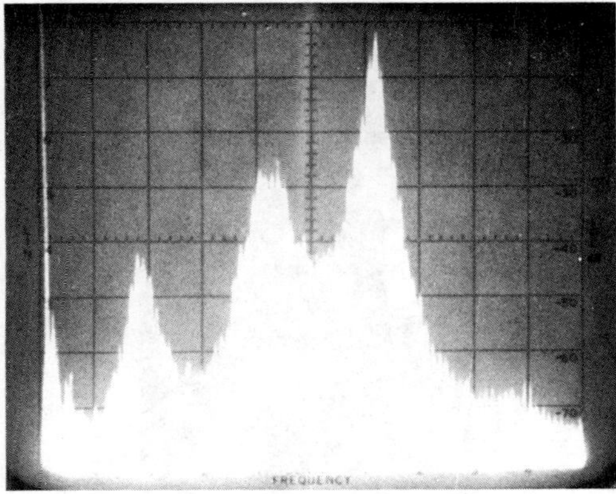

Figure 2 — Spectral analysis photograph of return echo from flat fused silica specimen. The upper and lower sides respectively show the results of a contact versus immersion test. Although the peak frequency remained at 6 MHz (1 MHz per division, linear amplitude presentation) the lower frequencies are significantly damped when the transducer is loaded by direct contact with the specimen.

(including the pulser front end limiters) are affected by coupling to the tranducer. The rest of the package may be affected by inherent design differences, line source problems, temperature fluctuations, etc., all of which may cause errors in the measurement.

As previously mentioned the many tolerance degradation modes which exist may be subdivided into two basic categories:

1) Modes related to the test technology which affects the quantities being measured; intensity, time, location.
2) Modes related to the accuracy and stability of the measuring instrument; the electronics package.

To avoid or at least minimize the first kind of error mode requires knowledgeable, competent, and experienced ultrasonic practitioners. The second kind requires a means to test the precision of the measurement instrument. To this end a device called the Electronic Test Block (ETB) (patent No. 3,351,977) has been developed at NRL. A previous report (1) discussed the overall capabilities of a device conceived to have a much broader application and the partial success of the preliminary laboratory model. This report presents the current status of a more limited in scope commercial version of the ETB, which is presently being evaluated at various field activities.

THE ELECTRONIC TEST BLOCK: OPERATING PRINCIPLES

The ETB is basically an electronic-controlled acoustic transponder (Fig. 3). It senses the acoustic signal from the instrument under test and returns pulses of known amplitude and time relationship. Form these known echoes a proper assessment can be made of the measurement accuracy of the instrument. The two basic qualities assessed are time and amplitude. Also, other operational tests such as resolution and dB attenuator accuracy can be made.

Figure 3 — Battery or line operated ETB built to Navy specifications. The front panel has been designed for ease of operation. Rugged case and electronic assembly allow its use under shipyard conditions.

As previously discussed there exists a two-way interaction between the transducer and electronics package. For this reason it is important to test the instrument and transducer combination. The ETB does this. By use of the built in transducers (Fig. 4) the calibrated electronic signals are linearly converted to the acoustic domain. At this point the known reference is introduced to the equipment being tested.

In practice the ETB returns two signals in synchronism with each incoming acoustic pulse (Fig. 5). While in the amplitude linearity mode, the amplitude of the second pulse is half that of the first. This ratio is maintained over a 40 dB range. For the resolution test mode, each pulser is independently controlled. While in these test modes, the timing of the two pulses is also controlled by the operator by means of thumb wheel switches on the front panel.

Figure 4 — A pair of transducers is provided per ETB. These units can adapt to immersion, through transmission and shear wave testing. The holding fixture for the tested unit allows the operator free use of both hands.

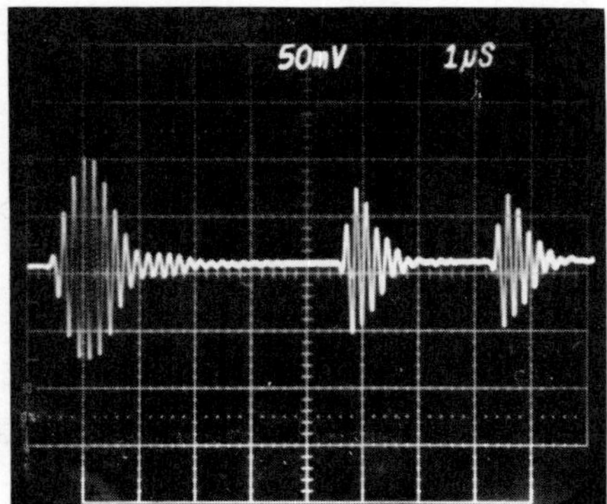

Figure 5 — Oscilloscope trace of incoming signal to ETB and two return echoes provided by ETB (from left to right).

Both the delay from the starting signal and the separation of the two pulses is so controlled. The principal reason for controlling the time and amplitude of the ETB is to be able to verify the ultrasonic instrument measurement precision under actual test conditions. Most instruments do not maintain the same accuracy throughout the entire dynamic range capability. For this reason the ETB was developed to make on-site assessments.

THE ELECTRONICS

There are four principal functional areas in the electronics package. These are the timing circuit, pulser network, operational logic board, and power supply.

The timing circuit comprises a hard start* 10 MHz pulsed oscillator and two sets of countdown counter chains. Upon receipt of an input pulse the oscillator is turned on. The counter chains in turn count down in 0.1 μs increments (1 cycle) from a preset number entered via the time control front panel switches. By this means an accurate time control is established in order to trigger the ETB's pulsers. The precision of the time measurement is principally dependent on the accuracy of the 10 MHz oscillator. This may be referenced, via established practices, to the National Bureau of Standards.

A unique pulser circuit is the heart of the ETB. The principal requirement is that the transducer be pulsed over a 40 dB amplitude range with no other changes occurring (i.e. frequency content, impedance, etc.). This is achieved by use of silicon-controlled rectifier (SCR) switches as shown in the simplified diagram (Fig 6). The main amplitude is controlled by the low-voltage amplifier

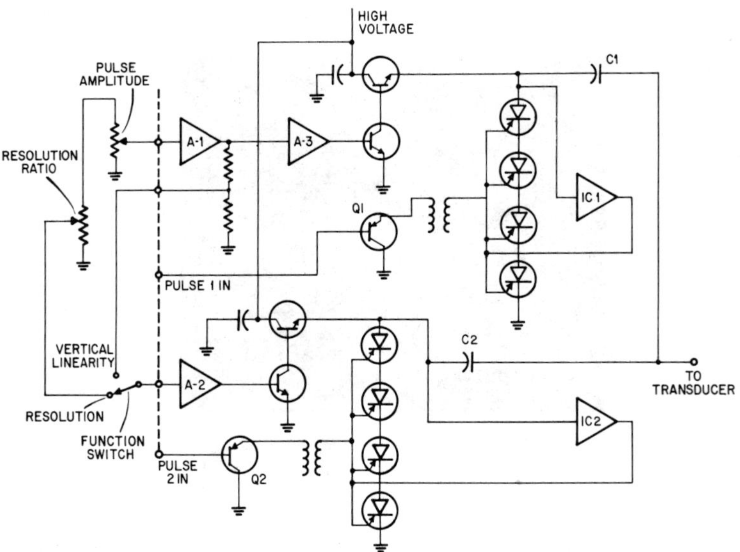

Figure 6 — Simplified pulse schematic. To the left of the dashed lines are the front panel controls.

*A proper current is kept flowing through the coil so that the turn on characteristics are controlled as to phase and amplitude. By this means the oscillator starting and running amplitudes are the same so no cycle count is missed.

A-1 via the front panel "pulse amplitude" potentiometer. The output of A-1 is independently amplified by each channel consisting of A-2 or A-3 and two transistors. By this means the pulse discharge capactitors C-1 and C-2 are charged to the desired voltage. When the SCR string is fired via trigger signals applied to Q-1 and Q-2, the transducer is pulsed by the discharge of the capacitors to ground through the SCR's. The SCR string is then temporarily turned off by means of a timed negative bias from IC-1 and IC-2 in order to prevent a refire or burn out. The second pulser's amplitude is independently adjustable by means of the front panel "resolution ratio" potentiometer when the "function" switch is set ot the "resolution" position.

The operational logic board consists of the input signal amplifiers and various flip-flop logic circuits. The received signal is amplified and, if above a preset level, an integrated circuit (IC) is triggered. This input port IC will start the 10 MHz oscillator and enable the first pre set timing counter chain (main delay). When the count reaches "0" the first pulser is activated and at the same time the second countdown chain is started. Upon reaching "0" count the second pulser is fired and, after a 10 μs delay, the two countdown chains are reloaded, the input port is enabled, and the clock is turned off. Power is supplied from internal lead acid batteries or an external ac supply. A special charger circuit provides the necessary current to maintain the battery. Either a 9 volt dc charge up to 1/4 ampere or a 10 volt dc charge limited to 1 ampere is supplied to the battery as conditions warrant. A special battery cutoff circuit is incorporated so as to prevent errors due to low voltage and to prevent damaging the batteries by complete discharge.

FIELD OPERATION

To assess the accuracy of the measurement to be made, the ultrasonic instrument is first adjusted by use of the proper reference standard for the test at hand. Next the instrument's transducer is coupled to the ETB's transducer. With the present model it is possible to test in the contact, immersion, and through-transmission modes of operation. The timing and the amplitude controls are then set on the ETB so that the two echoes appear on the the screen. If no pulses appear, then the ultrasonic instrument's transducer or pulser must not be strong enough to trigger the ETB. This is an arbitrary lower limit, but it does ferret out "weak" transducers and/or pulsers. If echoes appear, the test can now begin.

For amplitude linearity one merely observes the accuracy of the 2:1 ratio presented by the pulsers (Fig. 7) much in or the same manner as suggested in ASTM E-317 or MIL STD 271-E. However by using the ETB a significant improvement is offered. Instead of varying the ultrasonic instrument's gain setting (which in reality does not measure linearity*), the ETB's amplitude is varied

*Amplifier linearity is a measure of the linearity of the relationship between input and output at a constant gain. A change in the gain setting while observing the input to output relationship of two fixed input levels does not prove the linearity of the system (Fig. 8).

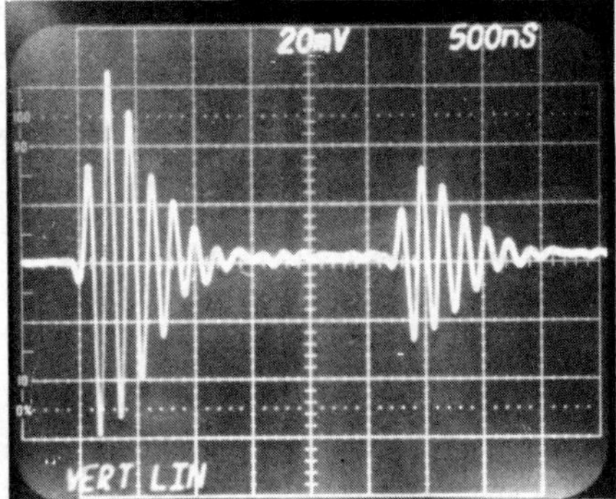

Figure 7 — Oscilloscope trace of the 2:1 ratio signal amplitude pulses returned by the ETB. This signal is taken from the ETB side of the transducer. A plot of dc voltage vs pulse output (no ETB transducer at pulses) or rf pulse output (ETB transducer at pulses) shows the relationship to be linear.

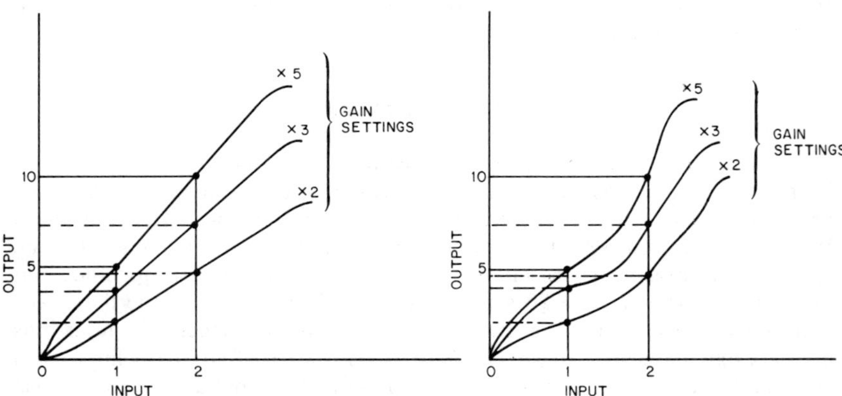

Figure 8 — A linear and nonlinear amplifier response is shown on the left and right drawings respectively. As shown, the two fixed pairs of input signals appear linear on the outputs for the various gain settings demonstrated. To properly verify the linearity of these amplifiers it is required that the two input signals (2:1 ratio) be varied (slide along the abscissa) over the desired testing range.

while maintaining the constant 2:1 ratio. By this simple means, a *true* linearity test is made. The time base linearity may also be examined by manually advancing the time domain controls. From experience it has been noted that most instruments do not maintain linearity for all combinations of transducers plus damping, gain, and attenuator settings. Conversely it is usually possible to find a combination of settings that will meet the test requirements. This is a very important point which unfortunately has not received enough attention. Such effects reinforce the futility of "testing" for linearity under the current practices. One can usually find a linear place, but will those be the same conditions for the actual test? In principle, if one needs to be assured of amplitude linearity then it must be measured on site under the actual test conditions.

The resolution is evaluated by observing the effects of moving the time domain having previously set the amplitudes of the pulses in a similar relationship as expected for that inspection (Fig. 9). The reason for this procedure has been demonstrated in a previous report (1). Simply stated, temporal resolution is a function of both time and amplitude. Current practice is to use a flat bottom hole (FBH) of known diameter and distance from the front surface of the specimen. Resolution is then described as the ability to display the echoes from the front surface and FBH to some specified criteria. The shorter a distance from the FBH to the specimen surface and or the smaller diameter FBH used is then a measure of "better" resolution. Unfortunately, a change in the relative amplitudes of the two signals may cause a change in the apparent resolution capability. A further point, less known but more vital, is the fact that the demonstration of a certain minimum resolution does not assure similar capacity for less stringent requirements (Fig. 10). Problems with both temporal and lateral resolution are really quite complex and have also not yet received the just attention deserved.

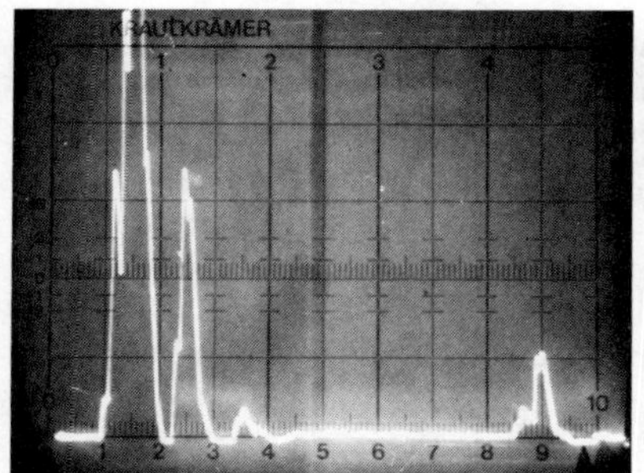

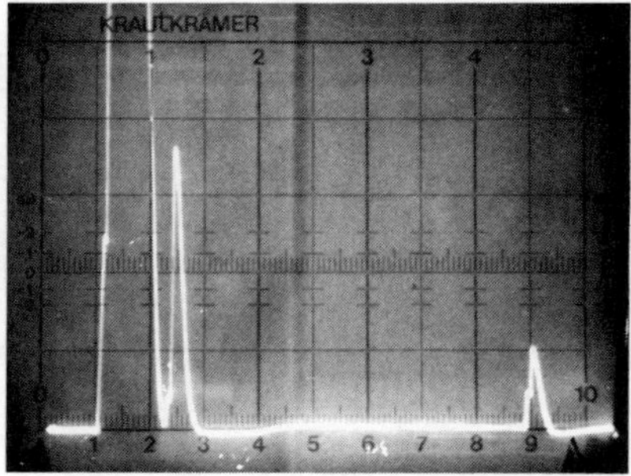

Figure 9 — At left, standard ultrasonic instrumentation resolving a 5 flat bottom hole near front surface during immersion test. At right the same result but emulated by ETB.

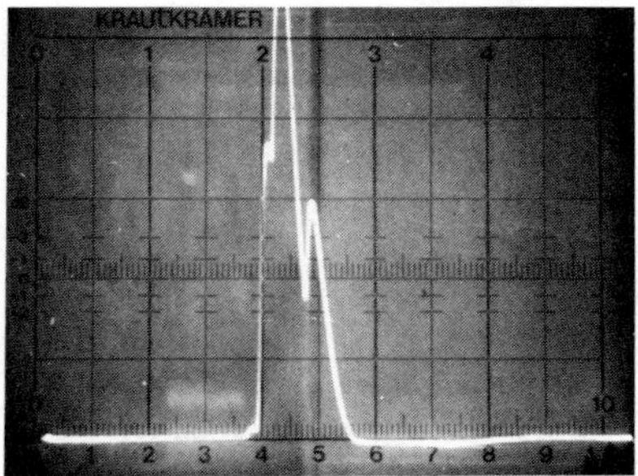

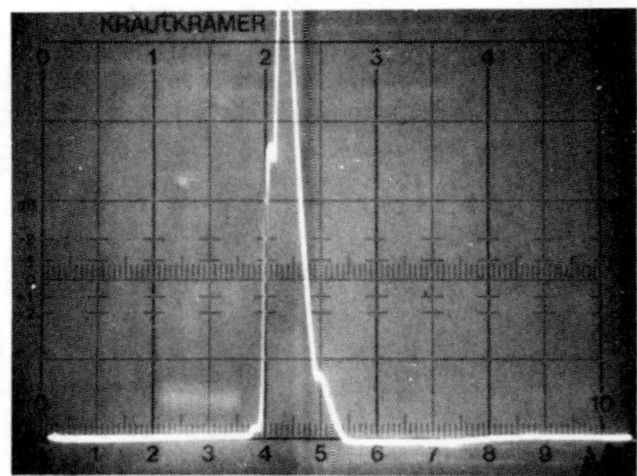

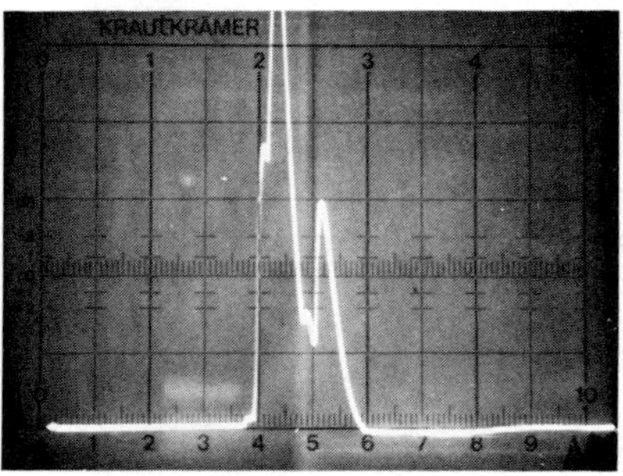

Figure 10 — ETB depicts resolution problems. Sequence from left to right shows the pulses separated by .6, .7 and .8 μsec. It should be noted that the relative amplitude of the signals will also determine the resolution capability.

CONCLUDING REMARKS

It has been shown that the accuracy of ultrasonic measurement may change due to two categorical variables: those related to the test technology and those related to the measuring instrument electronics. An Electronic Test Block has been developed as an aid in evaluating the tolerance degradation caused by the second kind of variable. With this device it is now possible to examine the precision of the measured quantities, time and amplitude. This may be accomplished in a very brief time and under actual test conditions. Such an evaluation will prove to be most beneficial for maintaining the reliability of ultrasonic inspection measurements.

REFERENCE

1. H.H. Chaskelis, "A Materials and Defect Simulator for Calibrating Ultrasonic Equipment used in Nondestructive Testing or Inspection," NRL Report 6984, Dec. 5, 1969.

CURRENT STATUS OF CODES AND STANDARDS FOR NUCLEAR STEAM SUPPLY SYSTEM COMPONENTS AND THEIR MATERIALS IN JAPAN (ACCEPTANCE STANDARD FOR NONDESTRUCTIVE EXAMINATIONS)

S. Onodera
The Japan Steel Works, Ltd.
Muroran, Japan

ABSTRACT

The current status and future trend of the Japanese Codes and Standards for nuclear steam supply system (NSSS) components and their materials are briefly reviewed herein from the points of the acceptance standard of nondestructive examinations (NDE) and the quality level of the products.

CODE FOR NSSS COMPONENTS AND THEIR MATERIALS IN JAPAN

In Japan, the Code for NSSS components and their materials has been issued as the national basis by the Ministry of International Trade and Industry (MITI) who has jurisdiction over the construction phase of nuclear power plants.

First governmental ordinance for NSSS in Japan was announced in 1965 by the Electric Utility Industry Law, and there are now two regulations which give specific technical requirements for NSSS, as below:

a) Ordinance No. 81 of MITI "Technical Requirements for Welding of Electrical Facilities (Nuclear and Nonnuclear)":
   Key points of the coverage are the requirements for welding and NDE for components.
b) Notification No. 501 of MITI "Technical Requirements for Construction of Pressure Vessels and Other Facilities (Nuclear only)":
   Key points of the coverage are the requirements for design, materials, fabrications and testing (including NDE) of components.

In the present paper, these two regulations are abbreviated as the MITI Code. The contents of the MITI Code are strongly influenced by the ASME Code.

ACCEPTANCE STANDARD OF NDE IN THE MITI CODE

Present Stipulation

The acceptance standard of the MITI Code is nearly the same as that of the ASME Code, except that the requirements for RT are more stringent than the ASME Code. The differences are comparatively shown in Tables 1 and 2.

Many occasions have been experienced in Japan where the utilities impose some additional requirements over the acceptance standard of the MITI Code to the fabricators and the material manufacturers. Examples are shown in Table 3. This is because that the utilities, final users of the materials, claim gradually severer standards than the previous ones through the experiences of the past material quality and that the materials supplied have been far better beyond the minimum requirements.

The quality levels of the large forgings and plates for RPV manufactured by the Japan Steel Works in recent five years (1973 to 1977) are shown in Figs. 1 and 2.

Trend in Future

The MITI Code is now in preparation of the third revision (the draft completed). As for NDE, the examination procedures and classification of defects detected by MT and PT which are specified in Japanese Industrial Standards (JIS) are adopted in this revision, although these acceptance standards are essentially the same as the present ones.

There remain still some areas to be studied. One is that the acceptance standards of RT are more strict than those in the ASME Code, while the acceptance standards of UT in both Codes are nearly the same. In order to establish a standard in which the same level of quality in both RT and UT is demanded, a committee composed of the Electric Power Generating Heat Engine Institute (one of the third party governmental inspection agencies approved by MITI), three utilities and four fabricators started investigations from April, 1977, with three years term.

On the other hand, no requirements for in-service inspection are included in the MITI Code. Though the Japan Electric Association (who provides the advices and suggestions for standards in electric industries as requested by MITI) made the draft of JEAC 4205 "In-service Inspection of Reactor Cooland Pressure Boundary" (dated March 1977), the official issue of the Rule is now suspended by MITI by the reason that this standard is much different from the MITI Code in the evaluation of the permissible defect size.

However, there remains no problem in the application of the in-service inspection procedure. The acceptance standard will therefore be handled by case-by-case basis, and JEAC 4205 is expected to be referred to in the MITI Code in the near future.

MATERIAL AND QUALITY REQUIREMENTS

As shown in Figs. 1 and 2, it can be said that the quality of the materials manufactured in Japan would be superior to the one required by the various Codes and Standards around the world.

Taking large ingot as an example, it is mainly due to the improvements of techniques in steel melting, ingot making such as multiple pouring method for reducing segregation, etc. Recently, the technique of VCD (vacuum carbon deoxidation), which would provide further improvement in the segregation of ingot, is becoming applicable for NSSS materials.

The requirements for material quality can be modified by the progress of steelmaking technology, which has been so remarkable in these two decades and would continue further. However, the material manufacturers are sometimes in perplexity because of inconsistency of NDE requirements for construction phase (pre-service) and operation phase (in-service). This inconsistency should be investigated and be put into a proper resolution.

Table 1 Acceptance standard for materials (UT)

		ASME Code, Sect.III	MITI Code, Notification No. 501
Forgings	Straight beam method	Frequency : 1-2.25 MHZ Sensitivity: Bl = 75 % Acceptance standard: Complete loss of back reflection (Bl≤5 %)	Same as ASME
Forgings	Angle beam method	Frequency : 1 MHZ Sensitivity: Notch reflection ... 75 % Depth of notch: 3 % of thickness, but 6.35 mm max. Acceptance standard: Over 100 % of reference curve	Frequency : 0.5-5 MHZ Sensitivity : Notch reflection ... 50 % Depth of notch: 3 % of thickness, but 9.5 mm max. Acceptance standard: Over 100 % of reference curve
Plates	Straight beam method.	Frequency : 2.25-5 MHZ Sensitivity: Bl = 75 % Acceptance standard: (a) Complete loss of back reflection which cannot be encompassed within a circle whose diameter is 76 mm or one-half of the plate thickness, whichever is greater. (b) Two or more defects smaller than described in (a) above shall be unacceptable unless separated by a minimum distance equal to the greater diameter of the larger defect, or unless they may be collectively encompassed by the circle described in (a).	Same as ASME
Plates	Angle beam method	To be applied for the plates 51 mm thick and less. Frequency : 2.25-5 MHZ Sensitivity: Notch reflection ... 75 % Depth of notch: 3 % of thickness Acceptance standard: Over 100 % of reference curve	Not to be applied.

Table 2 Acceptance standard for welds (RT)

ASME Code, Sect.III

The following defects are unacceptable.

1. Any type of crack, incomplete fusion or penetration.
2. Elongated indication which has a length greater than

t(thickness)	$t\leq 19$mm	$19<t\leq 57$mm	$57\text{mm}<t$
l(length)	6mm	1/3 t	19mm

3. Any group of indications in line that an aggregate length greater than t in a length of 12t except where the distance between the successive indications exceeds 6L, where L is the longest indication in the group.

4. Rounded indication in excess of that shown as porosity chart.

MITI Code, Ordinance No. 81

The following defects are unacceptable.

1. Any type of crack, incomplete fusion or penetration.
2. Elongated indication which has a length greater than

t(thickness)	$t\leq 12$mm	$12<t\leq 48$mm	$48\text{mm}<t$
l(length)	3mm	1/4 t	12mm

3. Any group of indications in line that an aggregate length greater than t in a length of 12t except where the distance between the successive indication exceed 6L, where L is the longest indication in the group.

4. Rounded indications in excess of the permissible sum of those specified in the following table.

Test field of vision (mm)	10x10	10x20			10x30
Thickness of base metal (mm)	$t\leq 10$	$10<t\leq 25$	$25<t\leq 50$	$50<t\leq 100$	$100<t$
Permissible sum	1	2	4	5	6

5. Rounded indication which has diameter greater than

D *	$D<25$mm	$25\text{mm}\leq D$
d(diameter)	0.2t (max. 3.2mm)	0.3t (max. 6.4mm)

* Distance between two indications

Table 3 Additional UT requirements for plates claimed by the customers

<table>
<tr><th></th><th>Straight beam method</th><th>Angle beam method</th></tr>
<tr><td>Customer A</td><td>Frequency : 3 MHZ
Sensitivity: Reflection from reference hole: 80 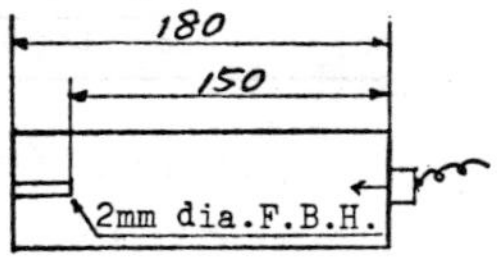 %
180
150
2mm dia.F.B.H.
Classification of defect shall be made by Fl to Bl ratio as follows;
<table>
<tr><td>Fl/Bl</td><td>50%</td><td>50-80%</td><td>80%</td><td>only Fl</td></tr>
<tr><td>Symbol mark</td><td>○</td><td>⊗</td><td>△</td><td>×</td></tr>
</table>
Acceptance shall be decided after discussion with the customer.</td><td>Frequency : 2 MHZ
Sensitivity: Reflection from notch ... 75 ± 5 %
Depth of notch: 2 % of thickness
Acceptance standard:
Over 100 % of reference line</td></tr>
<tr><td>Customer B</td><td>Frequency : 2.25 MHZ
Sensitivity: Primary reference level ... 75 %
Reference block: Reference block for welds of ASME Code Sect. III
Acceptance standard:
Any indication over 20 % of reference level shall be record and acceptance shall be discussed with the customer.</td><td>No additional requirement</td></tr>
</table>

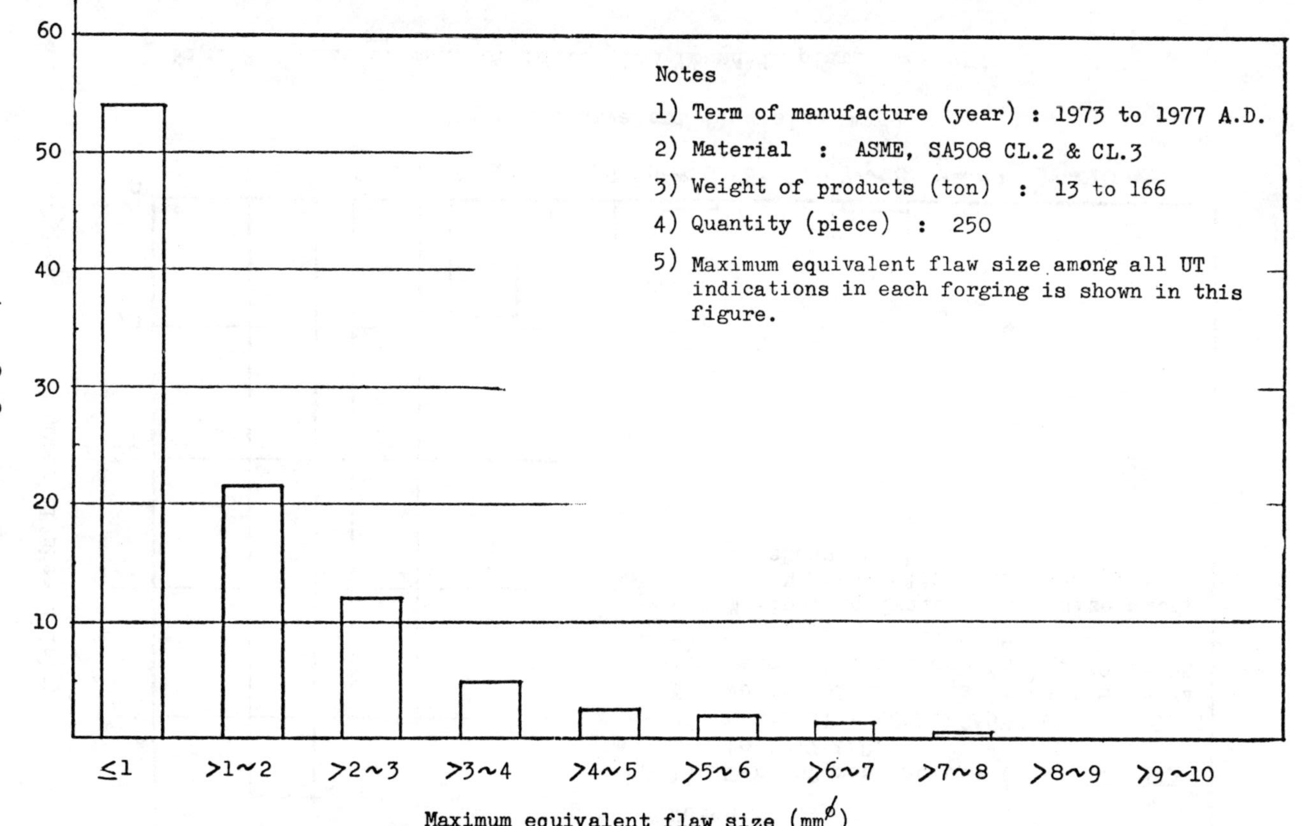

Fig. 1 Size of max. UT indication found in large forgings for RPV

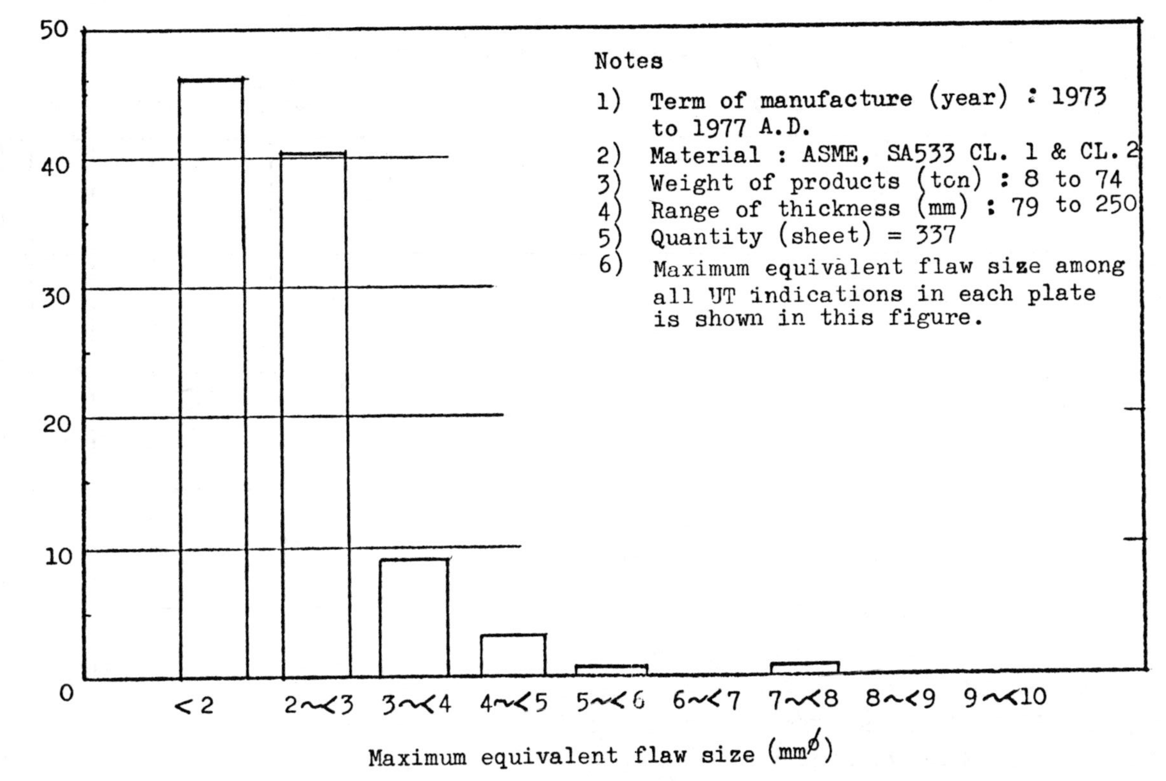

Fig. 2 Size of max. UT indication found in plates for RPV

A MULTIFREQUENCY EDDY CURRENT SYSTEM FOR INSPECTION OF STEAM GENERATOR TUBING (a)

T. J. Davis
Battelle-Northwest
Richland, Washington

ABSTRACT

Steam generators in nuclear power plants are sometimes subject to corrosion mechanisms in which the amount of degradation is difficult to accurately quantify with conventional techniques. This paper presents the results to date of a research program directed at using multifrequency eddy current inspection to improve assessment of material damage in steam generator tubing.

A prototype multifrequency test system has been developed and evaluated at Battelle-Northwest (BNW). The system can achieve discrimination against probe motion error signals and

---

(a) This report was prepared by Battelle-Northwest (BNW) as an account of work sponsored by the Electric Power Research Institute, Inc. (EPRI). Neither EPRI nor any member of EPRI, nor BNW, nor any person acting on behalf of either:

1. makes any warranty or representation, expressed or implied, with respect to the accuracy, completeness or usefulness of the information contained in this report, or that the use of any information, apparatus, method, or process disclosed in this report may not infringe privately owned rights;

2. assumes any liabilities with respect to the use of, or for damages resulting from the use of, any information, apparatus, method or process disclosed in this report.

against support plate indications when using two-frequency data analysis. The paper describes how this discrimination aids in:

- electronic assessment of flaw depth,
- detection of flaws in the U-bend transition region,
- inspection of tubes having ID variations, and
- detection of flaws in the support plate regions.

INTRODUCTION

This paper describes a four-frequency eddy current system which was developed by Battelle-Northwest (BNW) for the Electric Power Research Institute (EPRI). The overall objective of the program was to provide improved capabilities for inspection of steam generator tubes in nuclear power plants. This would enable utility operators to determine more quickly and with greater confidence which steam generator tubes should be plugged. Specifically, the objectives of this program were to: 1) improve the characterization of material damage presently achieved by the conventional single-frequency test, and 2) reduce the reliance on operator interpretation of signal information.

The paper deals primarily with the first phase of the research program which was directed towards evaluation of two-frequency testing combined with conventional differential probes.

SUMMARY

A multifrequency eddy current system based on the use of Walsh functions has been developed and has been evaluated on a steam generator mockup. The system records data from four test frequencies (100, 200, 300 and 400 kHz) and performs multiparameter signal processing on data from any selected pair of frequencies. The multiparameter processing permits electronic assessment of flaw depth between and directly under supports when using conventional probes. If a tight-fitting probe is employed to eliminate probe motion, the multiparameter separation capability which was used for probe wobble discrimination can be applied to total discrimination against supports. In this case, the system can both detect and size flaws located under the edges of supports. Such detection and sizing capabilities are not available with a single-frequency system.

Flaw size assessment capabilities of the system are shown in Figure 1. The recorded pulse heights are phase angles (depth information) of flaw signals produced as the probe was translated through a calibration tube which had machined flaws of the depths indicated. This type of real time electronic flaw characterization is not practical with the conventional single-frequency test because of probe wobble error signals. The multifrequency test has therefore demonstrated progress in removing visual data interpretation procedures from tubing inspection.

The system was evaluated in a steam generator mockup in July 1977. The evaluation was conducted by EPRI as part of a capability assessment of currently available inspection methods. Results indicated that in many cases multifrequency inspection had higher detection probability than a conventional single-frequency test at 400 kHz with ASME sensitivity, and that the system was capable of detecting defects in the U-bend transition region.

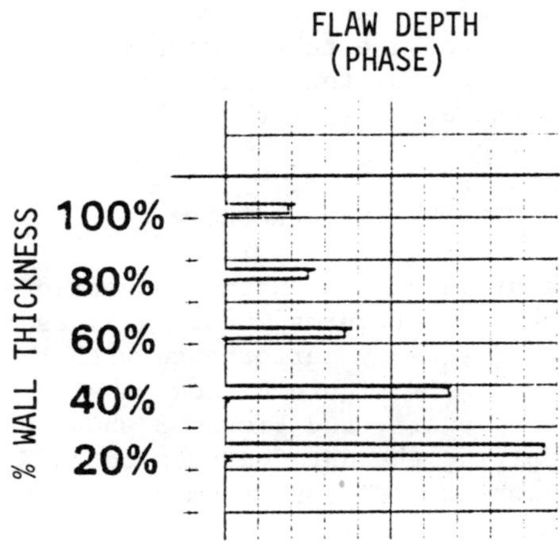

Fig. 1. Flaw Depth Output of System as Probe Was Translated Through a Calibration Tube with Flat Bottom Drill Holes of the Depths Indicated in Percent Wall Thickness.

Further development work on the system is being conducted under the continued EPRI program. This effort is directed at achieving the fullest practical utilization of the multifrequency method for inspecting steam generators. The work will center largely on the use of additional test frequencies and the development of specialized probes.

SYSTEM PERFORMANCE

In its present stage of development, the system can be operated by using either of two methods. The first of these employs a conventional differential bobbin probe of the type currently used for field inspection. The two-frequency multiparameter processing capabilities are employed to obtain discrimination against probe motion and/or pilgering (ID variations), and partial discrimination against supports. In this case, electronic assessment of flaw depth is performed between and directly under supports, while only presence or absence of a flaw can be detected at the edge of a support.

In the alternate method of operation, a tight-fitting or self-centering differential probe is used to reduce probe motion. All the capabilities of two-frequency multiparameter processing are then available to obtain total discrimination against supports, and sizing of flaws can be performed along the composite length of the tube. However, this method of operation does not provide the ability to discriminate against ID variations or the effects of the U-bend transition region.

PERFORMANCE WITH CONVENTIONAL PROBES

Performance of the system using conventional field inspection probes is shown in Figures 2 and 3. These data show a comparison of two-frequency performance with that of a single-frequency 400 kHz test. Figure 2 shows the data outputs as recorded on a strip chart, and Figure 3 shows the result obtained from applying the data outputs to an X-Y oscilloscope. Figure 3 also shows the response from a simulated tube support.

The major effect shown by the data is that the two-frequency test has eliminated probe motion indications so that all the X-Y Lissajous patterns originate from the same point. This permits the system to perform electronic measurement of the rotation angle (phase) of each figure-eight flaw pattern such that a flaw depth output is generated. In contrast, the

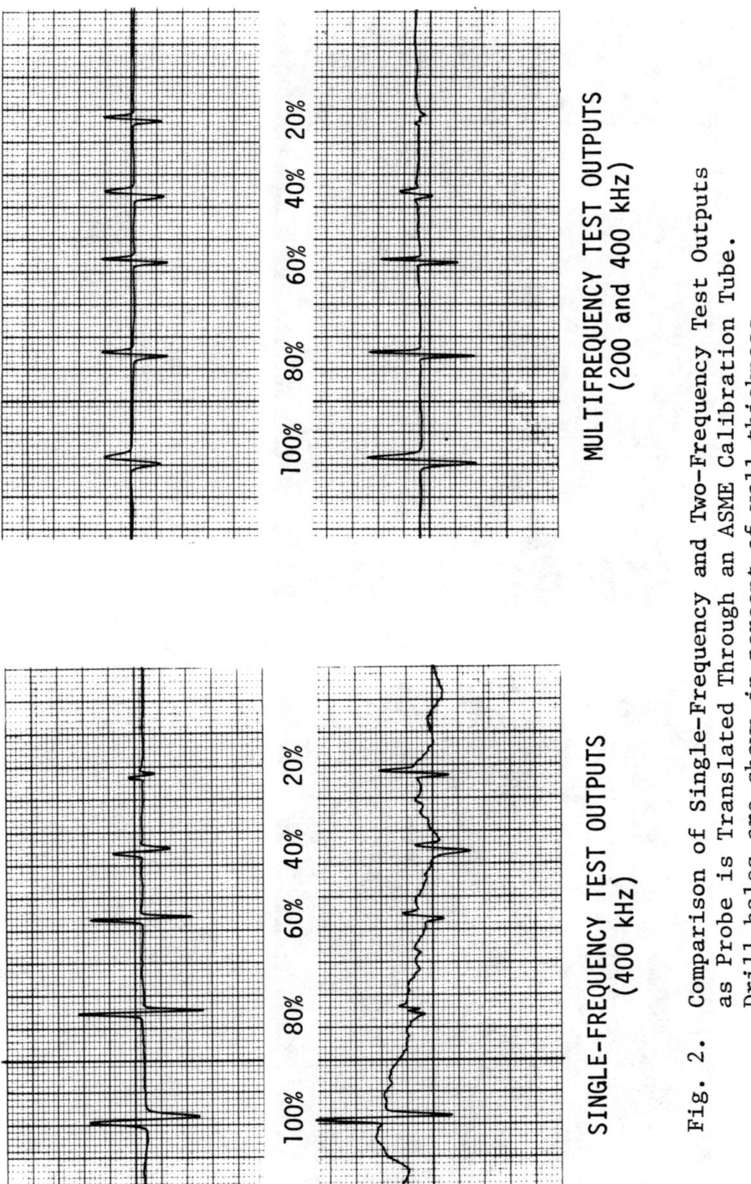

Fig. 2. Comparison of Single-Frequency and Two-Frequency Test Outputs as Probe is Translated Through an ASME Calibration Tube. Drill holes are shown in percent of wall thickness.

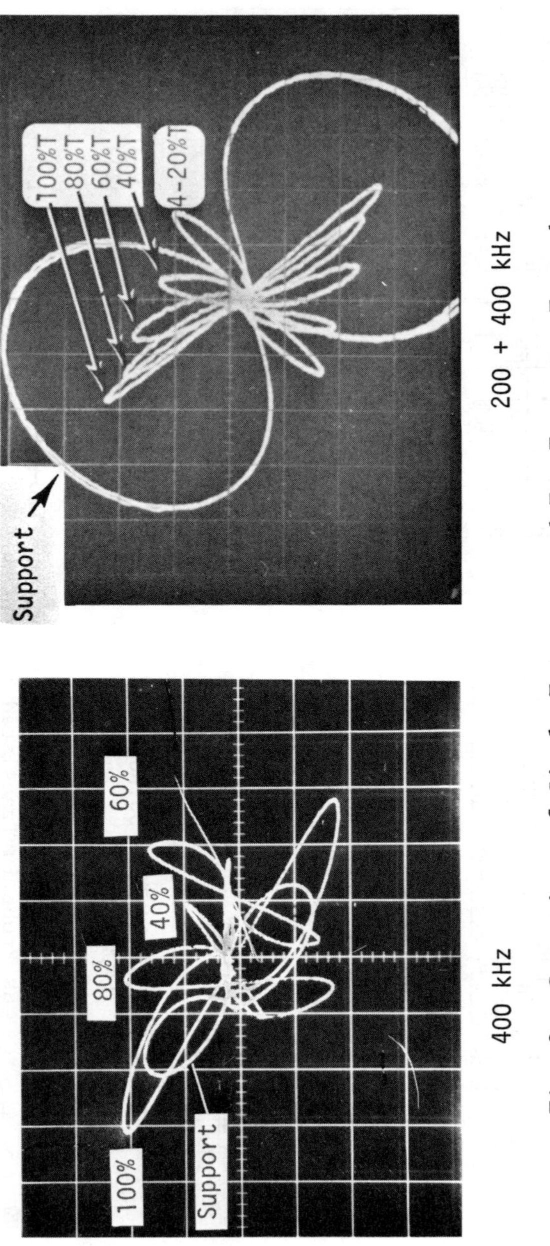

200 + 400 kHz

400 kHz

Fig. 3. Comparison of Single-Frequency and Two-Frequency Tests by X-Y Oscilloscope Presentation. Probe was translated through an ASME calibration tube with a simulated support. Defect depths are indicated in percent of wall thickness.

probe motion effects as shown in the single-frequency data preclude electronic measurement flaw depth. As a result, flaw depth assessment in a single-frequency test is done by first storing the figure-eight flaw pattern on the face of a memory oscilloscope and then measuring its phase by an electromechanical method.

Two multiparameter processing modes are peformed simultaneously when the system is set up to use conventional probes. The first of these, called the normal mode, is used between supports and directly under supports for electronic assessment of flaw depth. This mode is used in all cases except when a support edge signal is present at the system outputs. The normal mode processing electronically measures the phase angle represented at the peak amplitude of the wobble-free flaw signal and outputs this phase value for strip chart recording. (The flaw depth output for an ASME calibration tube is shown in Figure 1.) These values can be read directly off the strip chart for flaw depth assessment. Additionally, they can be printed by a digital printer, and they are suitable for acquisition by digital data processing systems.

System evaluation has shown that electronic flaw sizing is feasible for medium and large wastage flaws of the type most commonly found in steam generators. However, for small wastage flaws, such as an isolated stress corrosion crack, more accuracy is obtained by taking the peak amplitude value directly from the strip chart and calculating the arc-tangent to establish flaw depth. In either case, all flaw depth decisions can be made directly from strip chart data without having to pay recorded magnetic tape data into an X-Y oscilloscope for visual observation.

In the event that the system registers an indication from the edge of a support, an alternate multiparamter processing mode is used. This is called the support mode, and the corresponding processed outputs are shown in Figure 4. Under this processing mode, the support indication is reduced to a straight line located entirely within the horizontal channel, and the vertical channel is free of support indications. The vertical channel is then used to indicate presence or absence of a flaw under the support edge, while the horizontal channel automatically indicates presence or absence of a support edge signal. This information allows the data interpreter to ignore flaw depth indications if a support edge is present. This prevents a support edge from being interpreted as a flaw.

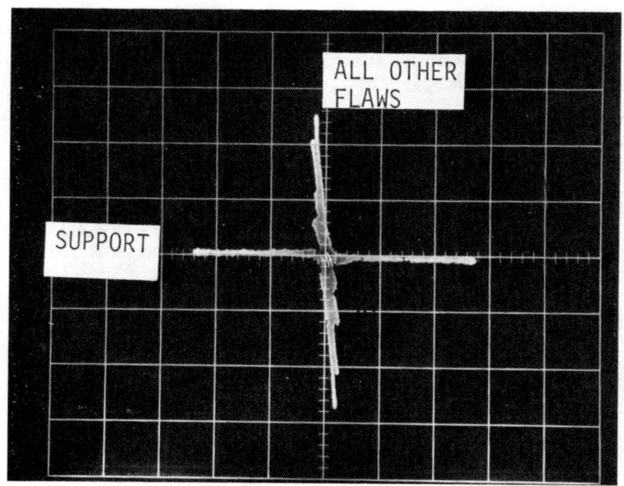

Fig. 4. Processed Signal Outputs for Support Mode when Inspecting an ASME Calibration Tube with Simulated Support.

PERFORMANCE WITH MODIFIED PROBES

Data are shown in Figure 5 for the alternate method, in which a tight-fitting probe is employed to reduce probe wobble. In this case, all of the two-frequency multiparameter capability can be employed for support discrimination and the support signal can be eliminated. It is then possible to size flaws electronically anywhere along the tube with the possible exception of the U-bend transition regions. The probe used for this evaluation is a conventional U-bend probe with shim material incorporated to provide centering of the probe inside the tube.

STEAM GENERATOR MOCKUP TRIALS

An evaluation of the system in its then-current state of development (conventional probes and two-frequency processing) was conducted in July 1977. The evaluation was performed by EPRI as part of a round-robin test to assess inspection capabilities currently available to the nuclear power industry.

Results of the evaluation indicated that in certain cases, the two-frequency test has a higher probability of detection for small wastage flaws than does a conventional single-frequency test at 400 kHz with ASME sensitivity. Results of other tests performed during the evaluation indicated that this

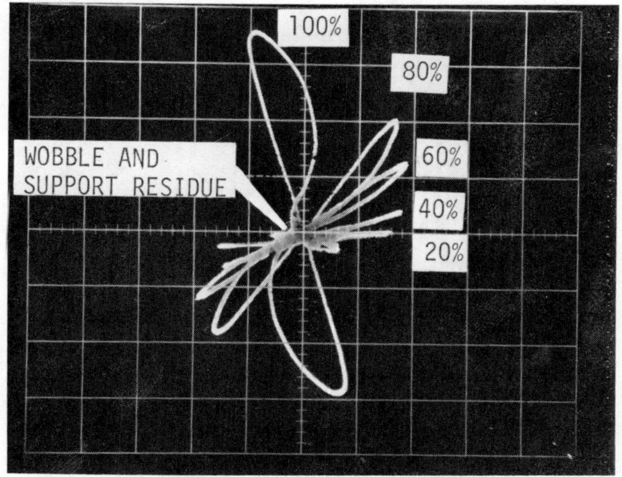

Fig. 5. Response of BNW-EPRI Eddy Current System as Adjusted for Total Support Discrimination Using a Tight-Fitting Probe. The probe was scanned past standard ASME calibration flaws and a simulated support using a combination of 200 and 400 kHz.

higher probability of detection could be due primarily to the added sensitivity to small OD flaws inherent in the 200-kHz inspection data. Probe wobble discrimination also aided in detecting and sizing these flaws. Detailed discussion of mockup test results will be the subject of separate EPRI reports (1,2).

Both intergranular stress corrosion OD cracking and defects in the U-bend transition region were detected and sized during the evaluation. Figure 6 shows detection probabilities for various flaws as calculated by the EPRI evaluation team.

SYSTEM DESCRIPTION

An abbreviated block diagram of the multifrequency inspection system is shown in Figure 7. The system consists of two instrument packages. One package is intended for in-containment acquisition of tubing data; the other is intended for post-analysis of the data. A summary description of the system is included in this paper. A more detailed description is contained in another report (3).

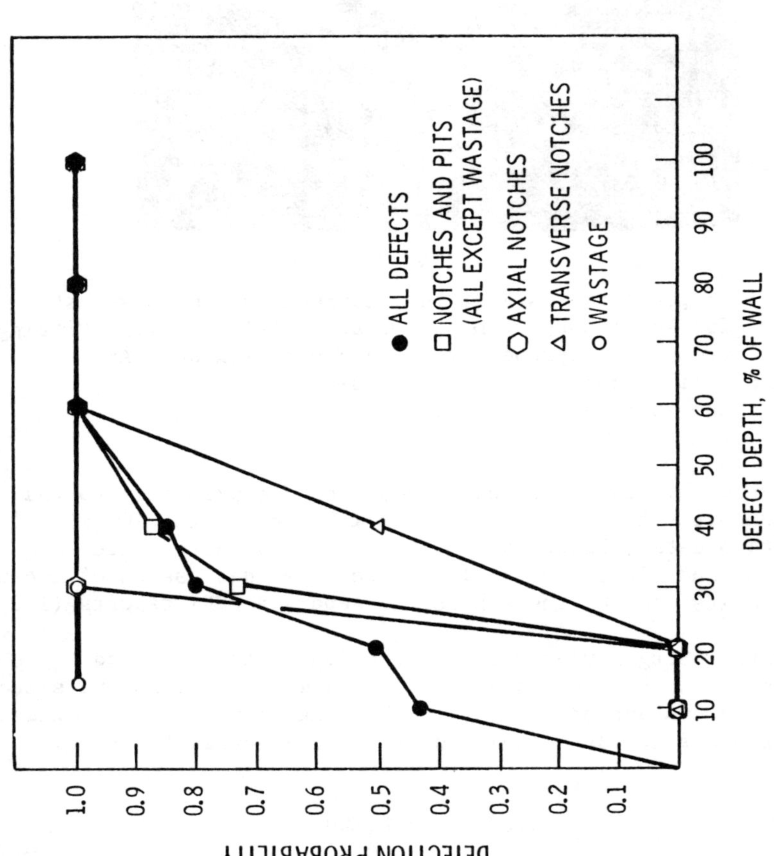

Fig. 6. Multifrequency Detection Probability Versus Defect Depth.

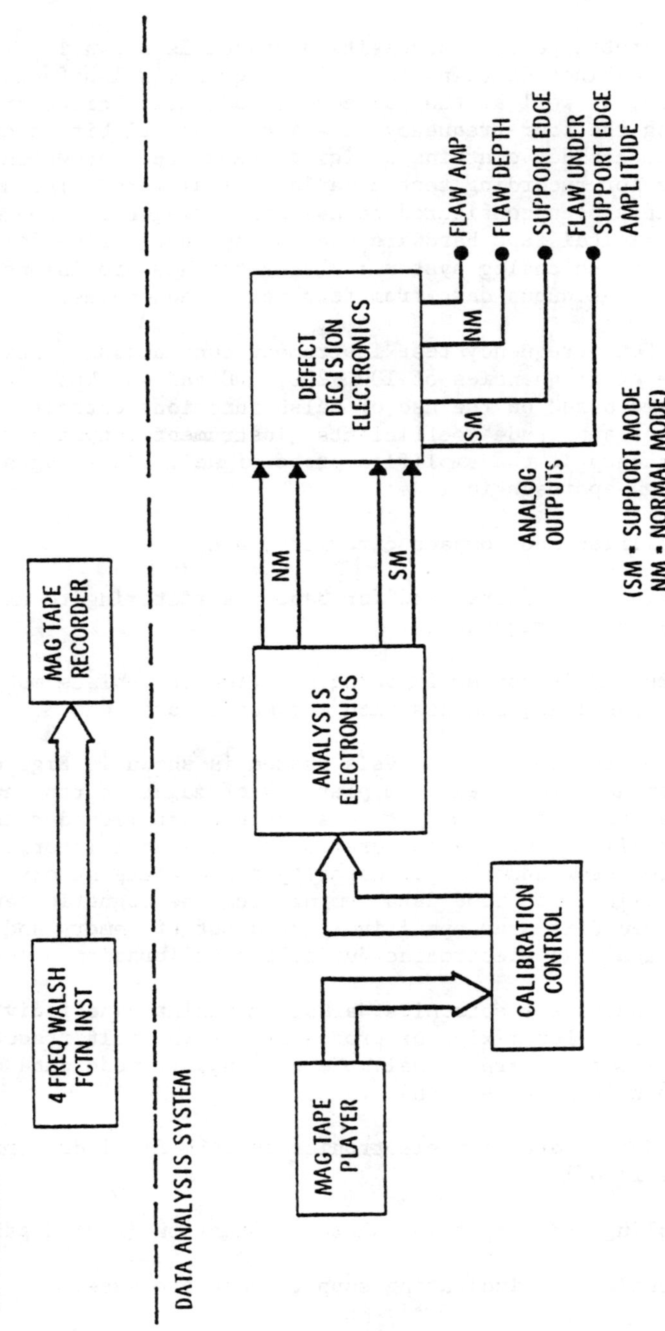

Fig. 7. Block Diagram of Multifrequency Inspection System.

The prototype data acquisition system is shown in Figure 8. The upper cabinet contains the multifrequency Walsh function instrument, as well as the pulse code modulation circuitry for digitizing the four-frequency data into a serial bit stream. The bottom cabinet contains a digital tape recorder which was evaluated for recording test data in digital form. The system has recently been configured to use analog magnetic tape in place of digital tape hardware due to the superior performance provided by the analog system. The system has provisions for recording continuous data from four test frequencies.

The four-frequency test instrument continuously excites the probe at frequencies of 100, 200, 300 and 400 kHz. A detection scheme based on the use of Walsh functions extracts the two Fourier amplitude coefficients (instrument outputs) from each frequency in the amplified probe signal. Advantages of this design approach include:

- simplified and compact circuitry, and
- elimination of the need for bandpass filtering to extract individual frequencies.

The instrument is currently being modified to replace most of the front panel adjustments with automatic functions.

The prototype data analysis system is shown in Figure 9. The system acquires data from playback of magnetic tape and processes the data for output to a strip chart recorder and monitor oscilloscope. A calibration control unit incorporates a microprocessor and a large capacity solid state memory for storing calibration tube data coming from the magnetic tape. These stored data are repetitively read out of memory and played into the analysis electronics during the calibration process.

The analysis electronics package contains nine individual phase rotators for mixing or processing of the multifrequency data. Its outputs are normal mode and support mode data of the types shown in Figures 3 and 4.

The defect decision electronics perform final data processing in terms of:

1. sampling and outputting phase information (flaw depth),
2. detecting and indicating support edge presence,
3. detecting absolute value of the flaw amplitude indications.

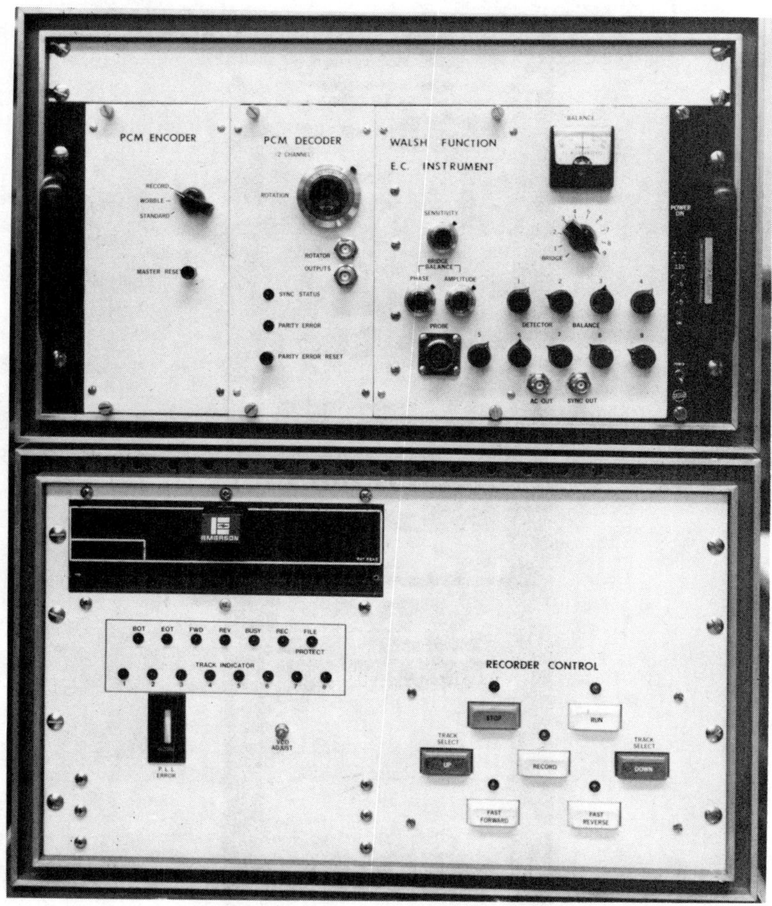

Fig. 8. Data Acquisition Portion of System.

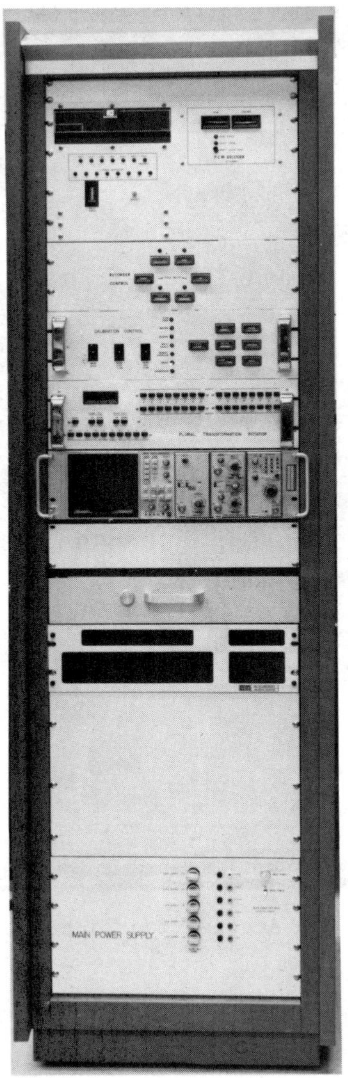

Fig. 9. Data Analysis Portion of System.

4. detecting support-free amplitude indications from flaws located under a support edge, and

5. providing automatic gain control for amplifying small indications and reducing large indications.

CONCLUSIONS

A prototype four-frequency eddy current instrument has been developed and evaluated for inspecting steam generator tubing using conventional probes and a selected pair of frequencies (200 and 400 kHz). Evaluations were conducted both on a laboratory scale and in a steam generator mockup. Major conclusions drawn by BNW from both the work reported in this paper and from the overall program are:

- Two-frequency inspection with conventional probes substantially reduces the complexity of data interpretation. Most data can be interpreted from strip chart observation, thereby minimizing the need for playback of magnetic tape data into an X-Y oscilloscope and for performing subsequent observation of flaw patterns.

- Automatic detection of flaw depth is practical for medium and large wastage flaws. Manual interpretation is required for small wastage flaws such as stress corrosion cracks, and for compound flaws or closely spaced multiple flaws.

- Multifrequency inspection has higher detection probability for most flaw geometries than does a single-frequency test.

- Both detection and sizing of flaws located under support edges can be performed if a tightly suspended or self-centering probe is used to eliminate probe motion indications (liftoff).

- Multiparameter compensation for probe motion plays an important part in reducing the effects of ID variations such as pilgering, and also aids in detecting flaws in the U-bend transition region.

- Detection of flaws in dents is marginal using conventional probes. New probe designs and/or three- and four-frequency testing will be required to inspect dented regions.

- Detection of circumferentially oriented flaws with conventional probes is marginal with either single-frequency or two-frequency testing. Alternate probes will be required to improve detection of these flaws.

- Certain types of muliparameter data processing result in compression of the phase angle versus flaw depth response and, hence, in a loss of accuracy over the single-frequency test in assessing flaw depth. (This compression of phase is prevalent in Figure 3.)

FURTHER DEVELOPMENT

Further work is being conducted under a continuation of the EPRI research program. This work is directed at obtaining the maximum possible utilization of multifrequency methods for inspecting steam generator tubing, simplification of system operating procedures, and evaluating multifrequency methods for solution of new industry problems as they occur. Particular emphasis will be placed on inspection of dented regions, adequate detection of circumferential cracks, sizing of ID flaws, and detection of cracks in the support plate. Primary technical areas being investigated are the use of three- and four-frequency data processing and the development of specialized probes.

REFERENCES

1. S. D. Brown and E. R. Reinhart, "An Evaluation of PWR Steam Generator Tubing Inspection Methods," Presented at the Second International Conference on Nondestructive Evaluation in the Nuclear Industry, Salt Lake City, Utah, February 13-15, 1978.

2. E. R. Reinhart, "Use of Round Robin Tests to Determine Eddy Current System Performance," Proceedings of the Workshop on Eddy Current Standards, National Bureau of Standards, Gaithersburg, Maryland, November 3-4, 1977.

3. T. J. Davis, Multifrequency Eddy Current System for Steam Generator Tubing Inspection, Vol. I, Progress Summary, EPRI NP-758. Electric Power Research Institute, Palo Alto, California, February 1978. This report may be requested from the EPRI Research Report Center, P.O. Box 10090, Palo Alto, California 94303.

A DC SATURATION EDDY CURRENT DEVICE FOR INSPECTION OF MAGNETIC MATERIALS WITH I.D. PROBE COILS

Robert A. Brooks
Donald N. Bugden
Magnetic Analysis Corporation

This presentation describes an eddy current technique for the inspection of magnetic tubing installed in heat exchangers, through the use of internal probe coils. Similar methods are, at present, extensively utilized throughout the nuclear industry for the in-service inspection of non-magnetic tubing. Examples are Inconel 600 in the steam generators, austenitic stainless steel in the feed water heaters and various copper alloys in the surface condensers. Problems associated with the extension of these techniques to magnetic material will be considered in addition to a description of the apparatus required. The application of this equipment to various magnetic materials has produced results which occasionally were predictable but also others that will require further analysis. These results will be presented and explained, and possible additional applications indicated by the results of this program will be discussed.

SATURATION, PERMEABILITY AND PARTIAL SATURATION

The application of saturation with an eddy current test system to magnetic tubing can easily provide a dramatic improvement in results. This initial improvement is attained with such relative ease and at such low field strengths, even in carbon steel, that one is left with the false impression that true saturation is really not necessary. Since 200 - 300 Oersteds is all that is required to reduce the noise, why bother applying more? This question is properly answered by considering what is meant by saturation in the first place.

There are three types of permeability; relative, differential and incremental, and all of them are merely slopes of a curve. They are values of proportionality between, B, the flux density, and H, the field intensity. Of these three, only the incremental permeability has any bearing in eddy current testing. That B is multi-valued with respect to H is illustrated in Figure 1 which shows the effect that is produced in magnetic material when a biasing field is added to the small alternating field of the test coil. The value of H is proportional to the ampere-turns/unit length of the coil and is the driving force that eventually causes the material to saturate. As the current is increased, the value of flux density will also increase following the magnetization curve of the material. This is shown by the line joining the small minor hysteresis loops. Should the current be turned off or reduced then the flux density will decrease, following the appropriate minor loop, to the new value. Should the current be repeatedly increased and then decreased it will continue to simply follow this minor loop. The slope of this loop is the incremental permeability where:

$$\mu_\Delta = \Delta B/\Delta H$$

Note that the slope of each loop decreases as the field intensity increases with the value approaching unity at saturation. This is one of the definitions of the term saturation.

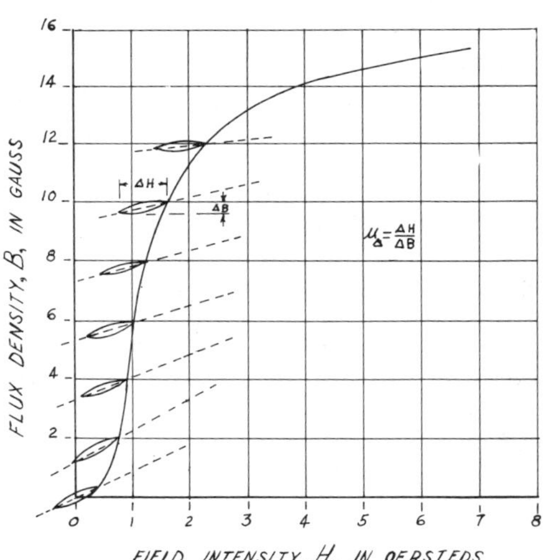

Fig. 1. Flux Density vs Field Intensity showing the change of Incremental Permeability.

Stresses within the material will cause a shift in the magnetization curve and thus alter the value of the incremental permeability. Since these stresses are not distributed evenly throughout the length of the material then there is a continual change of the permeability. A high noise level will be observed when it is tested with eddy currents. This condition is further complicated when a defect is introduced into the material.

The need to insure operation at, or near saturation is aptly illustrated by the next two figures which include four recordings of a 1" x .065 wall Ebrite (26-1) tube. Each recording was made at a different current level so that the complete series is a study of the effect of incremental permeability. This tube has four artificial defects: a 1/8" thru hole, a 1/16" thru hole and two filed notches .008" and .013" respectively. A comparison of Figure 2(A) recorded at 15 amps with Figure 2(B) recorded at 6 amps shows that the indication for the 1/16" hole is nearly eliminated.

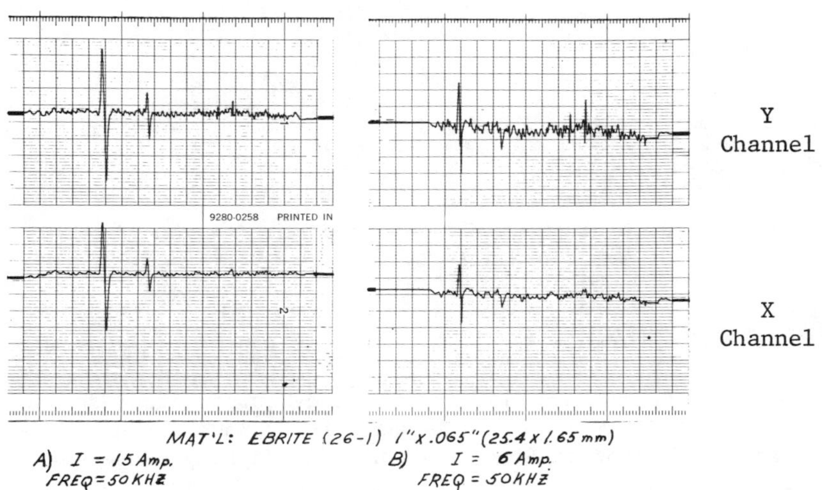

Fig. 2. Recordings of Ebrite (26-1) at Probe Currents of 15 and 16 Amps.

Further comparison with Figure 3(C) at 3.5 amps shows that the hole indication has once again increased but the notches which were just detectable at 15 amps now are similar to the 1/16" hole in amplitude. The noise level has also increased. Figure 3(D) at 2 amps shows still further change. Even the frequencies of the noise for the recordings at the two lower current levels are different. Since that much change has been demonstrated by variation of the degree of saturation, it would not be advisable to consider testing at less than 12 amps. In general, testing at less current than required for saturation should only be attempted after careful examination at other current levels. The effect upon performance can then be determined. Simply reducing the noise level is not enough.

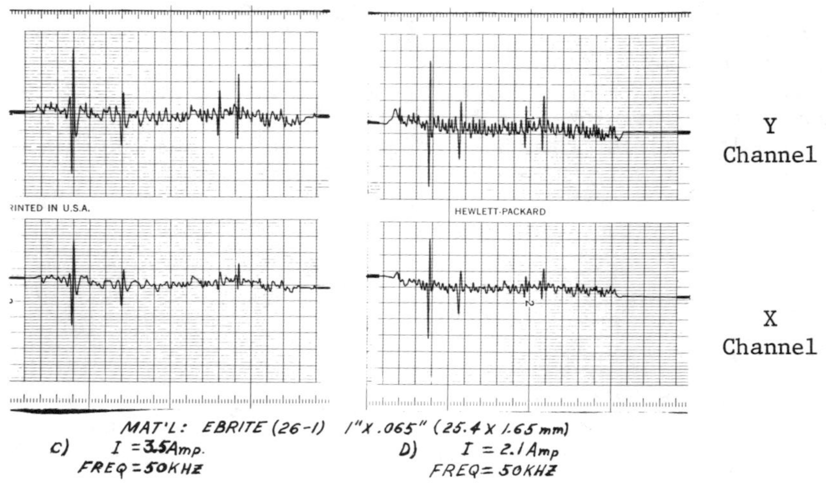

Fig. 3. Recordings of Ebrite (26-1) at Probe Currents of 3.5 and 2.1 Amps.

PROBE AND SYSTEM DESCRIPTION

It has been common practice to eddy current test carbon steel and other types of ferro-magnetic tubing, during manufacture, utilizing external magnetic fields to produce saturation. Essentially the same basic theory is used when applying this technique to testing this material with internal probe coils. However, saturation is much more difficult to attain, for in this case, the field must be produced by current flow in a coil that is limited to the internal cross section of the tube. This physical limitation, in addition to the tube's magnetic characteristics, ultimately determines the design of the probe. An effective means must then be provided to couple the magnetic flux lines from the coil of the probe to the material, while at the same time providing a sufficient number of flux lines to attain saturation.

In order to achieve good coupling, thereby concentrating the flux lines and reducing the leakage flux to a minimum, a core of Vanadium Permendur is provided as shown in Figure 4. The broken lines represent the lines of flux, with the mean path as indicated along the center line of the core through the pole pieces and then across the air gap into the material. If a flux density of 20-21 Kilogauss is to be maintained in the material, assuring saturation, then the air gap must be reduced to as small a value as possible. This requires a diameter difference of .010" - .020" for probes less than 1" O.D. and .020" - .040" for probes that are larger than 1" O.D. The cross hatched area indicates the location of the primary which supplies the magnetic field. Located above the primary are two detector coils.

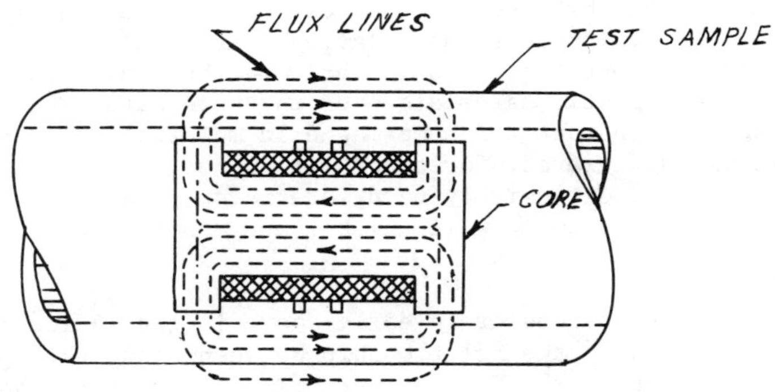

Fig. 4. Magnetic Circuit of Probe Core and Tube Sample.

The saturation of high permeability material such as 1010 carbon steel requires a magnetomotive force of 2500-5000 ampere turns with a corresponding current of 25-1000 amps. The actual value depends upon the diameter of probe required. In order to accommodate this amount of energy, a high power pulse is applied to the primary with sufficient magnitude and duration to saturate the material. Compressed air is provided for cooling. The duration and duty cycle of the pulse is dependent upon the the dimensions of the tube and its magnetic characteristics. The repetition rate is limited by the ability to cool the probe coil. Since this rate can vary from 10-80 pulses/second then the maximum speed of test will vary from 10-80 feet/minute. Sample and hold circuits are provided to insure stability regardless which pulse rate is chosen, and a vector presentation is supplied for readout. Just as the pulse rate limits the test speed, so too, does the force of attraction of the probe impede its movement. This is particularly true if continuous current is applied. Therefore, the liability in the pulsed system which restricts thruput speed, may be traded off against the advantage of being able to move the probe at all. The maximum attainable speed of a continuous saturation system may be substantially reduced because of the attraction between the magnetized probe and tube.

SYSTEM PERFORMANCE FOR NICKEL AND CARBON STEEL

The in-service inspection of steam generator tubing with eddy currents was greatly influenced by the apparent ability to measure the residual wall thickness of the Inconel. This was accomplished by observing the amount of phase shift produced by the O.D. defect and relating this value to is depth. Basic eddy current theory would indicate that this would also be true for magnetic materials once they are saturated. Actual tests have shown that for some magnetic materials this conclusion is valid, however, for others it is not. Unfortunately, one of the most extensively used materials, carbon steel, falls in the latter category, while nickel and monel behave as theory dictates.

These differences are clearly shown in the following two figures which show the relationship between phase shift and depth of defect at 25 KHz. The artificial defects in both the 3/4" x .085 carbon steel tube and the 1" x .065 nickel tube were flat bottom holes 1/8" diameter drilled to a depth of 80, 40 and 20% of the wall and one .060" thru hole. The familiar pattern of a succession or overlapping loops at various phases is demonstrated for the nickel sample in Figure 5. The .060" thru hole has been rotated in the direction of the y axis. The pattern for the carbon steel sample, however, consists simply of loops of varying amplitude and width, super imposed upon one another as shown in Figure 6. This pattern has been repeated over the entire frequency range of 10-100 KHz. One can only surmise that the reason for this difference is related to the cross-sectioned area and the fact that even near saturation the permeability factor dominates the test for carbon steel. This fact, although disappointing and troublesome, is still only a limitation. It does not nullify the capability of the test. The signal to noise ratio is still excellent for many applications.

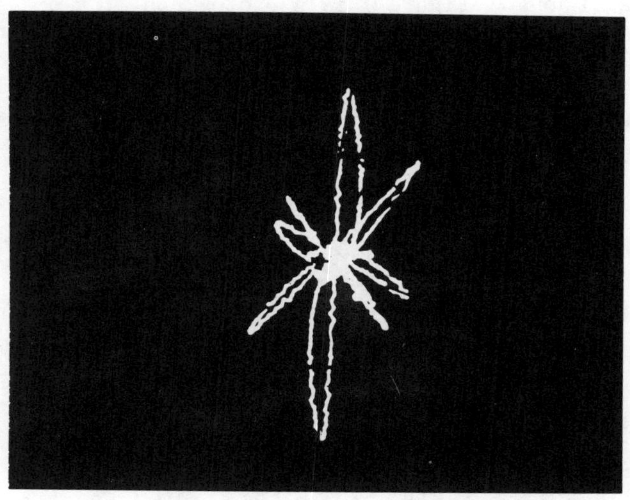

Fig. 5. Oscilloscope Pattern for a Nickel Tube (1" x .065) with O.D. Defects of Various Depths.

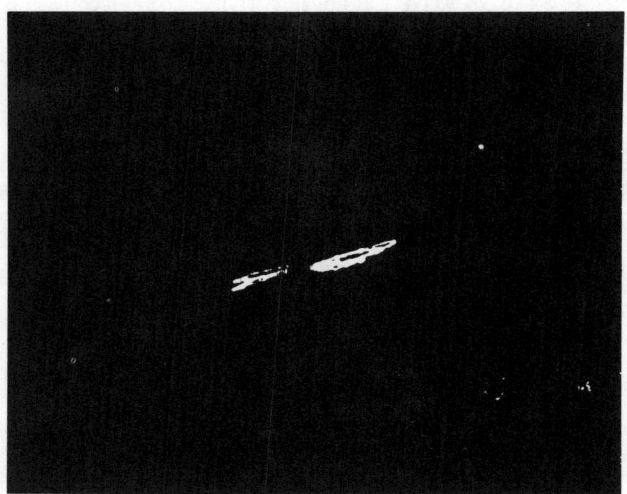

Fig. 6. Oscilloscope Pattern for a Carbon Steel Sample (3/4" x .085) with O.D. Defects of Various Depths.

Some of these applications will be demonstrated by the following figures. Figure 7 is a photograph of a 6" section of a 2-1/2" x .120 wall carbon steel tube which contained extensive O.D. corrosion. This defect extended ober the length of the sample and covered an area 1" to 1-1/2" wide. It was very deep and in some locations extended through the wall. Those small areas are noted by their very black color.

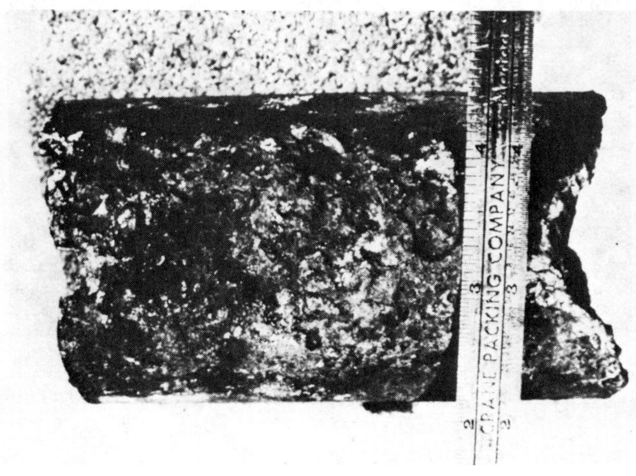

Fig. 7. Photograph of a 2-1/2" x .120 Carbon Steel Sample with Extensive O.D. Corrosion.

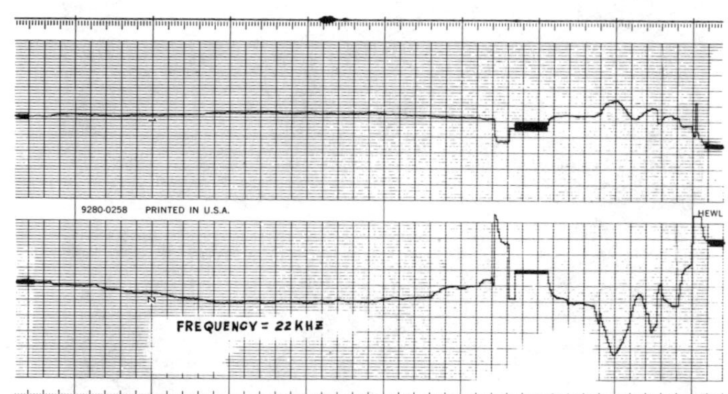

A) MAT'L: 1010 CARBON STEEL 2" x.120" (50.8 x 3.04 mm) 2' Long New Tube.

B) 6" Defective Tube OD Corrosion.

Fig. 8. Recordings of Two 2" x .120 Carbon Steel Samples showing the response for both Corroded and Noncorroded Sections.

A two channel recording of this tube sample in comparison with the test of a 2' section of a new tube is shown in Figure 8. The phase of the instrument was adjusted so that the indication for a 3/8" diameter hole 50% of the wall was rotated entirely in the X channel. Note that the corrosion caused a considerable phase shift.

An example of moderate O.D. pitting in a 1" x .093" wall carbon steel tube is shown in Figures 9 and 10. Two samples were provided for comparison. One of these samples was in relatively new condition, and the other was extensively corroded. Identical groupings of holes of various sizes had been drilled into the O.D. of both of them. The sizes and number of holes drilled in each group are noted below each recording. Some of the holes were through the wall and others only 50% deep. Figure 9 is a recording at a sensitivity of 66 while Figure 10 shows recordings 10 db higher. This was done to show the 1/32" hole and to accommodate overdrive of the recorder. Notice that the 1/16" hole can be detected in both tubes while the 1/32" diameter hole can only be detected in the new sample. The corrosion obscures the indication.

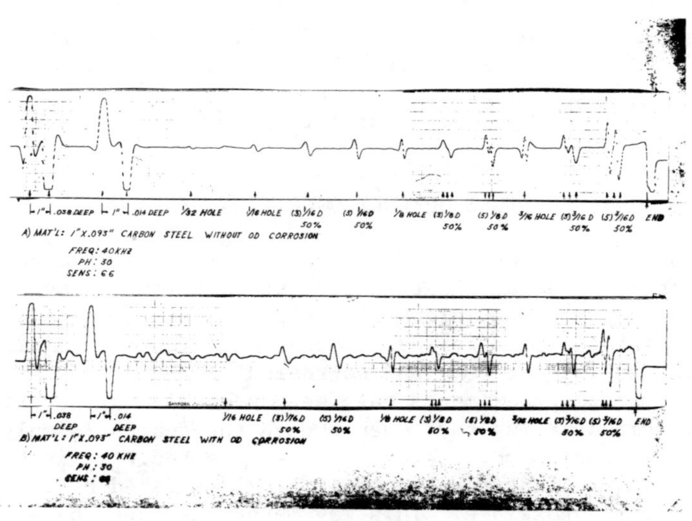

Fig. 9. System Response for Two Samples of 1" x .093 Carbon Steel at a Sensitivity of 66 db.
(a) Without O.D. Corrosion.
(b) With O.D. Corrosion.

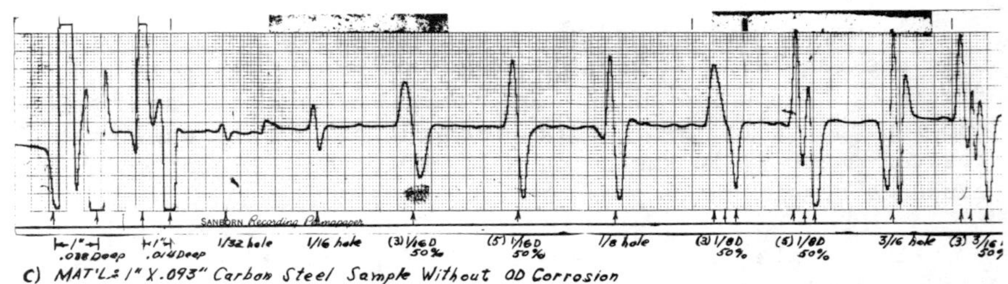

C) MAT'L: 1" X .093" Carbon Steel Sample Without OD Corrosion

Freq: 40 KHz.
Ph : 30°
Sens: 76

Recorder
5 mm/sec
5 mm = 0.5 Volts

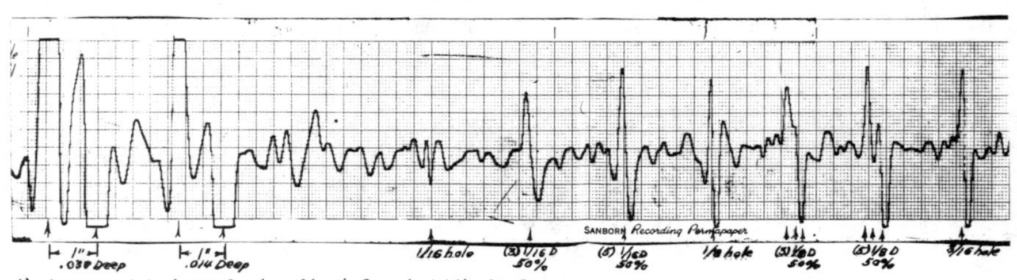

D) MAT'L: 1" X .093" Carbon Steel Sample With OD Corrosion

Freq: 40 KHz
Ph : 30°
Sens: 76

Recorder
5 mm/sec.
5 mm = 0.5 Volts

Fig. 10. System Response for Two Samples of 1" x .093 Carbon Steel at a Sensitivity of 76 db.
(c) Without O.D. Corrosion.
(d) With O.D. Corrosion.

TUBE SUPPORT CONDITIONS

Some of the more serious problems in any heat exchanger occur in the area of the tube supports. Since the supports will cause a redistribution of the flux lines, making it difficult to saturate the tubing, one can question the ability to test the tubing for wear and corrosion in their vicinity. However, numerous experiments have been performed with various supports, demonstrating that even though the tubing is only partially saturated, severe wear can be detected. It is necessary, however, that the amount of flux developed by the probe be as high as possible. This may be demonstrated by Figure 11 which shows the scope pattern for a 3/4" x .050 wall carbon steel tube and a 1" long tube support at current levels of 16, 8, 12 and 4 amperes respectively. The test frequency chosen was 5KHz and the tube clearance in the support was 1/16". Included with the support indication is a series of O.D. holes drilled to various depths. Note that both the holes and support indications become clearly defined between 12 and 16 amps.

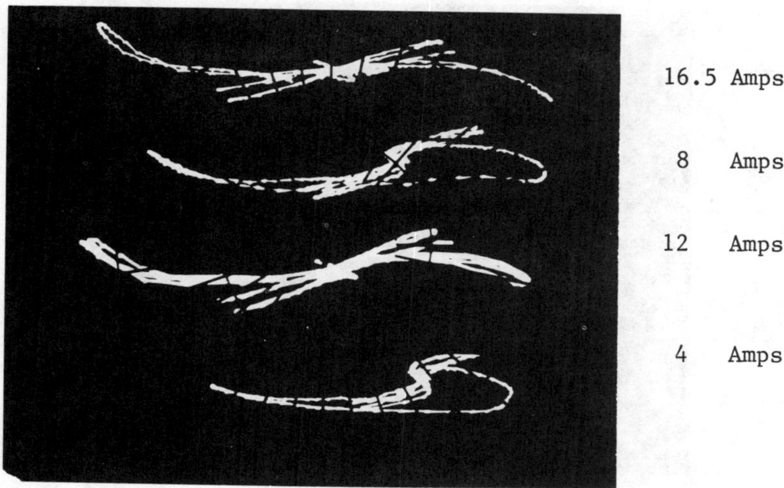

Fig. 11. Tube Support Indications at Four Values of Probe Currents. Tube: 3/4" x .050" Carbon Steel, F = 5 KHz

The next four photographs, Figures 12-15 show the response of a close fitting 1/2" support plate, with a 3/4" x .085" wall carbon steel sample at a test frequency of 10 KHz. In the four figures, the tube support indication was rotated to the horizontal axis. In Figure 12, the series of flat bottom holes are inclined about 20%.

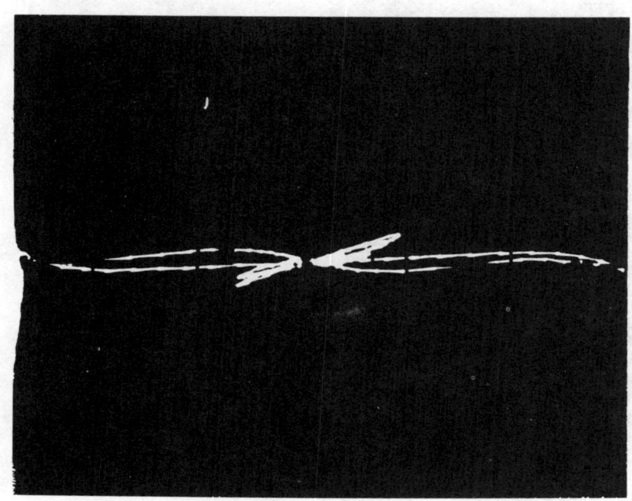

Fig. 12. Response for a 1/2" Tube Support and 3/4" x .085" Sample with Flat Bottom Holes, F = 10 KHz.

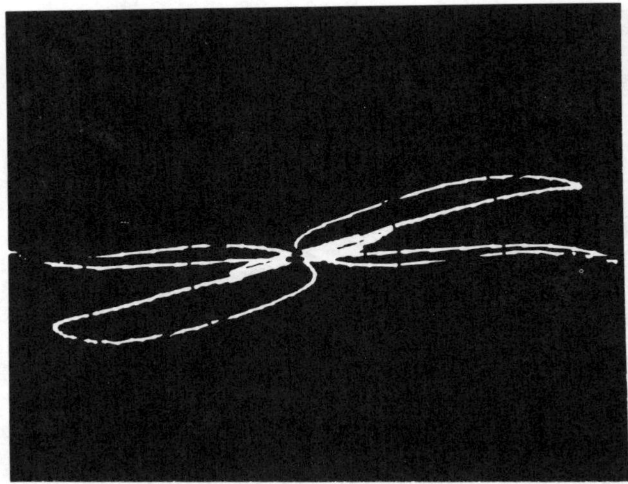

Fig. 13. Response for a 1/2" Tube Support and 3/4" x .085" Tube with Holes and a Flat .08" Deep, F = 10 KHz.

An additional artificial defect of .082 deep simulating wear by the 1/2" support is shown in Figure 13. The support has been displaced from the other defect. Note that the indication is in the same phase as the holes; however, due to its depth there is a large opening of the loop.

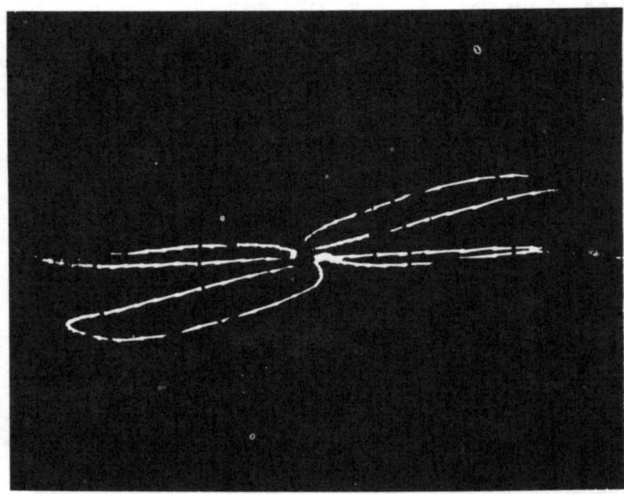

Fig. 14. Response for a 1/2" Support and 3/4" x .085" Tube with Wear .08" Deep, F = 10 KHz.

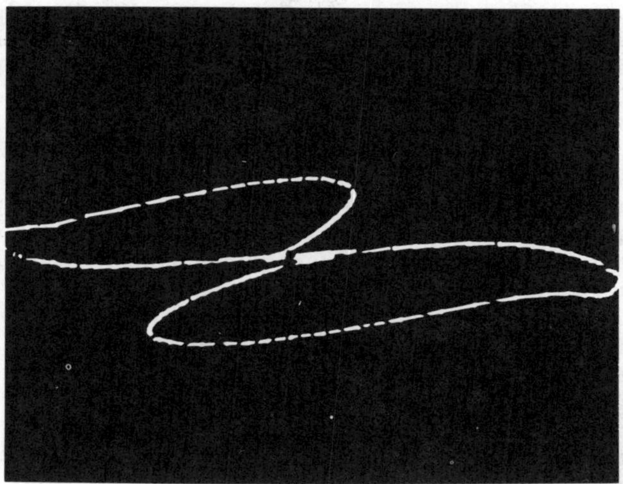

Fig. 15. Composite Response for the 1/2" Tube Support and Simulated Wear .08" Deep, F = 10 KHz.

Figure 14 shows only the support and the flat artificial defect comparing this scope pattern with that of Figure 15 in which the support was moved so that it covered the flat defective area. Note the change in the phase of the resultant composite trace. This is due to the fact that it is a vector addition with time as a factor, which is not indicated in Figure 14. In both figures the wear is clearly shown.

Consideration of the relationship between the supports and the tubing they retain and the fact that they are both part of the same magnetic circuit suggests that some measure may be made of the support themselves. In the previous figures the diameter of the holes in the support plates were nearly the same diameter as the tubing. This clearance, however, can change in the course of the heat exchanger's usefulness. An indication of the capability of the instrument to detect this condition is demonstrated by Figures 16-18, which are photographs of the responses for three separate tube sheets with enlarged diameters of 1/32" and 1/16". The test frequency for all three was 25 KHz and the test sample was 3/4" x .085 carbon steel. Observation of any one of the three figures shows that the response for the edge of the 13/16" diameter hole in the support is considerably smaller than the indication from the 3/4" hole. It should also be noted that the magnitude of each indication is related to the thickness of the support. That is the indication for the 13/16" diameter hole in the 1/4" plate is smaller than the indication from either the 1/2" or 1" thick plate. In addition, a comparison of Figure 17 with Figure 14, it can be easily seen that the amplitude for the support is decreasing as the frequency increases. A summary of these results indicates the following: (1) Major wear can be detected in the presence of the support plates; and (2) The signals from the support plates are a function of both frequency and distance. Observing these results we can infer that this system can also be applied to existing steam generator tubing having ferromagnetic tube supports subject to corrosion. In fact, preliminary work in this area has already begun and with promising results.

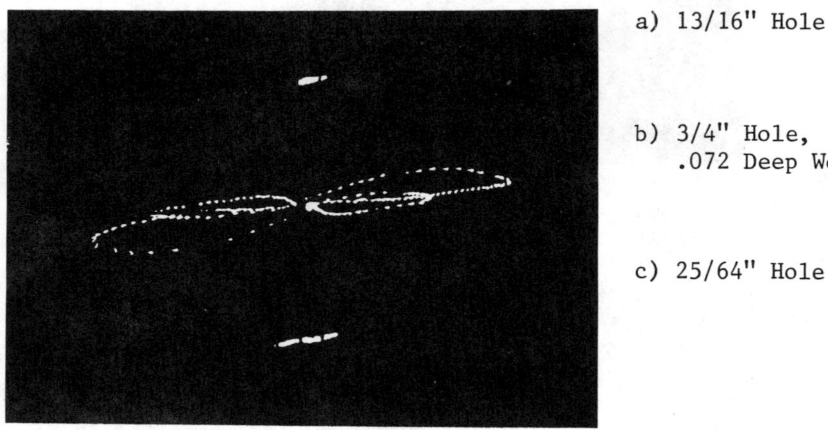

Fig. 16. Response for a 1/4" Support with a 3/4" x .085" Tube. Hole Diameters of a) 13/16", b) 3/4", c) 25/64", F = 25 KHz.

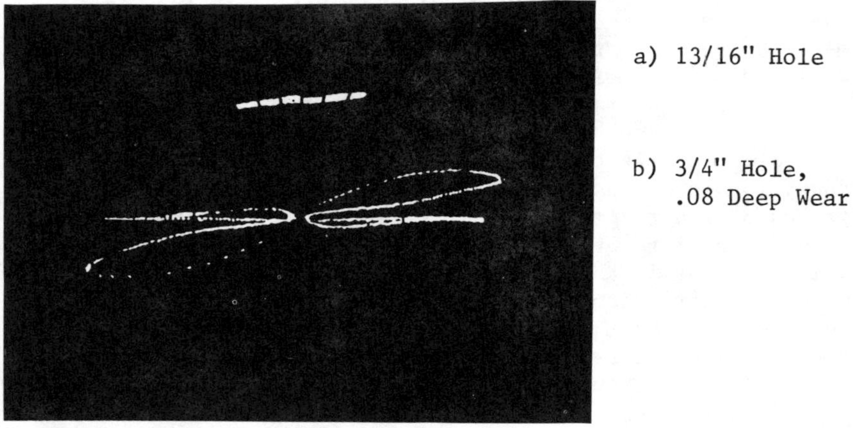

Fig. 17. Response for a 1/2" Support with a 3/4" x .085" Tube. Hole Diameters of a) 13/16", b) 3/4", F = 25 KHz.

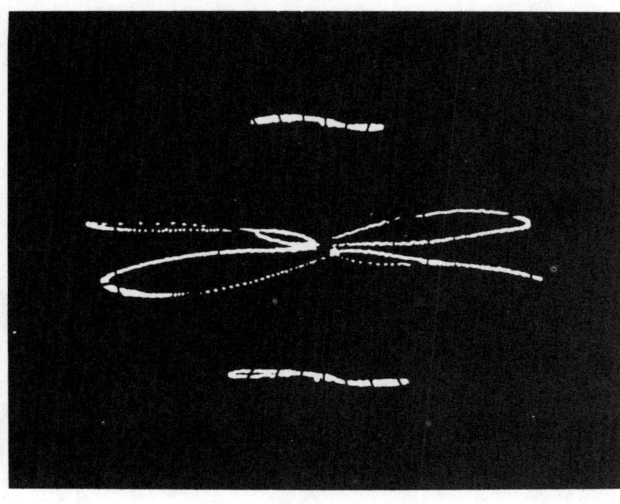

a) 13/16" Hole

b) 3/4" Hole,
.08 Deep Wear

c) 25/64" Hole

Fig. 18. Response for a 1" Support with a 3/4" x .085" Tube. Hole Diameters of a) 13/16", b) 3/4", c) 25/64", F = 25 KHz.

CONCLUSION

The major objective of this presentation was to discuss a practical system which has demonstrated the capability of probing magnetic materials without limiting the discussion to the specifics of the nuclear industry. Undoubtedly this new inspection tool will find useful application in the future as the need of the nuclear industry broadens. More and more, as we learn of the problems being encountered, this D.C. Saturation System can be adapted to satisfy some of these demands.

RADIOGRAPHIC WELD EXAMINATION IN RADIOACTIVE AREAS
(Carried out on behalf of and sponsored by THE SWEDISH NUCLEAR POWER INSPECTORATE)

Thomas Ulvsand
The Swedish Testing and Inspecting Company. STK

Jan Osterberg
The Swedish Plant Inspectorate. SA

The aim of this project was to investigate the possibility in using radiographic examination to detect weld defects, cracks for example, in nuclear power stations, which have been in operation.

The three main questions were:
1. How does the background radiation affect the possibility to detect the defects?
2. Can you avoid to remove the insulation before the examination? The removal of the insulation can give large radiation doses to the personnel.
3. Do you have to drain the pipes before the examination?

Of the questions mentioned above, number one was considered to be the most important.
Most of the work done in this project was to investigate how the visibility of cracks is changed. The reason of this change is the background radiation, caused by the radioactive waste products in pipes and vessels.This radiation will blacken the film, which is to be used in the radiographic testing.
The result of this investigation is that radiographic testing can be carried out at quite large background radiation levels.

THEORY

The detectibility of a defect in an object depends mainly on two things, the contrast and the sharpness of the image.
The contrast can be divided in two, object-contrast and image-contrast.
The object-contrast is the difference in radiation intensity caused by the defect. For a certain material it is governed by the energy of the radiation, an increase in radiation-energy gives a decrease in object-contrast.
When a film is used to detect the radiation, this object-contrast is the cause of an image-contrast. The filmgradient, the film-type and the processing of the film, decides the magnitude of the image-contrast. Large gradient, slow film and correct processing is important.
The total sharpness of an image depends on three factors namely a) unsharpness due to movement b) geometrical unsharpness and c) inherent unsharpness.

a) In radiographic testing you consider the unsharpness due to moment to be of no importance.
b) Geomtric unsharpness depends on the thickness of the object, the film-focus-distance and the focal spot-size. As long film-focus distance and small focal spot-size as possible is important to get small geometric unsharpness.
The acceptable geometric unsharpness in different codes lies within 0,1–0,4 mm.
c) The inherent unsharpness comes from the secondary radiation created in the film or the lead-foils by the absorption of the primary radiation. Up to 500 keV the inherent unsharpness lies beneath 0,15 mm.

IMAGE QUALITY INDICATORS (IQI)

There obviously are a number of factors that influence on the image quality. All codes demand the use of some sort of IQI at every exposure. Their purpose is to make it possible to decide afterwards, if the correct technique of examination was used.
In the beginning of the project we studied different types of IQIs. The purpose was to see which of them that responded best to the disturbance caused by the background radiation.
The IQIs we used were:
a) The IIW wire-indicator. There are seven steel wires of various diameters (0,1–0,4 mm) in a plastic cover. The image-quality is determined by the thinnest visible wire.
b) The AFNOR step-and-hole indicator. It consists of six steps with holes drilled in them. The diameter of the hole is the same as the thickness of the step in wich it is drilled. The quality is determined by the smallest visible hole.
c) The Cerl-indicator.
This indicator consists of two parts, one step-indicator in steel and one with parallel platinium wires of increasing thickness and spacing in plastic.
The visibility of the steps shows the contrast of the image. The sharpness of the image is shown by the two thinnest wires you can see apart.

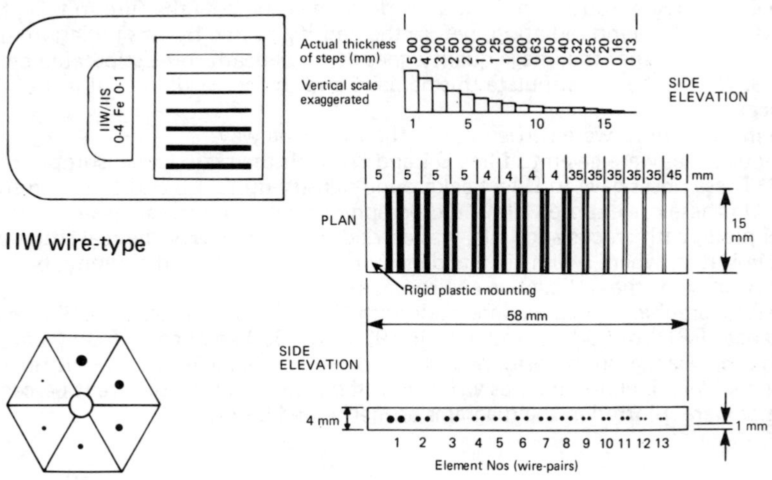

AFNOR step and hole-type

CERL step-type and double wire-type

TESTS

The film used in the investigation has been Agfa-Gevaert D7 and D4.

Choice of testing-plate

To be able to see and compare how the crack-visibility and the IQI visibility varies with different radiographic parameters, an object with cracks had to be used. These cracks had to be well defined and of a suitable size.
Some different testing-plates with cracks were radiographed to density 2 and were compared concerning the cracks. This made it possible to choose one of the plates in steel, 12 mm thick, with five fine crack details. It was these five cracks that were used in the proceedings, to determine how the crack-visibility varies with different radiographic parameters and background density.
The results of our test are shown in diagram-form. In these diagrams the varying parameter is on the horisontal and the different perceptibility-values on the vertical axis. The perceptibility-values are the number of wires, step, holes or cracks respectivily seen on the radiograph. The values put into the diagrams are a balancing of opinions of two, sometimes three persons. Trying to show some trend in the perceptibility, lines have been drawn in the diagrams. As it is a continuous course that is to be shown by a number of discret levels, these lines are permitted in some cases to be drawn one value above the point in the diagram. The difference between two perceptibility-values which in the diagram is one, for example between three and four, in reality can be just above zero. You may hardly see the fourth wire, but very well the third.
In the diagrames the scales for both of the CERL-indicators is half the scale in the other cases.

Investigation number one (steel plate)

The interesting question was how large dose one can allow the film to take, because of background radiation, before the density caused by this radiation, becomes so large that the image-quality goes down beneath some stipulated demand. We also had to stipulate this demand and to decide which IQI is useful to indicate it.
The investigations were carried out in the following way:
Various doses were given to films D4 and D7 with the radioactive isotope Ir-192. These doses gave to D4 a background-density up to 1,5 and to D7 up to 1,8. On these pre-exposed films, radiography in the conventional way of the steel plate, with cracks and IQIs, was carried out. In this way the real situation was imitated, where a film gets its density not only by the radiography, but also because of the radioactive environment.
The radiographic exposures were made with 170 kV high-tension and a film-focus-distance (FFD) of 100 cm and with Ir-192 (300–600 keV) and FFD 50 cm with each »background dose». Several exposures were made to different final density. This final density was written on the X-axis and the different perceptibilities were plotted in the diagrams as mentioned before.

Results (X-ray)

With Agfa-Gevaert D4 the results is shown in diagrams 1–5 where it can be seen that the perceptibility by the different IQIs and crack increases with density, which depends on the increase in film gradient. It also can be seen that of the different IQIs the IIW-indicator is the one that best follows the crack visibility.

Diagrams 1–5.

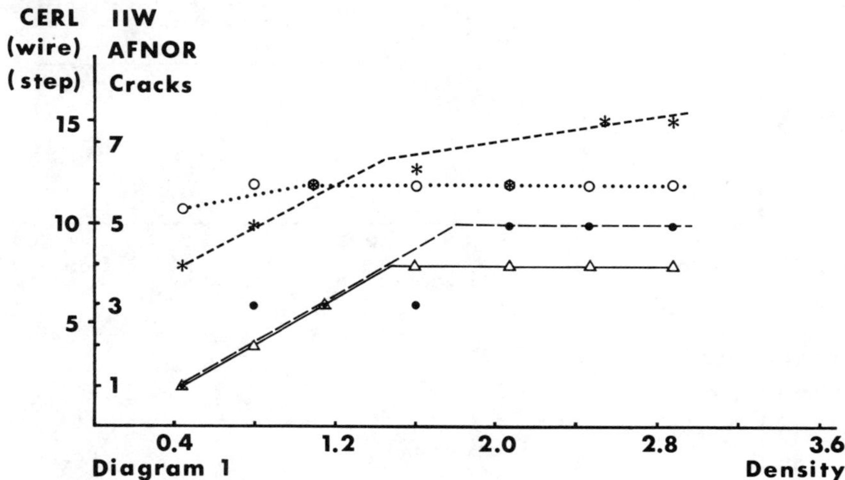

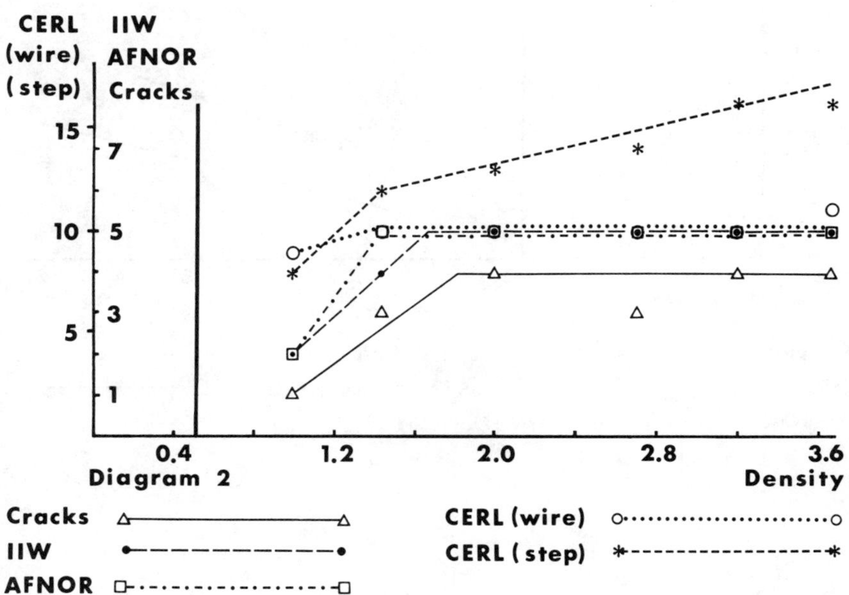

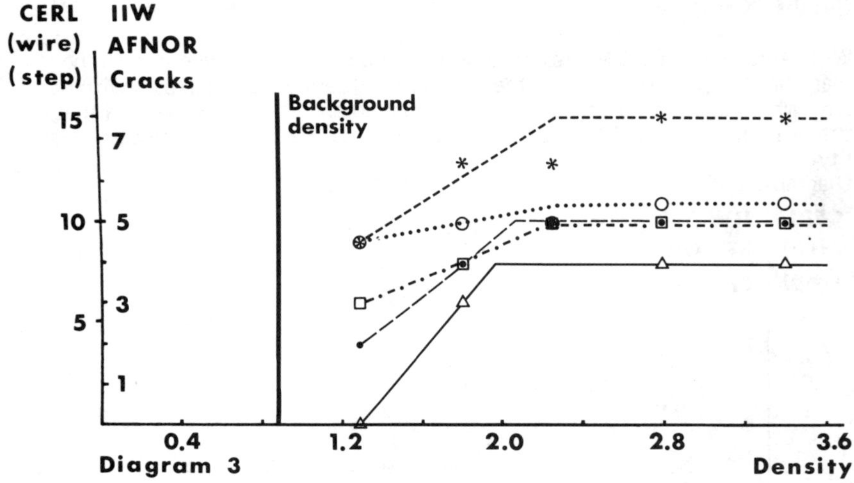

Diagram 3

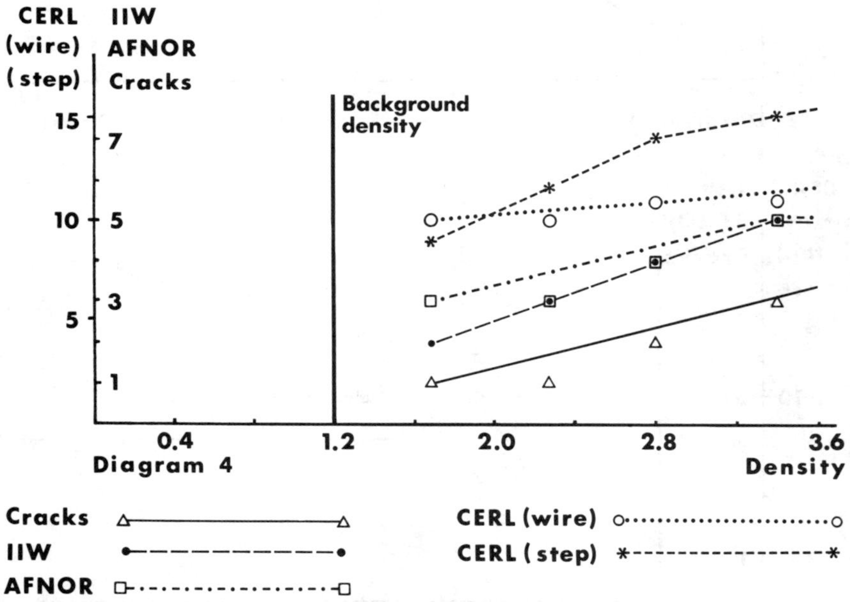

Diagram 4

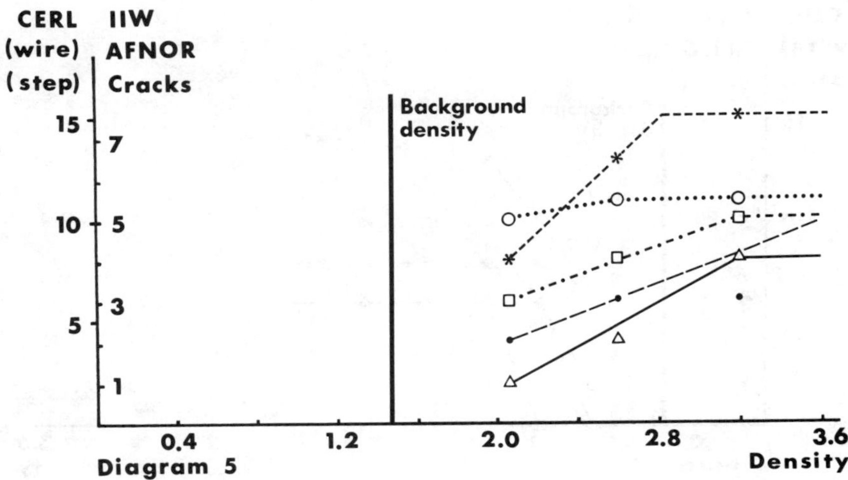

Diagram 5

With Agfa-Gevaert D7 the result is similar.
Diagrams 6—11.

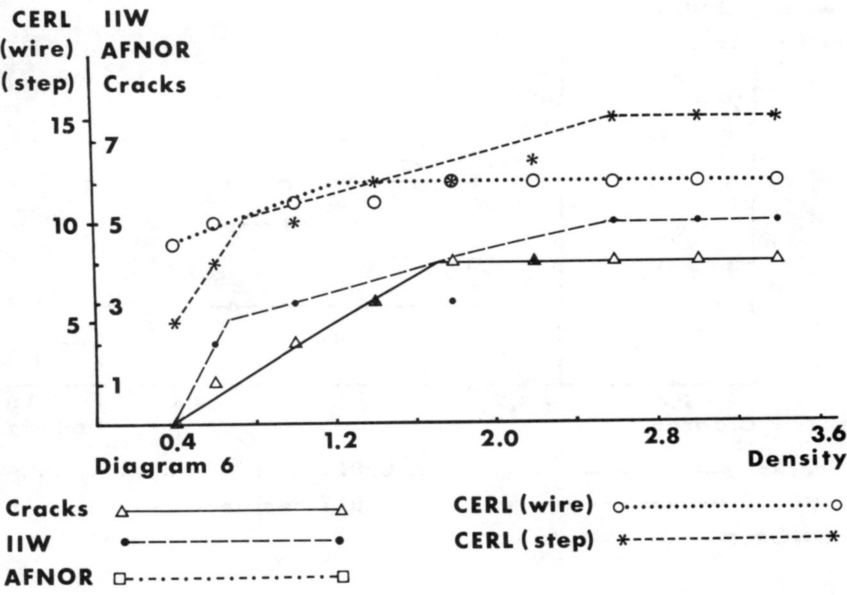

Diagram 6

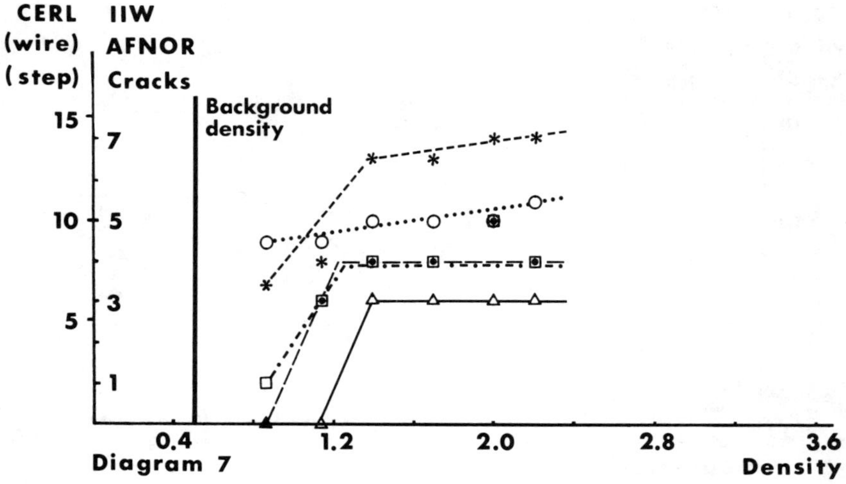

Diagram 7

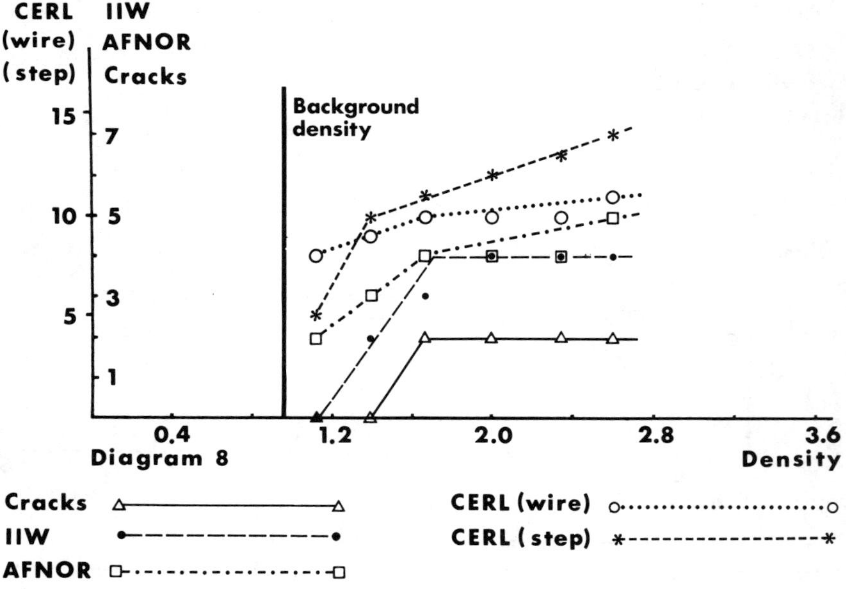

Diagram 8

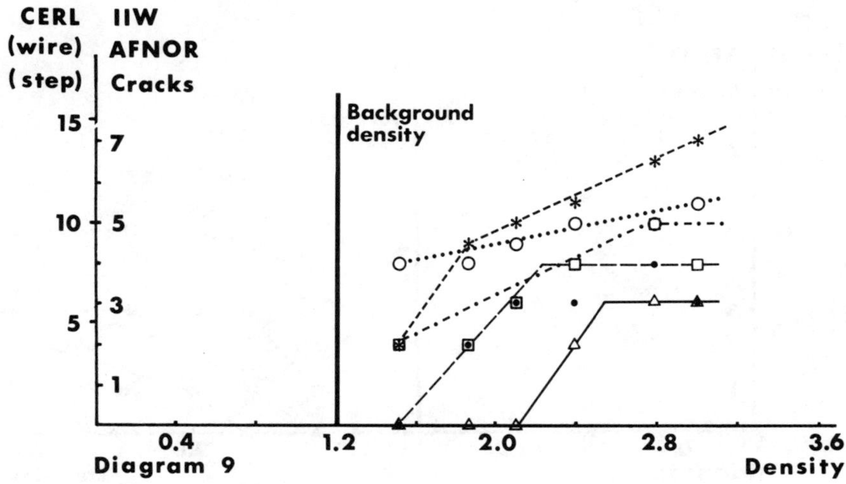

Diagram 9

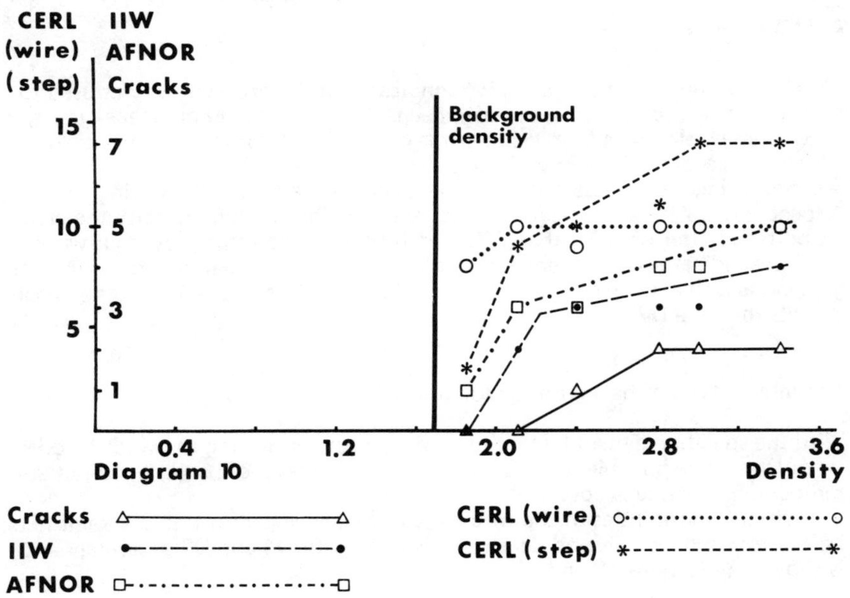

Diagram 10

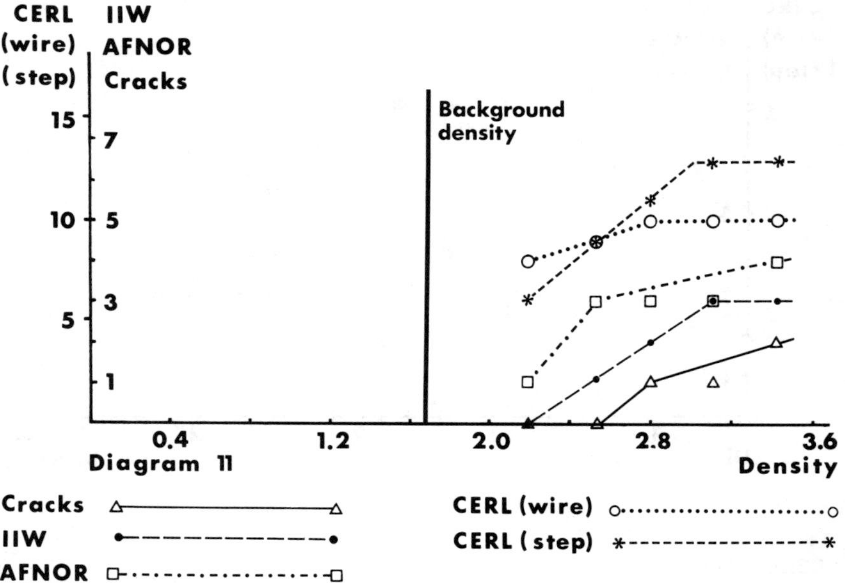

Diagram 11

These results implied that the IIW wire-indicator in the proceedings was used to indicate image-quality. This, in turn, means that the demand on image-quality now could be stated as four visible wires on this indicator (for 12 mm steel), which is according to Swedish Codes.

An interesting fact can be noted. If one compare the diagrams showing D4 respectively D7 it can be seen that with the same background-density the crack visibility is better with D4 than D7, even if both of the films show four wires. The wire-indicator is a rather rough indication on image-quality and the greater gradient and finer grain of the D4 makes it better to detect cracks or other small defects than the D7.

Acceptable level of background radiation

With the specified demand, mentioned previously, according to Swedish codes, it now should be possible to investigate up to which level of background-radiation radiographic testing is possible.

The different visibilities of the wire-indicator have been drawn in diagrams like before for each background density. It is done with D4 and D7 separately and is shown in diagrams 12 and 13.

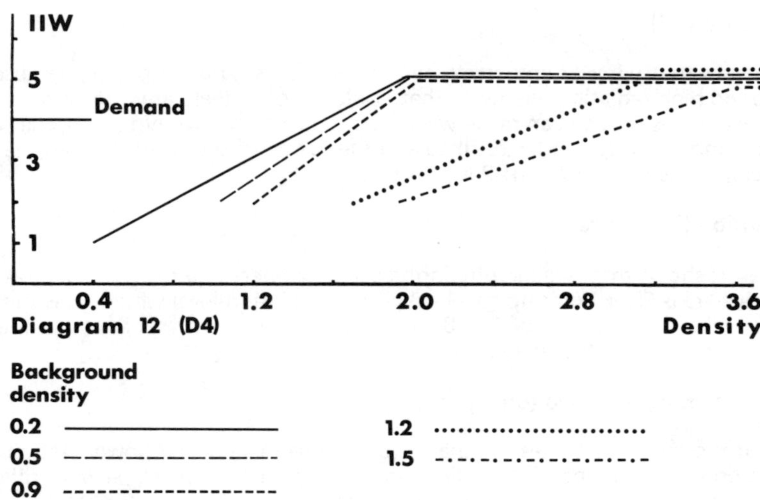

Diagram 12 shows that with D4 the four wires-demand is well fulfilled in all cases up to background density 1,5, inclusive, if the difference between the background density and total density is 2 or more.
The background density 1,5 corresponds to a dose to the film of about 2500 mrads in the energy range 300-600 keV.

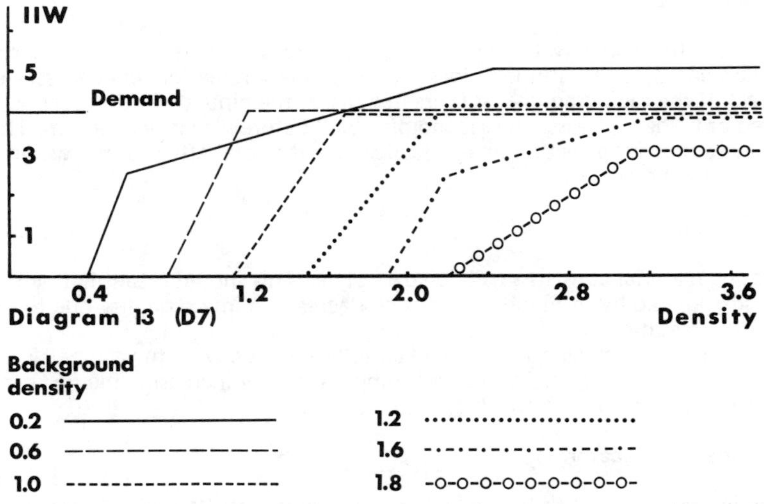

Diagram 13 shows that with D7 the demand is certainly fulfilled up to 1, inclusive, and with some uncertainly up to 1,5. The difference between background and total density must like before exceed 2.
With D7 the density 1,5 corresponds to a film dose of about 600 mrads.

Results (Ir-192)

In no cases four wires or any cracks can be seen. This is not surprising because of the too high radiation-energy to that thickness of a steelplate.
But the conclusion you can make with the support of the X-ray tests is, that if the wire-indicator demand is fulfilled and the net density exceeds 2, then the radiographic testing with Ir-192 is possible.

Conclusion (Test-plate)

The result shows that radiography in the usual manner is quite possible in radioactive areas. This holds up to a background radiation level which gives the film a background-density of 1,5. Because of larger gradient and finer grains a slower film is preferable to a faster.

Investigation number two (on pipes)

Most of the radiographic testing that is to be done in a nuclear power station will be on welds on pipes. These pipes are in some cases going to be waterfilled and/or surrounded with heat-insulating materials.
Because of what just have been said the next question to be answered was how the results of the testplate could be applicated on pipes and how waterfilling and/or isolation will influence.
With empty, unisolated pipes the previous results can be applicated. This means that if the radiography is carried out in a proper way, a background dose corresponding to density 1,5, can be accepted.

Pipes filled with water

When pipes filled with water is radiographed there is an increase in secondary radiation compared to unfilled pipes. This secondary radiation will decrease the contrast of the radiograph. As it is created inside the pipe it will to a part be filtered out when it passes the second pipe-wall before it reaches the film. This will have a positive effect on image-quality, but the total effect of the waterfilling should be negative.

Insulated pipes

Insulating material consists to a large part of air. This should mean that with pipes surrounded by insulation there is no decrease in image-quality due to secondary radiation or filtration.
But because the film cannot be placed close to the pipe-wall, the geometric unsharpness will increase. This can be compensated by increasing the FFD and should be done when it is possible.

Image quality indicator

As the investigations with the test plate shows that the IIW wire indicator follows the crack visibility best, it has been used in the following tests.

Description of the test

Unfilled, un-insulated pipes were radiographed. Then they were waterfilled or insulated with 3–5 cm insulating material, respectively, and radiographed again. By comparing the values of the wire-visibility, it was possible to see if there were any changes. The exposures were done both with X-ray and Ir-192. In some cases the high tension with unfilled pipes were little too high according to what is recommended by the IIW (International Institute of Welding). Swedish Codes are according to these recommendations.
The reason for this too high tension, was to get reasonable exposure time even with waterfilled pipes, without too big variations in the kV.

Tables

The number of visible wires is shown in Table 1 with X-ray and Table 2 with Ir-192. Beside this, these tables show the used kV, geometric unsharpness and demand of wire-visibility according to Swedish Codes.

X-ray radiography

The variation in geometric unsharpness lies within 0,05–0,3 mm. It can be shown that this variation does not have any influence on the wire-visibility. The high-tension varied between 130–150 kV with a few exceptions. (100, 120, 160 and 170 kV). For each separate dimension the variation is small or none. This makes it possible to conclude that the films concerning the wire indicators are comparable in spite of the variation in unsharpness and kV. Of course you shall always use as low high tension and small unsharpness as possible when you carry out radiographic testing. This is because cracks are more sensitive to change in contrast and unsharpness than wires are.

Results

From Table 1 it can bee seen that the visibility of the wire-indicator goes down with 20–50 % when the pipes are filled with water. Insulating material around the pipes does not affect the visibility.

Table 1

(X-radiography of pipes)

Dimension (mm)	High tension (kV)	Geometric unsharpness (mm)	Wires			
			Demand on percep-tibility	Seen in tests		
				No w No i	With w No i	No w With i
26,9x2,6	150	0,05	7	6	6	
42,4x3,2	150	0,08	6	6	3	
60,3x3,6	150	0,11	6	5	3	
88,9x5,5	150	0,16; 0,22	4	5		
114,3x8,56	130, 150	0,16; 0,3	5	5		5
141,3x6,55	130, 150	0,12; 0,25	6	6		5
168,3x10,97	170; 200	0,25	4		3	
205,0x2,5	120; 150	0,05; 0,18	7	7		7

w: water

i: insulation

Radiography with Ir-192

Table 2 shows that the visibility-demands never are fulfilled at these dimensions.

Table 2

(Ir-192 radiography of pipes)

Dimension (mm)	Wires			
	Demand on percep-tibility	Seen in tests		
		With w No i	No w With i	With w With i
114,3 x 8,56	5		2	2
141,3 x 6,55	6		3	1
168,3 x 10,93	4	1	3	0
205,0 x 2,5	7	2	4	1
218,0 x 12,7	4	0	2	0

w: water

i: insulation

THE RESULTS OF THE INVESTIGATION

Radiographic testing in radioactive areas can be carried out without loss in defect-visibility up to background radiation-levels corresponding to density 1,5. This means for the slower Agfa-Gevaert D4 2500 mrads and for the faster D7 600 mrads. The difference between total and background density have to be at least 2.

The slower film should be used for two reasons:

1. Finer grains and larger gradient make the defect-visibility better.
2. One can divide the influence of background radiation in two moments, one during the radiographic exposure and the other during application, removal and transportation of the film. The use of slow film makes the influence of background radiation smaller than with a fast film.

One finally can conclude that you have to drain the pipes, to achieve an acceptable result of the radiographic testing.

The insulating material does not have to be removed if it is not too inhomogenous. One can have to compensate the decrease in geometric unsharpness, due to the distance between film and object, by increasing the film-focus-distance. This can not always be done, but the geometric unsharpness shall be as small as possible.

Now when the restrictions caused by the background radiation and the normal requirements are known, it can be stated when it is possible to carry out radiographic testing in radioactive areas.

X-ray

Normally used high-tension gives with 15–20 mA-min a density of about 2. This means an exposure time of 3–4 minutes.

The time of application and removal of the film can be estimated to 3 minutes. This means that the total time the film is in the radioactive area becomes 6–7 minutes. With a maximum permissible dose to the film of 2500 mrads (Agfa-Gevaert D4) it is then possible to carry out testing in a radioactive area with 20–25 R/h. Variations up and down are possible, it depends on how fast the test can be done.

It should be noted that the dose-rate-levels mentioned shall be measured at the surface of the object being tested.

Isotope Radiography

Take as an example an object of 25 mm thickness to be radiographed with Ir-192 at 700 mm FFD. This means (from an exposure table) an exposure time of 28 minutes with 50 Ci. A handling time of 3 minutes makes 31 minutes for the film in the radioactive area. If the background-radiation level is beneath 5 R/h it is possible to carry out the testing.

Other objects or other radiation-sources will give other permissible radiation levels, but the way of calculating this level will be the same.

NONDESTRUCTIVE EVALUATION OF LWR SPENT FUEL SHIPPING CASKS*

D. W. Ballard
Nondestructive Testing Division 9351
Sandia Laboratories
Albuquerque, New Mexico 87115

ABSTRACT

This report summarizes an analysis of nondestructive testing (NDT) methods currently being used to evaluate the integrity of Light Water Reactor (LWR) spent fuel shipping casks. An assessment of anticipated NDT needs related to breeder reactor cask requirements is included. Specific R&D approaches to probable NDT problem areas such as the evaluation of austenitic stainless steel weldments are outlined. A comprehensive bibliography of current NDT methods for cask evaluation in the USA, Great Britain, Japan and West Germany was compiled for this study.

I. INTRODUCTION

This report was prepared to summarize the nondestructive testing (NDT) methods currently being used in the USA, Great Britain, Japan and West Germany to evaluate LWR spent fuel shipping casks. An assessment of their effectiveness on current casks was considered a necessary prerequisite to projections of NDT R&D needs to meet the more rigorous requirements to be imposed on future generation casks[1].

Conceptual breeder reactor cask designs from Sandia's materials and structures groups were reviewed for their inspectability using state-of-the-art NDT methods. Profitable development areas are indicated which could optimize specific NDT methods for Sandia's early conceptual designs of advanced technology casks.

---

* This work supported by the United States Department of Energy.

II. BACKGROUND

Nondestructive testing has played an important role in the development of a safe nuclear power industry from its first introduction[2]. Nondestructive evaluation (NDE), which includes NDT but goes beyond testing to include interpretation and interdisciplinary decisions on effect of interpretation on a product's end usefulness, can be expected to play an even greater role in the 1980's through improvements and extension of the use of the basic NDT methods and a more complete understanding of their indications[3,4]. The causes for this greater dependence on NDE methods include increased structural performance requirements, the use of defect-sensitive materials, changes in design philosophy based on fracture mechanics, more rigorous construction and safety codes, and a recognition of the need to determine the conditions and degradation of critical structures during their service life[5].

Spent fuel shipments, as are all radioactive material moves in the USA, are regulated by the Nuclear Regulatory Commission (NRC) and the Department of Transportation (DOT). International shipments in most cases are made consistent with the standards of the International Atomic Energy Agency (IAEA) with DOT serving as the USA competent authority[6].

Chapter 1 of Title 10 of the Code of Federal Regulations contains the rules related to transportation of radioactive materials. Specifically, Parts 20, 70, 71 and 73 of Chapter 1 deal with standards for protection against radiation, special nuclear materials, packaging of radioactive material for transport, transportation of radioactive material under certain conditions, and physical protection of certain radioactive material. Spent fuel casks contain quantities of radioactive materials in excess of Type B limits of 49 CFR 173.393 which defines the general packaging requirements for radioactive material. Acceptable release fractions apply only to the gaseous and volatile components of the spent fuel. It is considered that the remaining fission products are retained within the fuel elements or shipping cask.

There is currently little LWR spent fuel being shipped in the USA because of a lack of operational fuel reprocessing plants. Only one private plant was ever licensed to operate, and it was shut down in 1972 for modifications. Its owner, Nuclear Fuel Services Inc. of West Valley, New York, has since withdrawn their application to reopen. In Barnwell County, South Carolina, Allied-General Nuclear Services has completed the separation facilities of a reprocessing plant with a capacity of 1500 tons per year. The owners are awaiting a Nuclear Regulatory Commission license, which hinges on a government decision on waste storage and on rules governing the utilization of recovered plutonium. President Carter's interim decision to forego spent fuel processing has essentially stalled the final development of the Barnwell Facility. As a result, the spent fuel from the nation's 62 operating LWR or fission power reactors (another 72 are under construction and 61 more are in the planning stages) have been piling up at reactor sites with a current inventory of about 2500 metric tons[(7)]. Britain, France, USSR and other countries reprocess spent fuels from their nuclear power reactors in government facilities, but that is not the current policy in the USA.

Part 71 of Title 10 CFR Chapter 1[(8)] gives the following four basic safety requirements for transporting radioactive materials:

- Adequate containment;
- Adequate control of radiation;
- Safe dissipation of heat generation;
- Prevention of nuclear criticality.

Neither Part 71 or NRC/DOT regulations attempt to specify how these four basic safety requirements are to be met. This is left up to the cask designer who must qualify casks built to his design to a series of DOT/NRC-specified tests which represent the most credible accident conditions during the transportation cycle. In 1966 the Division of Reactor Development and Technology (RDT) requested that the Oak Ridge National Laboratory develop a Guide for good engineering standards of practice in the design, fabrication, testing, inspection, and maintenance of irradiated fuel shipping casks(9). It was decided that, initially, the information in the Guide would apply to lead-shielded spent fuel casks having steel inner and outer shells since this type of LWR cask is most widely used in the United States today.

The accident damage tests defined in 10 CFR 71 provide a means for reproducing in the laboratory or in the field the same general type and degree of damage a cask might reasonably be expected to sustain in a severe rail or truck accident. The damage test sequence includes: 1) a free fall from a height of 30 feet onto an unyielding surface with the cask landing in an orientation which should see the most damage; 2) a free fall from a height of 40 inches onto a 6-inch-diameter steel plunger long enough to do maximum damage; 3) heat input from exposure of 30 minutes to a fire or other radiant environment having a temperature of 1,475°F and an emissivity of 0.9; and 4) immersion of cask in water to a depth of 3 feet for 8 hours.

The designer must prepare a detailed Safety Analysis Report (SAR) prior to licensing. If his stress analysis and/or cask tests are adequate, an NRC-issued license will be granted before the cask can be used. Casks of varying sizes and mode of transportation have been licensed for either commercial transportation or on-site movement of LWR spent fuel elements or assemblies.

In addition to the damage test sequences required by 10 CFR 71, Sandia Laboratories has conducted over the past 18 months four full-scale crash tests of LWR casks for the Department of Energy(10). These tests fell within the "extra-severe or extreme" accident categories and have been estimated in risk assessment studies of transportation environments to be events of very low probability; no such accidents involving nuclear materials have ever occurred.

In the first two impact tests, a 22-ton cask was mounted on the beds of tractor-trailers, which were slammed into the concrete wall weighing 690-tons and backed with more than 1740 tons of earthfill. Impact speeds were 60 and 84-mph. For the third test, a tractor-trailer rig with a 28-ton cask secured to its bed was struck broadside by a 120-ton locomotive traveling at a speed of 81.5 mph. The integrity of the casks was not breached in any of these crashes.

The final test in the series involved the 81.4-mph impact into a massive concrete wall of a special railcar carrying a 74-ton fuel cask. The impact demolished the front of the railcar, but the cask received only minor external damage. The cask was subjected to a jet fuel fire (temperatures between 1800°F and 2100°F) for more than 90 minutes.

All crash results to date closely paralled predicted results based on 1/8 scale model crashes and computerized mathematical analyses[11]. The tests have helped confirm our ability to predict damage from extra-severe transportation accidents.

III. LWR SPENT FUEL CASK DESCRIPTION

Figures 1 and 2 show the general configurations of a typical spent fuel shipping cask for LWR fuels. The inner cavity containing the irradiated fuel is filled with water for cooling and shielding purposes. The inner and outer shells are usually made of stainless steel and the main shielding material for gamma radiation is lead or depleted uranium. The fins on the outer surface may provide some impact protection as well as adequate heat transfer. The cask travels on either a railcar or truck bed in a horizontal position resting on appropriate supports with tiedowns. To keep radiation doses and surface temperatures below the specified limits, a personnel shield may be used around the cask during transport.

CASK INFORMATION

LOCATION	SHIELDING LEAD	STEEL
COVER	18.1 cm	4.45 cm
BOTTOM	18.1 cm	4.45 cm
SIDES	21.3 cm	4.45 cm

CAVITY
33 cm DIA x 328 cm LG.

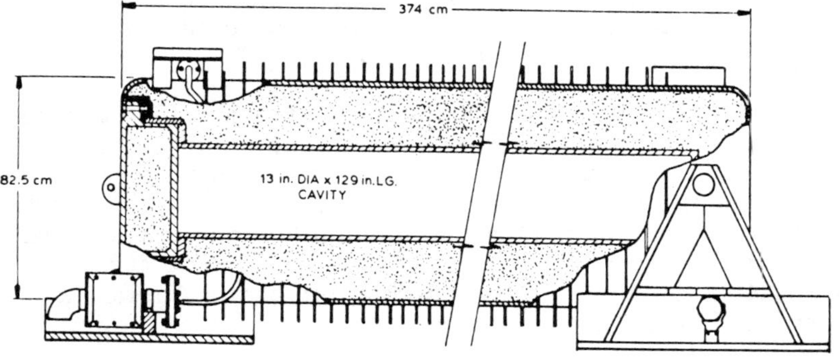

Fig. 1. Typical Cross-Section of a Large Spent-Nuclear-Fuel Shipping Cask

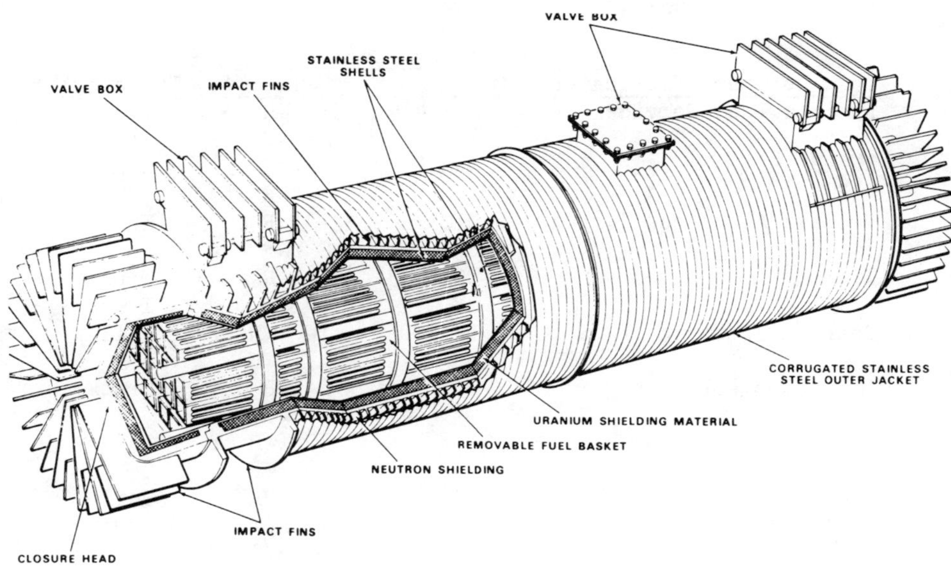

Fig. 2. Above, the shipping casks for spent fuel are built to withstand even a major accident

The largest and latest LWR cask licensed for use in the USA to date by NRC was built by NL Industries and is designated as NL 10/24. Each cask, weighing approximately 87 metric tons and mounted on a railroad flatcar, can carry up to 10 pressurized water or 24 boiling water reactor spent fuel assemblies. The cask has a 4.5 ton capacity for uranium, measures 5.3 m (17 feet) in length and 2.4 m (9 feet) in diameter, and uses helium rather than water as a coolant during shipments.

Other commercial LWR fuel casks licensed or still in design are listed in Table I.

TABLE I

Some Current LWR Casks

Cask Designer/Fabricator	Designator	Capacity	Transport
Nuclear Fuel Services	NFS 4	1 PWR or 2 BWR	Truck
	NFS 5	2 PWR or 4 BWR	Truck
Nuclear Assurance Corp.	NAC 1	1 PWR or 2 BWR	Truck
Transnuclear, Inc.	TN 8	3 PWR	Truck
	TN 9	7 BWR	Truck
General Electric	IF 300	10 PWR or 17 BWR	Rail
Exxon Nuclear		2 PWR or 4 BWR	Truck
		12 PWR or 26 BWR	Rail
Excellor, Ltd.[12]	EX-3	15 BWR	Rail
NL Industries	NL 1/2	1 PWR/ 2 BWR	Truck
	NL 10/24	10PWR/24BWR	Rail

Depending on the size and intended mode of transportation the weight fraction of fuel in a loaded LWR cask ranges from only 2 percent for small trucked casks (25 tons) to 6 percent for large railcar casks (100 tons). Lead shielding for gamma radiation is typically poured into an annulus around the fuel assemblies formed by two concentric shells of steel. Where necessary, hydrogenous material such as water is used for neutron shielding and heat transfer from the spent fuel assemblies. Some casks (i.e., TN 8/9) are shipped dry (no liquid coolants), while others use water or other coolants. Some casks (i.e., IF 300) substitute depleted uranium for lead in gamma shielding. Because of its high density (~ 19 g/cc), a cylindrical uranium shield weighs less than 75 percent of a lead shield and less than 50 percent of a steel shield having the same internal cavity dimensions and comparable radiation attenuation capabilities. Another variable is the cooling of the outside surface of the cask by using either natural convection (passive) or forced circulation (active). This cooling is required since a typical LWR rail cask may release as much as 70 KW of heat or 250,000 BTU/hr and DOT/NRC regulations require that accessible surfaces of LWR casks not exceed 180°F under normal operation conditions. (Breeder fuel assemblies will generate much higher heat fluxes and it may be necessary to raise allowable cask surface temperatures from 350 to 400°F, and to provide a separate personnel shield from the heated surfaces.)

IV. CURRENT QA REQUIREMENTS - LWR CASKS

The fabrication and inspection requirements for LWR shipping casks are not specifically covered by existing codes and standards. Experience to date indicates that LWR casks designed to meet IAEA and ASME Boiler and Pressure Vessel Codes (Class MC, Subsection NG) are more quickly and straightforwardly licensed than those built without resorting to code requirements. Nevertheless, general LWR cask design guides[9,13] devote several sections to the quality assurance (QA) programs expected from the designer, fabricator and user of current casks. Moreover, specific cask design specifications for the latest of LWR casks are quite detailed from a quality assurance standpoint. The following paragraphs summarize the major QA requirements gleaned from typical LWR cask specifications.

Designer's QA Responsibilities. The cask designer must furnish, in addition to complete materials/process/dimensional callouts, a cask design report to the licensing authorities and to the potential user which lists all design assumptions, cask limitations, and operational/maintenance procedures. Weldments, gaskets, seals and tiedowns are recognized as critical design and fabrication elements of all LWR casks. They are potentially "critical" points that require special quality assurance attention during the design phase. One of the provisions of 10 CFR 71 requires that the shipping cask licensee establish and maintain operating procedures and periodic inspection procedures to assure that the performance of the cask is in compliance with the regulations at all times. These cask inspections are imposed on the cask prior to its first use and prior to and following its use for transporting irradiated fuel assemblies.

Manufacturers QA Responsibilities. The nature of the materials being transported makes it imperative that the manufacture and inspection of LWR casks be in strict conformity with the design drawings and construction specifications. Further, a quality control program patterned after Appendix IX of Section III of the ASME Boiler and Pressure Vessel Code must be defined and carried out by the manufacturer covering such as the following:

- Rigid material conformance including mill test reports and marking;
- Rigorous control of critical fabrication processes such as heat treatment and welding;
- Maintenance of a complete Fabrication Record indicating compliance (or authorized deviations) with the design and process drawings and covering reports of all inspections and tests (including nondestructive tests) at the component and assembly levels;

- Provide full access to representatives of the design agency, cask purchaser and licensing authority.

Cask Purchaser or User's QA Responsibilities. To ensure that the fabrication requirements are met, an inspector(s), as a representative of the cask purchaser, must audit the manufacturer's procurement, fabrication, inspection and tests to determine compliance. The inspector must have the authority to stop work on a cask at any time when he feels the assurance of quality is being jeopardized. An Inspector's Log must be maintained in which he records the dates of his visits to the manufacturer's plant, status of work on the cask(s) on those dates, notations of materials accepted, notations of inspections and tests witnessed, difficulties encountered in fabrication and authorized deviations from design details and specifications. On completion of the cask fabrication, the Inspector's Log(s) must be incorporated into the Fabrication Record maintained by the cask manufacturer.

V. NONDESTRUCTIVE TESTING OF LWR CASKS

This section summarizes the widely varient nondestructive testing programs that have been imposed on first and second generation LWR spent fuel shipping casks. The information was compiled following a comprehensive study comprised of the following elements, not necessarily in chronological order:

- Literature search and compilation of bibliography of pertinent documents, journal articles and symposium proceedings dealing with nondestructive evaluation methods and procedures for current LWR casks. This bibliography(14) was distributed on December 17, 1976 and a reference file of hard copies or microfiche of this information is located in the NDT Information Center at Sandia Laboratories.
- Personal contacts with NDT and cask design and fabrication personnel involved in LWR shipping cask production. Both telephone and personal visits were used to secure up-to-date information from key installations on current cask evaluation techniques and their limitations. A listing of organizations and persons contacted is given in Appendix A.
- Attendance and participation in selected conferences and symposiums related to shipping cask evaluations. A list of these meetings attended during the period from July 1, 1975 to December 31, 1977 for this purpose is also given in Appendix A.
- Interdisciplinary discussions with Sandia Laboratories' members of the LMFBR Shipping Cask Technology Project

were held periodically during the past year. As strawman or conceptional designs were derived[15,16], we attempted to supply NDT input on methods, sensitivities and problems engendered by design/materials/fabrication considerations.

Nondestructive testing and evaluation practices vary widely on LWR casks. While there are some common denominators such as pressure and leak testing that are applicable to all casks, the degree of application of other NDT testing using such methods as radiography, ultrasonics, dye penetrants and radiation gauging varies widely. Commercial cask documentation of tests was quite thorough since the requirements for licensing by DOT/AEC (now NRC) impose such documentation.

The variation in NDT evaluation practices can be partially attributed to the lack of definitive NDT requirements on casks such as those imposed on pressure vessels and associated plumbing by ASME Codes and Standards. (Here again, we can expect this to change on next generation LWR Casks since both NRC and ANSI Cask Guides are expected to reference applicable NDT sections of the ASME Boiler and Pressure Vessel Code.)

It is difficult to understand the lack of imposition of state-of-the-art NDT methods on critical cask elements such as closure weldments and tie downs on some of the LWR casks. In most instances, the cask design agency and the cask fabricator had access to knowledgeable in-house NDT personnel and equipment that they used routinely to process ASME code vessels.

NDT methods are used most often as verification of basic construction requirements during cask fabrication. Application of the same methods on subsequent in-service or post-service examinations is understandably most difficult in view of the residual radioactivity on all inside cask surfaces and the excessive thickness of the overall case assembly. Even so, good baseline ultrasonic data on critical weldments following initial fabrication does not exist on some of the current casks and would be difficult, if not impossible, to obtain at this time.

Our analysis of NDT capabilities versus current LWR requirements for NDT application revealed only one major limitation. Volumetric NDT of austenitic stainless steels being used on the interior and exterior surfaces of most LWR casks is exceedingly difficult. This limitation is being studied widely in the USA and other countries[17] but results to date of "round robin tests" have merely confirmed the extent of the evaluation problem. This subject and weldments in general deserve more discussion; hence, a summary is given in Appendix B.

NDT of A508 or A533 structural steel weldments is being successfully applied to one or more of the current LWR casks without major difficulty. Other NDT requirements such as radiography or radiation gaging for LWR shielding defects are well within the capabilities of current nondestructive testing equipment[18,19,20].

Visual inspection is given a high degree of importance in cask evaluation both during fabrication and after use. In fact, the total NDT in-service practice on several casks is to visually inspect for shipping damage and corrosion (using special mirrors and floodlights or borescopes for remote viewing of radioactively hot cask interiors) followed by a pressure drop or leak test. More sophisticated NDT methods are only applied if damage is detected or leaks are evident.

The following NDT test plan is a composite of several different practices used on LWR casks. It represents a workable and meaningful NDE program that should carry over to many of the future generation cask designs - with augmentation of additional methods as needed for more rigorous design requirements.

A. During fabrication or prior to initial use:
   - Certification of all materials;
   - Ultrasonic scans of all precursor plate stock;
   - Radiography and/or ultrasonics of all weldments on subassemblies and final assembly;
   - Visual inspection of all surfaces;
   - Dye penetrant checks of all ID surfaces and OD weldments plus tie downs and lifting trunnions;
   - Radiation gauging of shielding integrity and uniformity;
   - Leak test of all seals and entire cask;
   - Radiography and/or ultrasonics of all lifting and handling members;
   - Over pressure test ($1\frac{1}{2}$ x MAWP) of sealed cask.

B. During use and periodic maintenance:
   - Visual checks of all surfaces, trunnions, and tie downs for shipping damage and corrosion products;
   - Leak test of entire cask before and after each use;
   - Penetrant test of any damaged areas;
   - Ultrasonic test of OD weldments if damage is visually indicated;
   - Replacement and testing of seals if not designed for multiple closure.

VI. BREEDER REACTOR CASKS AND THEIR NDT PROJECTIONS

The cask for breeder reactor spent fuel must attenuate gamma rays by 8-9 orders of magnitude and neutrons by 4-5 orders. It must keep the outer surfaces of the cask below 350°F while rejecting as much as 100 kW of heat. It must provide the same structural and radiation containment as LWR casks after subjection to the damage tests specified in 10 CFR 71.

Several conceptual cask designs for shipping breeder reactor spent fuel have been generated to date based on the best trade-offs commensurate with the current knowledge of FFTF fuel and expected reactor operating practices. ORNL analyzed a conceptual railcar design in 1972 based on transporting 6 fuel assemblies (3000 pounds) of 30-day cooled fuel irradiated to 100,000 Mwd/MT (megawatt days/metric ton) or approximately three times as hot as LWR fuels. Their design[21] incorporated sodium as the primary passive coolant, steel for gamma shielding, lithium hydroxide/sodium hydroxide for neutron shielding and triplicate sealing.

Aerojet Manufacturing Company described a conceptual railroad car cask for FFTF fuel held for 180 days after removal from the reactor[22]. Their cask would carry 9 breeder reactor fuel assemblies in a liquid filled cannister shielded in a steel/depleted uranium cylinder for gamma shielding plus an exterior neutron shield water jacket. The liquid coolant would be Dowtherm A (eutectic mixture of diphenyl and diphenyl oxide). Helium gas in a space between the cannister and shield would be provided to enhance heat transfer and to serve as a buffer between the two seals.

Figure 3 shows a cross-sectional view of a reference breeder reactor spent fuel reference cask with only 2 layers - the depleted uranium gamma shield and the neutron shield - as variable elements in the conceptual design. The thicknesses of these 2 layers were adjusted independently using computer code calculations until the dose rate 1.83 m (6 feet) from the outer surface was 5 mrem/hr, half of which would be attributable to neutrons and half to gamma rays.

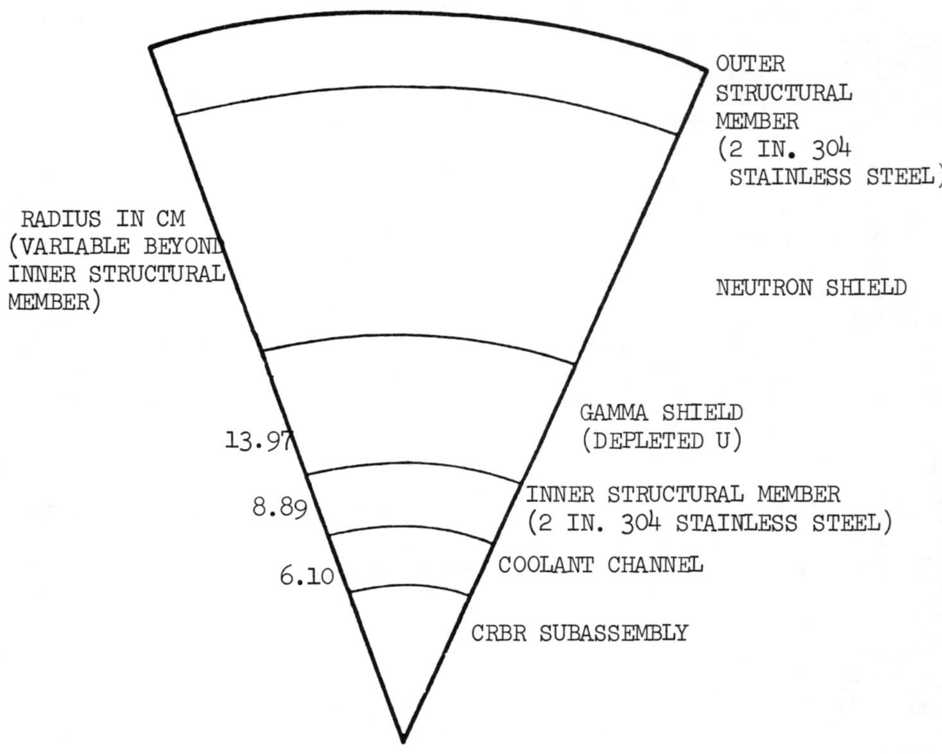

Fig. 3. Trial Reference Cask Configuration

Figure 4 presents the results of these calculations diagrammatically. Five different neutron shielding materials were used in the analysis - lithium hydroxide ($LiH_{0.8}$), didynium or misch metal hydride ($DiH_{2.5}$), beryllium, boron carbide (B_4C) and graphite (C). The latter three candidate neutron shield materials, although less efficient than the misch hydrides, are readily available, thermally stable and relatively chemically inoffensive.

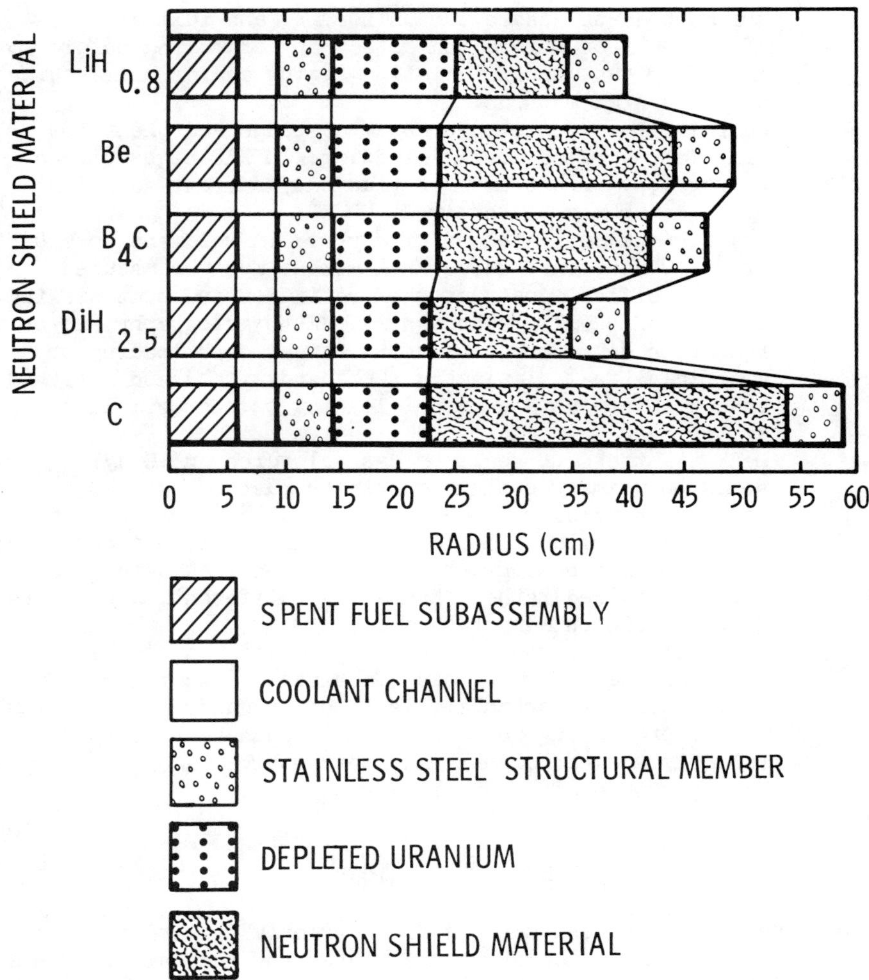

Fig. 4. Radial Dimensions of a Trial Reference Spent-Fuel Shipping Cask for One 30-day Cooled CRBR Spent Fuel Subassembly

Nondestructive testing requirements over and above those currently being applied to LWR cask have not been defined for the conceptual breeder casks. However, the following incomplete projections of specialized NDT needs appear valid at this date:

Improved weldment test methods for the austenetic steel structural members (SS 304). Focused and/or phased array ultrasonics may be necessary.

- . NDT methods to ensure the uniformity and full density of the neutron shield and detect any crumbling before and after cask use. Radiation gaging and/or ultrasonics may be useful methods.
- . Uniformity of depleted uranium castings and their 10 mil flame-sprayed copper cladding. LINAC high voltage radiography or ultrasonics plus eddy current scanning of cladding may be required.
- . Quality and completeness of copper fin weldment to O.D. of SS 304 external structural member must be assured.
- . Overall effectiveness of cask radiation and heat dissipation systems will be necessary. Autoradiography and/or radiation detector scans using actual cask loading or artificial fuel simulators (Co^{60} and Cf^{252}) and infrared scan with actual spent fuel loading or heat simulator are projected.
- . Leak tightness of seals on cask closures and Graylok seals on primary coolant cannister must be assured[25]. Worst case pressure rises of 10-30 atmospheres within the casks due to fuel element rupture have been estimated. Helium mass spectrographic and halogen leak detectors may be required rather than a PΔV (differential pressure drop) as run on many LWR casks.
- . Size/weight/thickness/processing considerations will make NDT checks during cask fabrication and prior to use of great importance. Tests of precusor materials before processing and after welding can be performed reliably and efficiently at subassembly stages, but may be impossible after final assembly.

VII. CONCLUSIONS

1. Nondestructive test methods are currently playing a major role in assuring the structural integrity and safety of LWR spent fuel shipping casks. However, NDT requirements vary considerably from cask to cask, but will become more definitive in the future.
2. Cask fabrication and users are not applying state-of-the art NDT methods which are used on other parts of the nuclear fuel cycle (i.e., on pressure vessels, fuel rods, generator tubing, valves, etc.). Lack of ASME code requirements[26,27,28] on LWR casks undoubtedly contribute to this generalized shortcoming.
3. Some basic improvements in NDT methods are desirable for LWR casks and will be mandatory for breeder reactor casks. Austenetic stainless steel weldment testing is the most deficient of current and projected NDT requirements.

VIII. REFERENCES

(1) L. B. Shappert, W. E. Unger, and J. M. Freedman, "Spent and Fresh Fuel Shipping Cask Considerations," Proceedings of the 23rd Conference on Remote Systems Technology, ANS (1975).

(2) G. J. Dau, "Thoughts on Emerging NDE Technology for the Nuclear Power Industry," Metals Progress, (June 1976), p. 35.

(3) R. W. McClung, "Into the Looking Glass: An Introspective View of Nondestructive Testing," Materials Evaluation, Vol. 33, No. 2, p. 16A.

(4) R. Halmshaw, "Potential Developments in NDT," British Journal of NDT, (January 1977), p. 21.

(5) R. W. McClung, "Needs for Development in Nondestructive Testing for LMFBR and Other Advanced Reactor Systems," ORNL-TM-5093, (November 1975).

(6) "Draft Environmental Statement on the Transportation of Radioactive Material by Air and Other Modes," U.S. Nuclear Regulatory Commission, NUREG-0034 (March 1976).

(7) W. P. Bebbington, "The Reprocessing of Nuclear Fuels," Scientific American, (December 1976), p. 30.

(8) "Packing of Radioactive Material for Transport and Transportation of Radioactive Material Under Certain Conditions," Part 71 of Title 10, Chapter 1, Code of Federal Regulations (April 30, 1975).

(9) L. B. Shappert, et al, "Cask Designer's Guide," ORNL-NSIC-68, (February 1970).

(10) H. R. Yoshimura, M. Huerta, "Full Scale Tests of Spent Nuclear-Fuel Shipping Systems," Transport Packaging for Radioactive Materials, IAEA-SR-10/17, Vienna, Austria (1976).

(11) R. M. Jefferson, H. R. Yoshimura, "Crash Testing of Nuclear Fuel Shipping Containers," SAND77-1462C, National Academy of Science Meeting of Transportation Research Board, January 1978.

(12) Kohtars Veki, "Spent Fuel Shipping Casks and Their Nondestructive Inspection Methods," Japan J. NDT 24, No. 6, p. 350 (1975), Translated for Sandia Laboratories, RS 3140/684.

(13) "Fabrication, Testing and Inspection of Shielded Shipping Casks for Irradiated Reactor Fuel Elements, Proposed ANSI Standard N14.8," (first draft dated August 1975).

(14) D. W. Ballard, "Bibliography of NDT Reports and Other References Pertinent to Sandia's Spent Fuel Shipping Cask Project," SAND76-0748 (December 1976).

(15) S. A. Dupree and H. J. Rack, "Status of Radiation Shield for LMFBR Shipping Cask Application," SAND76-0595, (September 1976).

(16) J. M. Freedman and R. B. Pope, "Program for the Development of Technology for the Safe Shipment of Spent Fuel for Commercial LMFBR in USA," Proceedings of IAEA Seminar on Transportation of Radioactive Materials, Vienna, Austria, August 23-27, 1976.

(17) J. C. Spanner, "Preliminary Development of In-Service Inspection Methods for LMFBR's," HEDL-SA-1092, Hanford Engineering Development Laboratory, (January 1976).

(18) "A Survey of Nondestructive Testing," Southwest Research Institute for NASA, NASA-SP-5113, (1973).

(19) R. L. Buckrop, "Nondestructive Testing Overview," Quality Progress, p. 26, (June 1976).

(20) "Practice Sensitivity Limits of Production NDT Methods in Aluminum and Steel," Boeing Commercial Plane Company for AFML, AFML-TR-74-241, (March 1975).

(21) A. R. Irvine, et al, "LMFBR Spent Fuel Transport: Conceptual Design and Partial Safety Analysis of a Sodium-Cooled Cask," ORNL-TM-3689, (February 1972).

(22) Aerojet Mfg. Company, "Conceptual System Design Description: LMFBR Spent Fuel Shipping Cask," Contract AT(04-3)-782, AMCO-05-M-01, February 1972.

(23) G. C. Allen, et al, "Spent Fuel Shipping Cask Designs for Breeder Reactors," Transactions for ANS Winter Meeting, November 27-December 2, 1977, SAND77-1111A.

(24) S. A. Dupree, "Radiation Shield Design for LMFBR Casks," Transactions for ANS Winter Meeting, November 27-December 2, 1977, SAND77-0971C.

(25) O. L. Burchett, J. H. Biffle, "A Program for Static Testing Cask Seals," SAND76-0287, (June 1976).

(26) "Rules for Inspection and Testing of Components of Liquid Metal Cooled Fast Breeder Reactors, Proposed Draft 7 for Division 3 of Section XI, ASME Code," M. S. Markowitz, Chairman.

(27) E. B. Branch, "The Impact of ASME Section III," Power, (October 1975), p. 23.

(28) R. R. Irving, "Why the ASME Code is in a Class by Itself," Iron Age (January 17, 1977), p. 29.

IX. APPENDICES

A. Acknowledgments and Key Organizational Contracts

A grateful acknowledgment is made of information from all of the organizations and individuals who were contacted specifically for contributions to this report either in person or by phone. Most of the key contacts are listed below, but there were others, both inside Sandia Laboratories and DOE, who also provided useful input to this cask NDT report. The organizations are listed below in alphabetical order, while the listing of conferences and symposiums attended is in chronological order and covers the period from July 1, 1975 through December 31, 1977.

Key Organizational Contacts.

A. Aerojet-Nuclear, Idaho Falls, Idaho
   - W. Francis, Materials Engineering
   - L. Lott, NDT

B. Babcock & Wilcox, Lynchburg, Virginia
   - A. Wehrmiester, NDT
   - A. Holt, NDT
   - J. R. Birchek, NDT

C. Battelle/NW Lab, Richland, Washington
   - G. J. Posakony, Manager of NDT
   - L. Becker, NDT

D. Boeing/Nuclear Power Division, Seattle, Washington
   - D. Martyr, Contract Administrator
   - W. R. Milne, Cask Design

E. Combustion Engineering, Chattanooga, Tennessee
   - R. M. Stone, NDT
   - W. A. Young, Nuclear Quality Engineering

F. DuPont/AE Division, Wilmington, Delaware
   - J. W. Langhaar, Cask Design
   - H. Brewer, Quality Control
   - J. Landrum, Quality Control

G. DuPont/Savannah River
   - L. E. Weisner, NDT
   - W. Joseph, NDT
   - R. Middleburg, Cask Design

H. Electric Power Research Institute, Palo Alto, California
   - G. T. Dau, Projects Manager for NDT

I. Foster-Wheeler, Livingston, New Jersey
   - I. Berman, NDT Liaison
   - R. Henschel, Solid Mechanics Department
   - M. F. Aiello, Solid Mechanics Department

J. Nuclear Assurance Corporation, Atlanta, Georgia
   - J. Viebrock, Cask Design
   - J. Rollins, Cask Fabrication

K. Nuclear Containers, Elizabethtown, Tennessee
   - W. R. Housebolder, Manager
   - N. L. Green, Vice President

L. Nuclear Regulatory Commission, Washington, D. C.
   - W. F. Anderson, Division of Engineering Standards
   - J. Muscara, Division of Reactor Safety Research
   - E. J. Brown, Division of Engineering Standards
   - V. S. Goel, Division of Engineering Standards

M. Oak Ridge National Laboratory, Oak Ridge, Tennessee
   - R. W. McClung, NDT R&D
   - J. N. Robinson, NDT Inspection
   - L. B. Shappert, Cask Technology
   - C. Brinkman, Reactor Metallurgy
   - J. Hammond, Technical Contact with Y-12

N. Philadelphia Electric, Philadelphia, Pennsylvania
   R. H. Zong, Quality Manager
O. Southwest Research Institute, San Antonio, Texas
   C. E. Lautzenheiser, Vice President
   S. A. Wenk, NDT, R&D
P. Union Carbide Y-12 Plant, Oak Ridge, Tennessee
   V. C. Jackson, NDT Production
   L. E. Burkhart, NDT R&D
   D. L. Mason, NDT R&D
   R. Oakes, Mechanical Properties
   W. Moyer, Mechanical Properties
Q. Westinghouse/Hanford, Richland, Washington
   J. C. Spanner, Materials Application
   D. Green, NDT

Conference/Symposium Participation and/or Attendance.

A. Materials/Design Forum, San Francisco, CA, July 9-11, 1975, J. C. Bushnell
B. Review of Quantitative NDE, Thousand Oaks, CA, July 15-17, 1975, A. G. Beattie
C. Failure Prevention & Reliability Conf., Washington, D. C., September 17-19, 1975, D. W. Ballard
D. Ultrasonics Symposium, Los Angeles, CA, September 22-24, 1975, A. G. Beattie
E. Third Annual Boiling Water Reactor Mtg., Gaithersburg, MD, September 29-30, 1975, O. J. Burchett
F. Eddy Current Methods Group (ASME), Denver, CO, October 9-11, 1975, J. C. Bushnell
G. Quality Testing Conference (ASNT), Atlanta, GA, October 13-16, 1975, D. W. Ballard
H. NDE in Nuclear Industry, Denver, CO, December 1-3, 1975, J. H. Gieske and D. W. Ballard
I. Acoustic Emission Working Group, Ft. Lauderdale, FL, January 19-23, 1976, A. G. Beattie
J. Critical Materials Availability Conf., Reston, VA February 24-26, 1976, D. W. Ballard
K. Spring ASNT Conference, Los Angeles, CA, March 9-13, 1976
L. Review of ARPA-Sponsored NDT Programs, San Diego, CA, July 8-10, 1976, D. W. Ballard and J. C. Bushnell
M. Fourth Henniker Conference, Henniker, NH, August 9-13, 1976
N. Second International Conference on Materials Behavior, Boston, MA, August 16-20, 1976, D. W. Ballard
O. Review of Quantitative NDE, Alsimar, CA, August 30-31, 1976, O. J. Burchett
P. Fall ASNT Conference, Houston, TX, September 27-30, 1976, R. W. Mottern
Q. IEEE Acoustical Imaging & Holography Symposium, September 29-October 1, 1976, J. H. Gieske

R. Acoustic Emission Working Group, Williamsburg, VA, October 17-20, 1976, A. G. Beattie
S. WANTO NDT Conference, Livermore, CA, December 6-9, 1976, D. W. Ballard, R. W. Mottern
T. ASME Code Meeting, Ad Hoc Acoustic Emission Group, Ft. Lauderdale, FL, January 23-26, 1977, A. G. Beattie
U. QUINTECH-77, Golden Gate Welding & Materials Conference, San Francisco, CA, January 26-28, 1977, D. W. Ballard and J. H. Gieske
V. ASNT Spring Conference, Phoenix, AZ, March 26-April 1, 1977, D. W. Ballard
W. ASME Code Meeting, Ad Hoc Acoustic Emission Group, Livermore, CA, June 15-17, 1977, A. G. Beattie
X. ARPA/AFML Review of Progress in Quantitative NDE, Ithica, NY, June 13-17, 1977, D. W. Ballard
Y. ASNT Conference on Innovative and Advanced Radiography, Wilmington, DE, August 1-3, 1977, F. A. Hasenkamp
Z. Energy Technology Conference, Houston, TX, September 19-22, 1977, J. H. Gieske
AA. Ninth National SAMPE Conference, Atlanta, GA, October 3-6, 1977, A. G. Beattie

B. Nondestructive Testing of Weldments

Welding is the specified method of joining that is common to all LWR shipping casks as well as most of the other components of LWR power plants. It is a highly reliable process that is well understood and produces consistently high strength joints when properly controlled. Weldment testing has been developed and refined over the past decade and a variety of useful NDT methods are available. Unfortunately, a combination of factors still make weld joints the weak links in most structures inside and outside the nuclear field.

Currently, the major problem that is plaguing LWR plants is the difficulty of testing, with a high degree of reliability, weldments in austenitic stainless steels (Type 300 series, commonly called 18:8 types). Austenitic steels are not heat treatable since they retain an essentially austenitic microstructure at all temperatures. Moreover, they have high ultimate strength, ductility and toughness, retaining notch-tough even at very low temperatures. At elevated temperatures they are relatively strong as compared to iron/chrome alloys, ferritic stainless and martensitic stainless (400 series). For these reasons, austenitic steels such as types 304, 308, and 316 are widely used for nuclear applications.

ASME Code Case 1594 requires double volumetric examination of critical weldments, allowing in some cases (i.e., in austenitic stainless steels because of problems with ultrasonics) two radiographic techniques at different angles to satisfy this requirement. This two-view radiography is questionable as to its validity and efficiency when compared with ultrasonics which is successfully used on standard nuclear structural steels such as A508 and A533.

Coarse grained austenitic stainless steel weldments are difficult to interrogate with standard ultrasonic methods because of the large and variable attenuation and the high level of noise encountered during an examination. False defect indications and masking of real defects in the weldments and adjacent heat-affected zone have raised real doubts as to the credibility of an ultrasonic method for such weldments. The problem is being studied extensively in the USA and abroad[5,17] and round-robin experiments are underway.

Potential solutions to excess noise from austenitic weldments include the use of focused transducers, use of frequency analysis, phased arrays of transducers, advanced data processing to improve the signal-to-noise ratios, artificial aperture scanning techniques similar to acoustic holography, and use of non-contact or electro-magnetic transducers for ultrasonic testing. Acoustic emission monitoring during and following welding and electro-thermal methods have also been proposed as alternates to conventional ultrasonics.

As initially stated, weldments can be highly reliable and are the preferred method of joining in cask assemblies. Factors requiring close attention to consistently achieve joints as strong as the parent metal include:

1. Optimum weld joint design considering local applications, wall thicknesses, materials, etc.
2. Optimum welding process selection considering materials, dissimilar metals, filler rods or electrodes, number of passes required, pre-heat and post-heat treatments, etc.
3. Qualification of both welders and NDT personnel involved in weld examinations. Section IX of the ASME Code is the preferred guide for the qualification procedures.
4. Rigorous weld rejection and repair criteria must be documented and followed.
5. All weld external surfaces, including the heat and affected zone, must be inspected for cracks using either liquid penetrants per procedure A-2 or B-3 of ASTM E-165 or with magnetic particles using ASTM E-109.

NEEDS FOR DEVELOPMENT IN NONDESTRUCTIVE TESTING FOR ADVANCED REACTOR SYSTEMS*

R. W. McClung
Metals and Ceramics Division
Oak Ridge National Laboratory
Oak Ridge, Tennessee 37830

ABSTRACT

The needs for development of nondestructive testing (NDT) techniques and equipment were surveyed and analyzed relative to problem areas for the Liquid-Metal Fast Breeder Reactor, the Molten-Salt Breeder Reactor, and the Advanced Gas-Cooled Reactor. The paper first discusses the developmental needs that are broad-based requirements in nondestructive testing, and the respective methods applicable, in general, to all components and reactor systems. Next, the requirements of generic materials and components that are common to all advanced reactor systems are examined.

Generally, nondestructive techniques should be improved to provide better reliability and quantitativeness, improved flaw characterization, and more efficient data processing. Specific recommendations relative to such methods as ultrasonics, eddy currents, acoustic emission, radiography, etc., are made. NDT needs common to all reactors include those related to materials properties and degradation, welds, fuels, piping, steam generators, etc. The scope of applicability ranges from initial design and material development stages through process control and manufacturing inspection to in-service examination.

---

*Research sponsored by the Division of Reactor Research and Technology, United States Department of Energy, under contract W-7405-eng-26 with Union Carbide Corporation.

INTRODUCTION

Because of recognition of the important role that nondestructive testing has in assuring the quality and integrity of nuclear reactor systems, a report was prepared to delineate problems and indicate areas in which development should be accomplished. The report is a summation based on (1) the knowledge and experience of the author and the NDT development group at Oak Ridge National Laboratory (ORNL); (2) an extensive survey conducted among knowledgeable personnel (designers, fabricators, and NDT experts) from government, industry, and other sources; (3) previous reports that discuss NDT needs. This paper is an abbreviation of portions of that report.

Although emphasis was given to the Liquid-Metal Fast Breeder Reactor (LMFBR), attention was also given to the High-Temperature Gas-Cooled Reactor, and the Molten Salt Reactor. The general organization for this paper includes (1) general needs for improved NDT methods, techniques, and equipment; and (2) needs for components that are common or similar for different reactors. Fuel elements will not be included. This does not imply that fuel assay and safeguards technology are not part of the overall family of nondestructive testing.

METHODS-ORIENTED PROBLEMS AND NEEDS FOR NDT DEVELOPMENT

Needs Common to All Methods

Many of the recognized problems (or shortcomings) of the current state of technology in nondestructive testing are independent of any given reactor system or component. The needs are shared by many or all of the reactor systems and components. Therefore successful solutions to these problems will accrue directly to the benefit of any reactor system under consideration.

A Brief Look at Administrative Needs. Although the principal thrust of this paper is toward necessary research and development for nondestructive testing, this aspect does not cover all that is considered to be vital to a realistic and optimum use of the technology. Although no significant elaboration will be provided on the "administrative" or "software" side of NDT, brief mention will be made to provide a more complete coverage of the overall subject. For example, NDT has traditionally been limited by many to the detection of discontinuities in manufactured goods and, more recently, to the in-service examination of assembled products as a function of maintenance. The consideration of NDT should begin with the first stage of design to assure that the manufactured item can be inspected, both from the standpoint of its configuration and the relative accessibility in the assembled system.

Any necessary development of NDT methods or techniques should begin early enough on critical-path schedules to assure that the tools are in hand to be used in other developmental phases such as fabrication, new materials, etc. By having proper dialogue in these early planning stages among designers, materials people, and NDT personnel, the correct questions can be asked (and answered) about what properties need to be measured; e.g., is it density, presence of porosity, cracks or other discontinuities, dimensions, modulus, or some other. The early development and application of NDT is necessary for the establishment of correlations among the signals obtained, conditions of material producing the signals, and the effects of these conditions on the materials performance in service. Thus, this will not only provide useful data to further the development of materials or manufacturing processes, but will also provide the guidance for practical acceptance limits on the finished goods. Major benefits can be obtained through greater application of NDT for process control; e.g., the number of acceptable items will increase per unit of manufacturing, decreasing costs and improving delivery schedules.

Another administrative problem that plagues many organizations and fabricators is the lack of competent, qualified personnel. Since most of the NDT technology is people-oriented (and will continue to be for some time), the results of examination are heavily dependent upon the personal abilities, judgment, and interpretation of the operator. With expanding demands for NDT throughout all industry and limited educational opportunities for development of new personnel (primarily short courses for engineers and a few vocational schools for technicians), it is becoming a greater problem to assure that inspection organizations have the qualified people who can meet the certification requirements of most nuclear standards. This can, of course, affect the reliability of the NDT being performed. Among the several solutions to this problem are increased education and training to provide upgrading and a greater supply of competent personnel, and development of improved equipment, techniques, and technology to reduce the reliance on the skill and subjective reasoning of the operator.

Need for Improved Reliability and Quantitativeness. Frequently, as nondestructive examinations are repeated, different results are obtained within an organization or between separate organizations, e.g., between the seller and purchaser of material. This variability creates confusion, debate, commercial problems, and questions about safety of components and the reliability of the examination method. Since NDT is a system embodying people, materials, procedures, techniques, and equipment, the solution to the problems of reliability may take several forms. We have already noted the need for qualified people. The operators require written procedures that will assure an unambiguous understanding, which, when properly followed, will result in more consistent performance. Too often existing documents are not adequately specific and definitive and need improvement. We need an increased knowledge of the many variables in both techniques and equipment that can affect examination results so corrective action can be taken. To cite just one example in this section, the frequency spectrum in an ultrasonic pulse can vary significantly in an examination, and with inadequate knowledge of both the changes and the frequency response of allegedly similar transducers and instruments, variable results can occur. Studies should be performed to gain a better understanding of the statistical variability and reliability of various NDT techniques. For example, the Air Force has sponsored work on the ability to detect various size flaws using both penetrant and magnetic particles; a few other limited studies have been conducted. The task at HEDL on FFTF fuel cladding, in which statistical limits and control on ultrasonic testing were established, is an example of such work, which should be accomplished for numerous other applications. To enable fracture mechanics or other analytical tools to be used as a design method, one needs to know with higher confidence the lowest level of detection (smallest flaw that will *always* be detected) on a reliable basis (and if necessary determine what needs to be done to achieve higher sensitivity). In some instances it may be sufficient to have numerical confidence in the smallest flaw that will be detected 95% of the time, or 90% or even

50% of the time. Quantitative statements about reliability of present systems can be made only after statistically controlled studies; necessary improvements to less-than-adequate reliability will require appropriately directed development on the elements of the test.

A related consideration is the increasing necessity for more quantitative data from NDT. The technology has been directed toward the qualitative detection and location of discontinuities. However, there is an increasing need for upgraded capability to achieve quantitative measurement and characterization of discontinuities or other conditions. Some of the conditions contributing to this need include the advent of fracture mechanics to calculate the significance of flaws to serviceability; but successful application of fracture mechanics depends on a confident awareness of the exact size and orientation of a flaw that is detected (or a quantitative awareness of the size of the largest flaw that could go undetected).

Need for Improved Flaw Characterization. The NDT techniques are frequently comparisons of response between a known artificially produced artifact such as a drilled hole or machined notch and the detected unknown condition (and the responses are usually described in terms of that from the artificial rather than that of the actual flaw). As discussed in the previous section on quantitativeness, there is need for new and improved NDT capabilities to identify and interpret the responses in terms of the exact character of the internal condition. For example, when a flaw is detected during in-service inspection several immediate questions must be answered with confidence: What is it? Is is new? If old, has it grown? What is its exact through-the-wall dimension? How close is it to a surface? These are questions to which accurate answers must be provided, and technological development must be accomplished to achieve the desired status.

Need for Increased Computerization of Data. Another problem with broad-based application to many NDT methods and reactor components is that for increased need of computers or other advanced instrumentation for processing and analyzing data. Some of the technology may be available in other instrumentation or communications technology.

There are several aspects and time scales for such need and application. The earliest applications (and embryonic work has begun) is for simple data collection, processing (possibly enhancement), and analysis of voluminous data in production-type applications for repetitive work and to relieve the human operator. For fully successful application of computers to such go no-go evaluations, data collection and storage, the previously described needs for improved reliability and quantitativeness in techniques and equipment are obviously paramount.

The next stage of computer needs and application will include the extraction of more basic data from NDT signals (e.g., frequency spectra and phase of ultrasonic pulses) to allow more interpretation capability or pattern recognition. However, achieving this stage will require additional development of knowledge within the respective disciplines of ultrasonics, eddy currents, etc. Successful completion could, of course, do much to solve the aforementioned problem for improved defect characterization.

A third stage of computer development and integration in NDT that is ultimately needed is for the development of basic NDT equipment designed for direct computer-coupling (e.g., with full digital electronics) rather than simply attaching a computer to receive signals tailored for human interpretation. This design philosophy would allow more distortion-free data to be available for automatic, real-time processing, analysis, and interpretation. To cite just one benefit, in manufacturing the feedback in process control would do much to assure that only acceptable parts are fabricated.

Need for Improved Shop, Field, or Commercial Application. As techniques and equipment are developed in laboratory studies, these should be completely followed through so that the technology has ample opportunity to reach its full application on the production line, in the field, or in other appropriate areas. This will not only require extensive effort beyond the feasibility or demonstration stage, but will also require consideration for simplicity of operation or performance, for system reliability, and for economics.

Individual Methods

The previous sections on needs common to all methods of NDT should be kept in mind as the following sections on individual methods are presented. Although occasional brief inference may be made to previously cited needs, to avoid undue repetition no concerted effort will be made to assure citation on every method to which the general needs are applicable.

Penetrating Radiation. This method covers all application of x rays, gamma rays, and neutrons for qualitative imaging (e.g., radiography) and quantitative measuring (e.g., radiation attenuation). The most common application, radiography, has been widely used for years without significant changes. A major problem is the reliance on the human operator to make a subjective evaluation of the radiographic image of both the specimen and the image quality indicator (IQI), or penetrameter, used to assure the basic radiographic quality. Work is needed to reduce the human element and make the method more reliable and the results less subject to controversy. Potential approaches for accomplishing this are studies to improve the IQI and work on radiographic image enhancement. The American standard IQI is a thin shim or plaque containing cylindrical holes. It has decreased applicability for thin (less than 0.25-in.) sections and for assuring that narrow linear indications are visible. The European standard penetrameters use small-diameter wire to demonstrate image quality. These, too, have limitations. A fresh look should be taken at each one of these (and other) concepts in an attempt to improve the IQI. Radiographic image enhancement is a special case of the NDT data enhancement that was mentioned earlier. Sophisticated work has been accomplished using large computers as an outgrowth of the space program and photographic enhancement. Most of these efforts deal with reducing distortion, reducing noise, and sharpening edges. Beneficial results have been obtained, but large, expensive computer facilities are required. Less expensive approaches are needed that can be within the price range of most radiographic laboratories. The limited work with closed-circuit television for image enhancement needs to be expanded to achieve better image (not just the edge) enhancement, and further to automatic processing to simplify interpretation. Another enhancement approach that should be investigated is Fourier optical processing to accomplish frequency filtering similar to that performed with computers.

There is a need for improved techniques and equipment for performing radiography as a part of in-service inspection in a radioactive background. With the necessity to examine through both walls of pipe (frequently containing fluid) this could require higher energy sources than are currently available for field application. An alternative for "hot" radiography would be to develop techniques incorporating specimen shielding and slit scanning similar to that which has been used for hot-cell application.

Among other problems are needs for (1) improved techniques for dimensional measurements, (2) crack detection, (3) better guidelines for choice of radioisotopes or x rays, and (4) improved ability to ignore surface irregularities (such as weld roughness).

As cited in an earlier section, there is a need for more quantitative data and improved flaw characterization. Several developmental approaches that could help to achieve this with penetrating radiation include the aforementioned radiographic image enhancement, layer radiography (e.g., laminagraphy or tomography), advanced film densitometry, and direct measurement of the transmitted radiation with quantifying detectors (such as scintillators or semiconductors).

Ultrasonics. Many of the needs cited in the section covering problems common to all methods such as improved reliability, quantitativeness, flaw characterization, signal enhancement, and computerization of data are directly applicable to ultrasonics, and detailed discussion will not be repeated here. In addition to these general needs (whose solution would significantly upgrade the applicability and and capability of ultrasound) there are a number of specific needs, which in some particulars may be also related to the general needs. The recognized needs will not be presented in order of importance or priority.

Improved calibration techniques are needed. This is a multifaceted problem including the transducers, the electronic circuitry of the instrumentation, the physical material used as reference and transfer standards (test blocks containing mechanical flaws), and the assembled system with the attendant inspection procedures. The calibration procedures should include both the in-laboratory or bench-type techniques for preinstallation and periodic maintenance and the on-line, rapid recalibration of an inspection system to assure continued consistent performance.

There is a need for higher confidence and control on the size, shape, and direction of the ultrasound beam emitted from the transducer and projected into the specimen. Successful solution would increase the confidence in flaw location, in resolution of closely spaced reflectors, and in defect characterization. One approach that appears to have merit, based on the results of several investigators, is the use of multielement transducers with adjustable excitation for beam direction and control. However, further study on the concept is needed.

A better understanding of ultrasonic propagation and reflection is needed. One method of gaining an improved practical awareness is through the use of ultrasonic schlieren techniques. Studies should be performed with advanced schlieren systems to investigate the behavior of new-design transducers, and to study the interaction of sound with discontinuities in both transparent (e.g., glass) and opaque (i.e., metallic) specimens, thus gaining new insight about transducer design, ultrasonic focussing, transducer placement, and technique development.

Most current ultrasonic techniques use primarily the amplitude of the signal and time of arrival for interpretation, with only nominal attention given to the frequency and no attention to the phase of the signals. Investigations have shown that the spectral content of the ultrasonic pulses can vary and have a significant effect on the performance. The phase of reflected signals will be affected by the acoustic properties at the interfaces. Major effort should be directed toward gaining a better understanding of the role of frequency spectra and phase in ultrasonic nondestructive testing. The expected benefits would be the gaining of significantly greater information from the ultrasonic signal about the character, size, and orientation of reflectors, and improvement in the reliability of ultrasonic technology. Preliminary studies have already demonstrated the feasibility of such endeavors.

The successful utilization of more data (e.g., for production, in-service inspection, or the above-mentioned frequency and phase) would increase the need for improved instrumentation to simplify processing of the signal (perhaps with computers) without the distortion that frequently occurs with present-day devices. New instrumentation should probably be based on digital circuitry to enhance the possibility that the real signals are computer compatible for processing, analysis, and storage.

Some of the previously cited problems of quantitative measurement, characterization, and location of flaws and test reliability are particularly troublesome in thick sections (e.g., greater than 4–6 in.). For example, based on in-service inspection studies in light-water reactor pressure vessels, one of the major problems is confident measurement of the through-the-thickness dimension of detected flaws. Work is needed to overcome these limitations.

A new technological approach to ultrasonic testing is acoustical holography, a technique for qualitative scanned recording of phase and amplitude information to allow reconstruction of an acoustic image with three-dimensional information. This allows qualitative visualization of a flaw (or other reflector) within an opaque specimen. Beneficial studies with the method should include increasing the examination speed and application to a large number of specimens containing actual flaws.

Acoustic Emission. In recent years there has been a tremendous upsurge of interest in and application of acoustic emission (AE) technology to monitor deformation and crack propagation or other stress-wave-emitting phenomena in a variety of materials and configurations. The several manufacturers of commercial equipment provide systems for qualitative detection and location of emitting sources. However, study is needed to obtain a better understanding of the the significance of the emission and an improved capability for quantitative interpretation of the signals in terms of, for example, flaw growth, deformation, or other material conditions. One current limitation on the interpretation capability is that the signals are frequently modified by the frequency response of the detector so that the actual characteristics of the event are obscured. A better understanding is needed of this as well as a determination of optimum processing of the received data (i.e., event counting, energy integration, etc.). In common with most NDT methods improved calibration procedures are needed to check the operation of the system. Improved sensors and mounting techniques are needed that have been demonstrated to be capable of operating for years at elevated temperature and in a radiation environment. As an in-service inspection

there must be knowledge developed about the effect of material degradation (e.g., embrittling, corrosion) on the initiation and propagation of AE events. Acoustic events have been demonstrated to have benefit during the hydrostatic testing of pressure vessels, but significant application to field-type installation is needed to gain experience and confidence. There is controversy over the applicability (or generation of signals) in austenitic stainless steels at about 500°F. For example, these reservations coupled with the low operating pressures raise questions about the benefit for monitoring breeder reactor vessels. Investigations should be made on stainless steel to resolve this question. Signal enhancement techniques are needed to improve the signal-to-noise ratio, and permit more quantitative data. Extensive application of AE on reactor components should be made to improve the quantitative and qualitative aspects and to gain a fuller appreciation of the limitations and impediments to the method. Application studies should be made with various welding processes because of the potential to serve as a real-time process control monitor, detecting flaws as they are created.

Eddy Currents. As with ultrasonics many of the common needs are also applicable to eddy-current technology; e.g., needs for reliability, quantitativeness, flaw characterization, signal enhancement, and data computerization. The large number of parameters that can affect a simple examination lead to the need for techniques that can allow discrimination and separation of unwanted variables from desired signals in both simple and complex structures. Directions for solutions include (1) development of improved understanding of the quantitative theory, (2) development and improvement of design technologies applying the theory of the electromagnetic induction process, and (3) the development of practical instrumentation capable of performing simultaneous interrogations at several frequencies (by multifrequency or pulsed techniques) and concurrently processing the multiparameter signals to isolate the desired data. Other needs are for improved measurement of flaws and characterization (and identification) of various discontinuities, including intergranular attack, tubing wastage, etc. With many reactor components being constructed of ferromagnetic material, there is a need for technique development toward eddy-current examination of such materials (despite the high permeability) for surface and near-surface flaws. For in-service examination there is a need for development to allow performance at elevated temperatures and in the presence of radiation. To enhance the application of eddy-current techniques both for manufacturing and in-service inspection, there is a need for simpler, more reliable instruments to be developed as well as improved calibration and procedures to assure proper functioning of this inspection system.

Liquid Penetrants. The developmental needs for the relatively simple penetrant techniques are significantly less than for the more sophisticated methods, but some of the needs are similar. For example, there is a need for improved techniques for checking the performance and sensitivity of a penetrant system. Since a requisite for the examination is that penetrant flow into the discontinuities, there is a need to assure that the flaws are not already filled with another substance (perhaps even the cleaning solution). There is a need for penetrant materials that can be readily used at high temperatures (e.g., 400–500°F) for welds that must be maintained at a preheat temperature. A study should be performed to determine if residual penetrant (even containing limited halogens) after an examination can have an adverse effect on subsequent welding or service.

Magnetic Particle Examination. The principal needs of magnetic particle techniques are for calibration procedures and standards to confirm the sensitivity and performance of the system, and development to achieve quantitative results in addition to the qualitative response currently obtained.

New Technology. The preceding sections have dealt with needed improvements to methods that are currently recognized and used for NDT. There needs to be continued, long-range study on new methods that have the potential for problem solving as a supplement to existing technology. To cite only a few, consideration should be given to exoelectron emission for stress and fatigue studies, Barkhausen noise, magnetoabsorption, Mössbauer techniques, optical holography, and microwaves. Some of these have had success in laboratory studies but limited success on real world samples because of a lack of preservice characterization or history on the tested items. With the increased emphasis for quality assurance, archive samples, and material traceability, the potential is now much greater for preservice characterization and subsequent detection of change due to deformation, degradation, flaw initiation, corrosion, or other undesirable condition.

Thermal techniques have also shown considerable potential for materials evaluation, but additional work is needed for improved techniques. In particular, study is needed to alleviate the problems due to variable emissivity without the necessity of painting the object black (the current practice). Tests at multiple wavelengths may resolve this problem. Thermal surface impedance techniques also show considerable promise for extracting additional data from a thermal examination (an analog to eddy-current technique), and further study should be performed. Because of the dependence on subjective interpretation of magnetic particle examination (see earlier section) there is a need for investigation of other magnetic techniques for applications to surface and near-surface examination of ferromagnetic materials. The techniques should include instrumented detection techniques for leakage flux to provide more quantitative outputs for flaw evaluation. Other magnetic techniques should use measurement of coercivity, permeability, or other magnetic properties as a tool for characterizing the material for such properties as hardness, heat treatment, and dimensions.

NDT NEEDS FOR MATERIALS AND COMPONENTS COMMON TO MOST REACTORS

This section is divided into two major parts: (1) general NDT needs that are related to the material of construction and may be applicable to several components and (2) the needs that are more particularly applicable to specific components. In either case the needs are generic to several (or all?) reactor systems.

NDT Development Needs Oriented to Materials for Reactor Usage

This section will include discussions of needs for techniques for measuring material properties, monitoring degradation, welds, and alloy identification.

Materials Properties and Degradation. There is a general need for development and application of NDT technology for the measurement of materials properties. There are at least three broad aspects of this need: (1) as an aid to materials characterization (e.g., in heat-to-heat or heat treatment variations), measurement of properties on actual structures to be used, minimizing destructive analysis; (2) determination of the degradation or loss of desired properties in service to enable a decision for continued service; and (3) determination of the effect of the change in properties or degradation on the performance of the NDT for overall integrity. Among the properties that could be measured are modulus, density, strength, (e.g., with appropriate correlations in graphite, concrete, or other brittle material), electrical conductivity (and with appropriate correlation the hardness of metals), and magnetic permeability. Included in conditions contributing to degradation of material properties are embrittlement (due to creep or irradiation), sensitization (making the material susceptible to intergranular attack), corrosion, deformation or other strain, creep, onset of third-stage creep, swelling, carburization, and decarburization. Each of these undesirable conditions needs to be detected (and measured) to assure that premature or unexpected failure does not occur. NDT techniques need to be developed to assist in-service detection and measurement of these conditions. In some instances it may be necessary to nondestructively detect conditions leading to degradation. For example, unknown residual stresses can contribute to premature failure, creep, or deformation. Surface contamination and subsequent elevated temperatures can lead to intergranular attack or stress-corrosion cracking. Appropriate development of NDT techniques should be conducted to detect the problem before the adverse effects occur.

Welds. Welding is a method of joining that is common to most components of most reactor systems. The several needs for NDT development are therefore common. One of the problems centers around the occasional use of the wrong welding electrodes. The solution is probably a combination of improved NDT technology to verify the identity of the welding rod at the vendor, in the fabrication shop, and in the field as well as improved quality assurance (QA) procedures and control to prevent the weld being made with the improper electrodes, leading to inadequate strength and cracking. It has been suggested that approximately one-half of the bad welds can be traced to inadequate QA control. A common examination technique for thick multipass welds is to use liquid penetrants at various stages of the fabrication. But there is concern that the residual penetrant material can be a contributor to faulty welds in subsequent passes. This should be investigated and, if true, supplementary development is needed to allow performance of a noncontaminating examination at the desired intervals. Since some welded materials (for example, high-chromium steels) are not permitted to be cooled before postweld heat treatment, there is a need for NDT techniques (e.g., a high-temperature dye penetrant, ultrasonic, or other) that can be performed at the elevated temperature so defects and other undesirable conditions can be detected before the heat treatment. In addition, techniques (e.g., thermal NDT techniques) are needed that can provide accurate knowledge of the preheat temperatures and cooling rate for those high-chromium (9—12%) steels. The ASME Code Case 1594 requires double volumetric examination of the welds, allowing in some instances (e.g., in austenitic stainless steels because of problems with ultrasonics) two radiographic techniques at different angles to satisfy this requirement. Because of the disagreement over the validity and efficacy of the special radiographic requirement, a developmental study should be performed to base the requirement on demonstrated facts or to recommend changes.

As mentioned in the discussion on acoustic emission, promising studies have been conducted in which the welding process was monitored. In many instances the signals received were correlated with the overall integrity of the joint and the presence of discrete discontinuities. In different materials the responses were different (i.e., high emission rate in some materials indicated bad welds, in others good welds). Additional work is needed to establish useful process control instrumentation to increase the possibility that only good welds will be made (or corrective action initiated early for out-of-control conditions).

Because of the importance of the nondestructive examination of austenitic stainless steel weldments, this problem will be considered in an independent section.

<u>Austenitic Stainless Steel Welds.</u> A current and widely recognized problem is the volumetric examination of coarse-grained stainless steels, particularly welds. A principal reason for the problem is the difficulty in applying ultrasonic techniques because of the large, variable attenuation and the high level of noise encountered during an examination. There are widely divergent views ranging from optimism that existing techniques are applicable to pessimism whether ultrasound will ever be useful for stainless steel welds. However, the widespread use of such materials, and the need

for techniques other than radiography (because of the limited capacity of radiography for crack detection and the difficulties of application to in-service inspection) make this an important problem for investigation. The fact that occasional welds can apparently be ultrasonically examined lends confidence that development should be beneficial. Because of the multifaceted nature of the problems, a successful solution will probably require a well coordinated program with approaches from several directions. For example, since some (a few) stainless steel welds can be examined, there is a need for a joint effort with welding development to determine the welding practice, microstructure, or other conditions that can optimize service and inspectability. There is an attendant need for techniques to measure grain size and orientation in weldments since these can affect both the inspection and performance of the weld. The higher-than-average ultrasonic attenuation in austenitic welds fosters a need for techniques that can overcome this problem. The use of angle-beam longitudinal wave ultrasound has enjoyed limited success because of longer wavelengths and decreased sensitivity to the anisotropy in the grains. Other potential approaches (not to be considered exhaustive) include the use of lower frequencies (development will be necessary to overcome usual decrease of sensitivity and resolution with increased wavelength), development of techniques for increased pulse power, and the use of amplitude-insensitive frequency analysis techniques. If the attenuation continues to be a variable despite metallurgical studies, there is a need for techniques to recognize the change so that appropriate adjustments in calibration can be made during the examination.

The problem of excess noise from austenitic weldments also needs study to reach a solution. Potential avenues (again not exhaustive) include the study of focussing of ultrasonic beams, the use of frequency analysis, advanced data processing to improve the signal-to-noise ratio, artificial aperture scanning techniques that integrate the signals over an area of scanning (i.e., similar to acoustic holographic methods), and advanced imaging techniques.

Multiparameter eddy-current techniques should also be studied for application to base metal and welds to detect and measure near-surface flaws, ferrite, and heat treatment condition.

As mentioned in the discussion on acoustic emission, there is question about whether the ductile stainless steel welds will emit signals during deformation or failure. Sufficient study should be performed to satisfy this question because of the high current interest (or push) to use acoustic emission techniques for in-service inspection.

Alloy Identification. An occasional problem when more than one material may be present on a site is the inadvertent mix-up or loss of identification. In such cases there is a need for techniques to provide identification or verification of the proper alloys. This was briefly mentioned as a need for welding electrodes in the welding section, but it is common to all materials. Partial solutions have been achieved through the use of eddy currents (measuring or comparing characteristic electrical conductivities), x-ray fluorescence, thermoelectric properties, and chemical spot tests. The most difficulty is normally encountered in austenitic stainless steels because of the range of allowable chemical compositions and (for eddy currents) overlapping ranges of electrical conductivity. A coordinated effort is needed to develop uniform high-confidence, field applicable techniques and equipment to provide greater assurance that the correct materials are always used.

NDT Development Needs Oriented to Components for Reactors in General

This section is intended to address those needs for NDT development that are the same (or similar) for most advanced reactors.

Design and System Monitoring. As mentioned earlier under Administrative Needs, there must be closer cooperation and coordination between the designers and the NDT personnel. Among the several reasons are the assuring of adequate accessibility to the portion of the component or assembly requiring examination (both in manufacturing and in later in-service inspection) and the enhanced probability that a simple (or existing) examination technique can be applied. Both of these benefits could decrease the need for NDT development applied to specific hardware. Too often the consideration of nondestructive examination is too late (if at all), and expensive development of techniques and mechanical devices is required (or the benefits of NDT are sacrificed). For those items that must be designed and constructed so certain areas needing examination are truly inaccessible, consideration should be given to the design and implantation of fixed sensors (e.g., ultrasonic transducers) which with associated circuitry can be tailored to the specific interrogation. This will require the development of high-reliability transducers and associated devices that can withstand the environment for the life of the components. Rather than have the broad flexibility and versatility that are common to most NDT equipment, these systems could be designed with a "one-track mind;" e.g., concerned only with recognizing a specific signal pattern on a go no-go basis.

In a continuation of this line of thought, more on-line interrogation is needed for the proper functioning of entire components or systems; e.g., as a part of preventive maintenance using signature analysis techniques (thermal, mechanical, acoustic, electromagnetic) to monitor the continued health and proper functioning of items such as pumps and fan bearings. This is in addition to the discrete interrogation using more conventional methods of inspection to detect flaws or other degradation in the structural materials. The performance monitoring would confirm design data and monitor changes in service that could lead to failure. For continuous in-service monitoring capability (as opposed to periodic examination), there is a need for long-life sensors, environmental resistance, and high-confidence calibration techniques to assure factual monitoring.

In-Service Inspection. It is widely recognized that in-service inspection is required to assure the continued integrity of the reactor system. Many of the needs may be peculiar to specific components or reactor systems but a few are common to all. Because of the frequent requirement to perform the examination remotely due to the temperature, irradiation, or other adverse condition, improved devices and techniques are needed to minimize the manpower exposure and to increase the confidence, reliability, and significance of the examination. Heavy reliance is being placed on ultrasonic and acoustic techniques. Investigations and development should be accomplished to allow the extensive use of other methods (e.g., penetrants, eddy currents, and radiography) in the remote adverse environments.

Tubing. Much tubing is used in a reactor system for steam generators, heat exchangers, fuel cladding, etc., and there must be assurance that high integrity is achieved during manufacturing and maintained during service. Good inspection techniques are available for tubing inspection, but improved capability and conversion of laboratory technology into production procedures are needed. As discussed under general needs for NDT methods, there must be improved reliability and quantitativeness in the applied techniques. Developments are needed to allow more reliance on eddy-current techniques for high-speed examination of tubing. As part of this development there must be improved capability for (1) measuring flaw size, (2) characterizing defects, and (3) isolating noise (signals from insignificant variables) that can arise, e.g., as permeability variations caused by localized cold work during straightening of stainless steel tubing. Improvements are also needed in production inspection techniques using ultrasonics to improve the capability for (1) detecting small rounded flaws that are occasionally being overlooked, and (2) determining actual size and orientation of the flaws.

Most tubing specifications establish acceptance (or rejection) levels based on a finite size of flaw regardless of the relative wall thickness at the position of the flaws. Yet studies on the significance of flaws in tubing have indicated a first-order effect that tube strength in ductile materials is based on the residual wall thickness. Thus with tube wall eccentricity a flaw in the thick section could be innocuous, while the same size flaw in a thin section could be cause for failure. For more realistic specifications and to decrease unnecessary rejections, there should be an investigation and, if feasible, a development of inspection techniques that would determine residual wall thickness.

In addition to the usual application of NDT to detect flaws, development should be accomplished to allow nondestructive characterization of tubing materials for manufacturing variability — e.g., measurement of cold work, anisotropy, and grain size. Further comments on the needs for NDT of tubing are in the section on steam generators.

Valves and Pumps. Allegedly pumps and valves have been the major cause of aggravation, problems, and downtime in operating reactor systems. This should generate wariness for advanced reactor systems. Improved volumetric examination procedures (e.g., radiographic and ultrasonic) are needed on the thick-section bodies and impellers for both manufacturing and in-service inspection. Also important (and difficult to examine) are the heavy welds that join the units to the piping system. In-service examination for internal components such as impellers and valve seats may require development of unique NDT devices and techniques to allow valid application in awkward (almost inaccessible) locations. The thickness and bonding of hardfaced valve seats needs to be monitored for both manufacturing and in-service inspection.

Steam Generators. The history of steam generators in power generating systems indicates that leaks between the primary and secondary fluid are almost an accepted fact of life. When water is the common fluid the problem of leaks creates inefficiency, increases down-time at maintenance periods to find and plug the leaks, and requires initial overdesign to allow tubes to be decommissioned without reducing the capacity of the unit below that required. The leaks are aggravating, at times expensive, but in general there are no drastic side effects. However, when different fluids are used in the primary and secondary circuits (e.g., sodium and H_2O, helium and H_2O, molten salts and H_2O) the leak and subsequent mixing can have much more severe consequences as a result of interaction between the fluids (for example, a sodium-water reaction) or as a consequence of the leaking fluid attacking other portions of the system [for example, (1) the sodium-water reaction products leading to caustic corrosion and stress-corrosion cracking in the steam generator, or (2) steam leakage in a gas-cooled reactor causing attack on graphite components]. With the tubing being the minimum membrane separating the two fluids, the greatest concern and need for improved nondestructive examination methods is for tubing and the tube-to-tubesheet joints. With recent recognition that very tiny leaks (below the normal detection sensitivity for tubing inspection) can rapidly grow in service, there is now a new need for higher sensitivity examinations for short, deep flaws (leaks or incipient leaks) in critical tubing; and the development of the higher sensitivity techniques must include the capability for discrimination from long, shallow, acceptable flaws and high-speed performance to keep the costs down. For incipient leaks, the prime candidates would include advanced ultrasonic and eddy-current techniques; for existing leaks, additional approaches include tracer gases and acoustic techniques. The tube-to-tubesheet joints are also important as potential leak sites. For the commonly used face-side welds, the current technology and application provide only marginal examination (usually visual or penetrant techniques to detect flaws on the outer surfaces and occasionally helium leak testing to find leaks that are open at the time of examination). Although the orientation of the fusion zones and the configuration of the weld reduce the potential for further examination, improved techniques can and should be developed to increase the confidence in the integrity of the joints. Depending upon the exact details of a joint design, beneficial advanced techniques could be developed using radiography (with a miniature radiation source in the bore of the tube), high-resolution eddy-currents, or finely focussed ultrasonics.

A major need is for improved technology for the in-service examination of the tubing and tube-to-tubesheet joints to detect incipient failure or, after a leak, to identify and and characterize the leak source for isolation of the tube.

Let's consider an active leak first. There are several instrumental approaches for monitoring the fluids for evidence of a leak (e.g., detectors of moisture, hydrogen, oxygen, and helium), and such "sniffing" systems will not be included in this paper. However, acoustic noise detection should be considered and developed as a complementary mode of leak detection. The active leak will produce acoustic signals (1) as a function of fluid motion through an orifice (the leak) and (2) as bubbles of leaking gas form and collapse in a liquid (either as a simple mechanical action or as a result of chemical reaction). For use as an active, on-line monitoring system to detect and localize (by triangulation) the initiation of a leak, it will be necessary to develop a system that can function despite the noise associated with an operating steam generator and will filter or otherwise discriminate between leak signals and normal background noise. If the latter is proven to be impractical, the developed acoustic system could have potential for localizing a detected leak in a static system by using a pressurized inert fluid in one side of the system to generate noise as it is forced through the leak. Techniques should also be developed for examining localized tubes individually (for example, with tracer gases, acoustic techniques, high-resolution ultrasonics, or eddy currents) to locate and characterize the leak — determining if it were caused by a crack, hole, general weld thinning, or other cause — as a aid to assessing the probability of reoccurrence. Of course, examination of surrounding tubes for damage is also necessary.

For both leaks and incipient leaks (ultrasonics and eddy-current development are recommended for the latter), development is needed to assure the capability to perform high-sensitivity high-resolution examinations despite the presence (and inspection handicaps) of external tubing supports and baffles, thick tubesheets, limited accessibility, and the frequent curves or bends in the tubing. The use of ferromagnetic materials and duplex tubing increases the inspection problems requiring solution. As discussed previously in the section on Methods-Oriented Needs, development of techniques for signal enhancement, quantitative reliable defect characterization, and computerized processing and storage of the data will be necessary not only because of the voluminous data accumulated during an examination but also because of the need for comparison between successive examinations.

Piping. The principal needs for NDT development recognized for piping systems include improved techniques for installation welding, in-service inspection of both pipes and welds, and — when remote maintenance and replacement of piping or components are required — the remote inspection of the replacement welds.

With the trend toward automatic pipe welds, there should be adequate development to allow the same type of automated mechanism to be used for high-confidence nondestructive examination of the joint using ultrasonic and eddy-current techniques immediately after the welding process.

For in-service examination more sensitive techniques are needed for the detection of stress-corrosion cracking in stainless steel piping, particularly at the welds. A current inspection problem (in addition to those cited earlier) with stainless steel welds is that the rough inner surface at the weld creates noise and impedes interpretation. More discriminating ultrasonic techniques are needed and improved capability to perform radiography with improved sensitivity in the field on radioactive specimens would be beneficial. The detection of propagating cracks is extremely important, even with materials that allow the consideration of the "leak before break" philosophy.

For remote inspection of remotely repaired welds, one needs automated examination systems that may be similar to the mechanized automatic system discussed for installation welding or in-service inspection, but probably with a greater tolerance for the environment (temperature, radiation, etc.) that makes the remote operation mandatory.

Pressure Vessel. Because of the diverse nature of pressure vessels for different types of reactors, discussion of needs should be limited to the specific reactor. However, there are some generic problems. For example, the pressure vessels are inherently very large units. Their size creates problems and needs for improved technology to perform volumetric examination. Residual stresses can be an important factor to the vessel, but the current examination technology is inadequate to detect and measure such stresses. Despite the very large size of the vessels, certain areas may need examination with extremely fine detail. Improved large-scale examination or screening techniques are needed to assure that the correct localized areas are identified.

SUMMARY AND CONCLUSIONS

An extensive (but not exhaustive) survey has been conducted of needs for NDT to be developed for advanced reactor systems. The findings have been divided into several major sections: (1) needs for upgrading of nondestructive testing and the respective methods that are commonly applicable without regard to specific reactor systems or components, and (2) needs that are related to materials or components for reactors in general and are not peculiar to any reactor.

Although this paper is a resultant from a multifaceted survey, the author recognizes that continued effort would probably uncover additional needs that have not been cited. Repeated conversations with the same interviewees would probably disclose additional thoughts as fresh experiences (and problems) occur. Further, it is recognized that many of the NDT problems have not yet surfaced because of the preliminary stage of design or planning for many of the components. For these reasons, this paper should be recognized as a first effort, and as a viable document with a continual need for review and an occasional need for updating. It is hoped, with a better definition of the needs, that practical solutions can be obtained, affording significant improvements in the safety, integrity, performance, and availability of advanced reactor systems.

ULTRASONIC INSPECTION TECHNIQUES FOR TWO WELD CLOSURES PROPOSED FOR RSSF WASTE STORAGE CASKS

C. J. Morris
S. R. Doctor
G. J. Posakony
Pacific Northwest Laboratory
Operated by Battelle Memorial Institute

One of the plans being considered for intermediate storage of high level radioactive waste materials is to place these materials in large sealed stainless steel canisters and subsequently store these canisters in a second sealed steel storage cask. Weld procedures are proposed as the closure or seal for these vessels. Inspection of these closures to assure initial and eventual integrity of the closure welds presents a challenge to nondestructive testing. The environment is thermally (500 - 1000°F) and radioactively (10^5 R/hr) hot, necessitating remote inspection procedures.

Under canister development and demonstration programs sponsored by the Department of Energy (DOE) at Rockwell Hanford Operations (Rockwell) and Pacific Northwest Laboratory (PNL), research was performed to develop an ultrasonic test method that could be employed in the environment. As a result of the work, ultrasonic test techniques were developed for inspecting the final weld closure of both the canister and cask vessels. This article discusses the engineering aspects of the weld design required to develop an effective test procedure. Special transducers, coupling techniques and fixturing were developed and demonstrated. Discussions include a description of the ultrasonic tests and results obtained from laboratory and mockup evaluations.

INTRODUCTION

The 2-in. full penetration weld closure technique is the fourth in the series of proposed Retrievable Surface Storage Facility (RSSF) waste cask closure concepts. In this study, the 2-in. full penetration weld closure, shown in Figure 1, was inspected with ultrasound. Before this research was carried out, ultrasonic inspection techniques were developed for three other weld closures. They were a full penetration butt joint closure, a multi-plate fillet closure, and a threaded plug with seal closure weld (1).

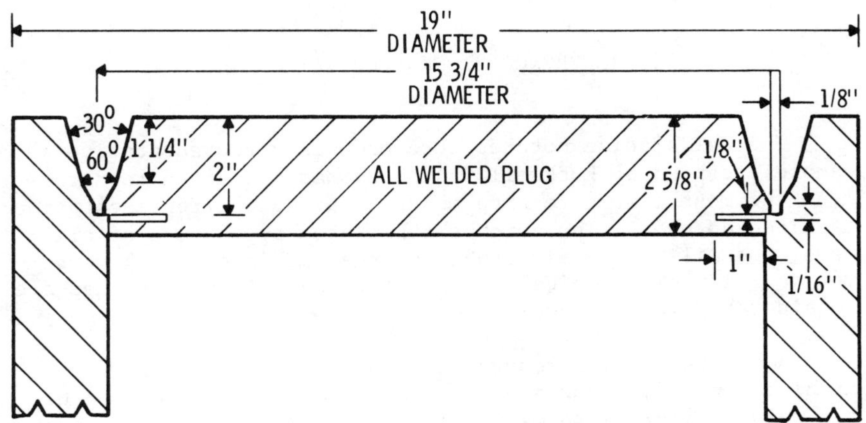

Fig. 1. Full Penetration Weld for RSSF Cask.

The ultrasonic examination of the butt joint, fillet, and seal closure welds was limited to laboratory contact testing at room ambient using a representative closure specimen. In addition to the laboratory testing of the 2-in. full penetration weld, Pacific Northwest Laboratory (PNL) was contracted to develop an ultrasonic inspection system capable of remote operation at temperatures between 400 and 500°F. This would be the anticipated cask temperature in air if the cask were filled with high-level radioactive waste. To accomplish the ultrasonic inspection PNL fabricated special high-temperature transducers and a special mechanical fixture which positioned and coupled the transducers to the RSSF cask. Upon completion of the remote high-temperature inspection, the RSSF cask was drop tested and again inspected to evaluate closure integrity. Two drop tests were conducted by Rockwell Hanford Operations (Rockwell) with the 2-in. full penetration weld and the threaded plug with seal closure weld inspected after each drop. (One closure type was on each end of the cask.)

This report presents a discussion of the work effort for the development of the remote high-temperature ultrasonic inspection system. Areas of concern pertaining to this development effort were:

- reference specimen,
- ultrasonic test technique,
- high-temperature transducer fabrication,
- remote mechanical fixture,
- ultrasonic system operation, and
- ultrasonic data resulting from the testing program.

REFERENCE BLOCK PREPARATION

The calibration notches were made in a representative weld sample provided by Rockwell. A study was conducted to determine the most cost-effective means of producing representative machined notches. The notches fabricated by the use of slit saws, end mills, and electro-discharge machining (EDM) were very similar ultrasonically; however, the slit saw notches are considerably less expensive to machine.

A series of five representative slit saw notches were machined into the test specimen at the crown and the root of the weld. Figure 2 shows the location and orientation of the notches. The orientation of the machined notches was selected on the metallurgical basis that flaws and weld defects are likely to occur along the heat-affected zone (i.e., weld interface). As a result of cask closure geometry, the heat-affected zone is at an angular orientation of ±15° at the crown (referenced to a 0° weld centerline) and ±30° at the root of the weld. In addition to flaws occurring in the heat-affected transition zone, fracture mechanics analysis has shown that flaws of equal

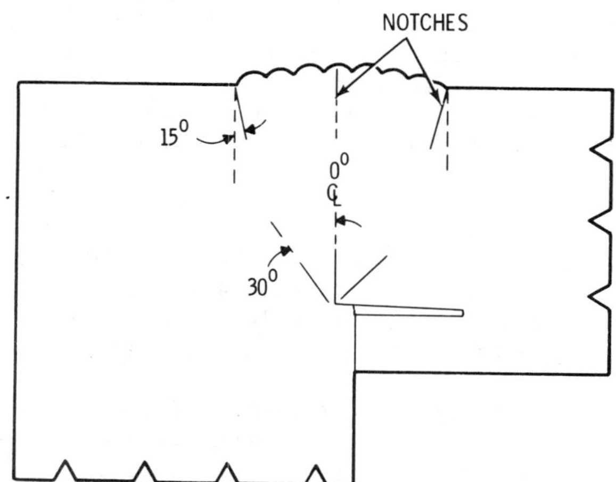

Fig. 2. Location and Orientation of Slit Saw Notches in the Reference Weld Sample.

severity could result at the center of the weld crown and root areas. With the information derived from the fracture mechanics analysis, PNL constructed a representative reference block (Figure 3). The reference block contained slit saw notches at angle orientations of ±15° and ±30° at a depth of 0.3 in. at the crown and root of the weld, respectively. In addition, three machined notches, at depths of 0.1, 0.3 and 0.5 in. along the 0° centerline position, were added to the reference block to provide defect sizing information and a level of system sensitivity.

LABORATORY TESTING

Using the developed reference block, tests were conducted to determine the optimum inspection technique for crown and root examination. Both longitudinal and shear mode techniques were investigated using 2.25 MHz and 5 MHz transducers of the sizes listed in Table 1. A Sperry 771 Reflectoscope with a 10N P/R was used as the ultrasonic pulser/receiver for the laboratory testing.

As a result of the laboratory testing program, an optimum inspection technique was established which requires a transducer positioned on the side of the cask and on top of the cover plate. The inspection of the weld crown is performed using a 0.5-in. diameter transducer, at 2.25 MHz, with a longitudinal wave propagating at an angle of 15° into the crown of the weld. This examination is accomplished with a transducer located on the side of the cask. The root inspection is achieved using a 0.25- by 1.0-in. transducer, at 2.25 MHz, with a shear wave propagating at an angle of 60° into the root of the weld. This examination is accomplished with a transducer located on top of the cover plate. The results of the longitudinal and shear tests on the reference block are presented in Figure 4.

Figure 4 shows a plot of the relative amplitude response versus notch depth. The amplitude response to 0.3-in. notches is 60% or greater regardless of angular orientation. For 0.5-in. notches a minimum reponse of 80% is achievable. Notches in the range of 0.1 in. are detectable, and serve to evaluate test sensitivity. However, a 0.1-in. flaw has a very low probability of growth that might threaten the integrity of the cask. The data show a nonlinear response that assures a high degree of test confidence for detecting the larger flaws of interest.

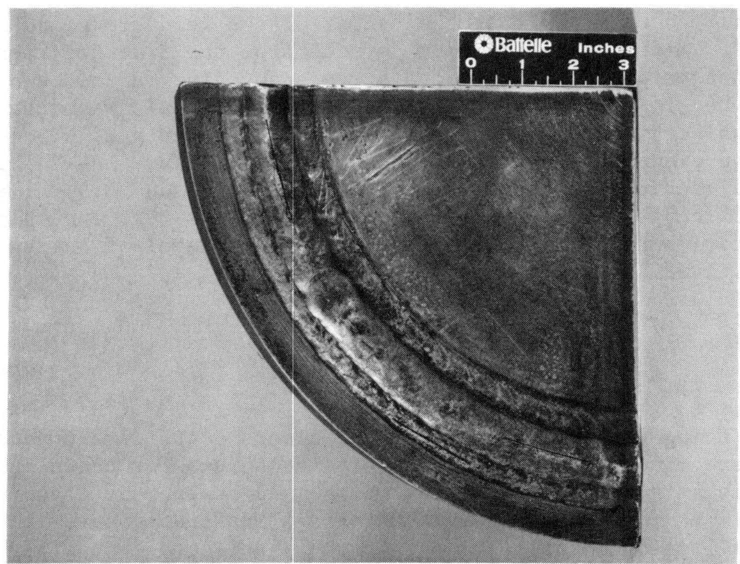

A. Crown Notches

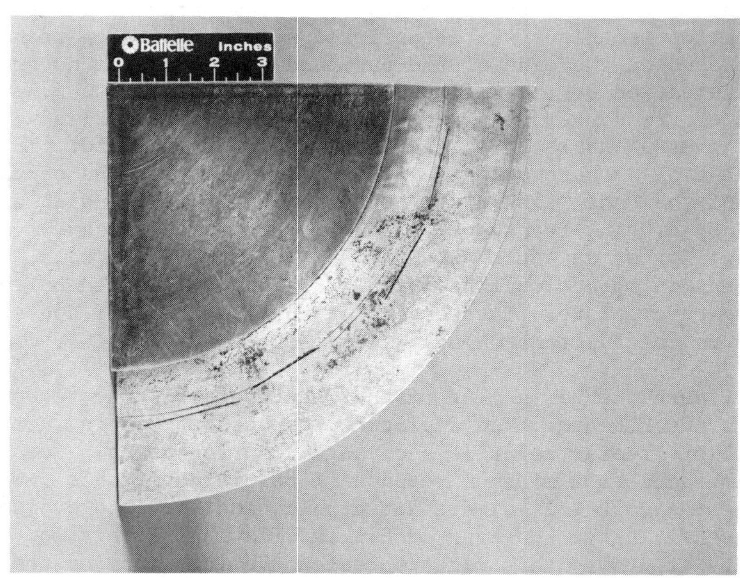

B. Root Notches

Fig. 3. Ultrasonic Reference Block.

Table 1. Size and Frequency of Piezoelectric Elements Investigated.

	Propagation Angle	Size	Frequency	Wedge Material
Longitudinal Technique	43°	0.5 in. x 1.0 in.	2.25 MHz	Macor®
	45°	0.75 in. x 1.0 in.	2.25 MHz	plastic
	15°	0.25 in. x 1.0 in.	2.25 MHz	plastic
	15°	0.25 in. x 0.3125 in.	2.25 MHz	plastic
	15°	0.5-in. diameter	5.0 MHz	plastic
Shear Mode Technique	45°	0.75 in. x 1.0 in.	2.25 MHz	plastic and vespel
	45°	0.5-in. diameter	2.25 MHz	plastic
	35°	0.5-in. diameter	2.25 MHz	plastic
	23°	0.75 in. x 1.0 in.	2.25 MHz	plastic
	45°	0.25 in. x 1.0 in.	2.25 MHz	plastic
	35°	0.25 in. x 1.0 in.	2.25 MHz	plastic
	60°	0.5-in. diameter	2.25 MHz	plastic
	60°	0.25 in. x 1.0 in.	2.25 MHz	plastic

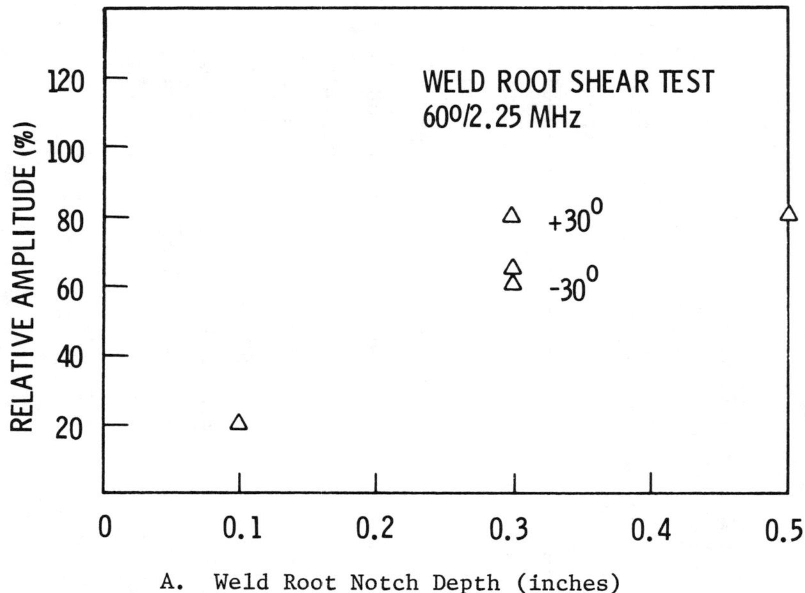

A. Weld Root Notch Depth (inches)

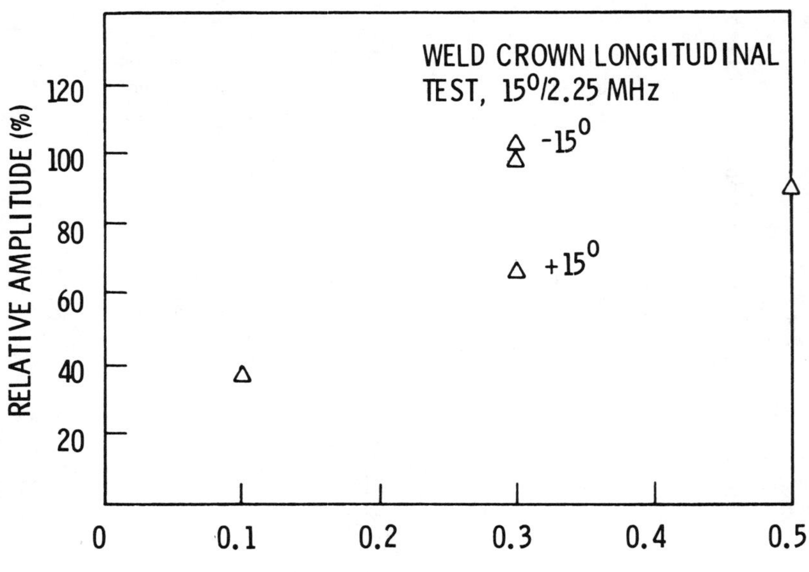

B. Weld Crown Notch Depth (inches)

Fig. 4. Ultrasonic Response to Slit Saw Notch Depth for the 2-in. Full Penetration Weld.

TRANSDUCER DEVELOPMENT

High temperature transducers were designed to withstand thermal surface temperatures between 400 and 500°F. Special bonding agents were used to bond the piezoelectric elements to the selected wedge material. In addition, special transducer backing and damping material was used to achieve optimum transducer performance.

The transducer wedges were fabricated from high density graphite, Macor®[(a)] (a ceramic), and Aerotherm®.[(b)] The graphite exhibits the best high temperature performance with essentially no change in ultrasonic velocity up to 600°F. The high density graphite was selected as the primary wedge material for the demonstration test with backup transducer wedges constructed using Macor® and Aerotherm®.

MECHANICAL FIXTURE DEVELOPMENT

A mechanical fixture was developed which was used to position and hold the transducers as the waste cask rotated. The fixture was designed with the knowledge that horizontal, vertical, and angular positioning was available if the fixture could be attached to the weld head assembly system. A dovetail bracket was constructed to adapt the fixture to the weld head assembly. Figure 5 shows the mechanical fixture with associated transducers. Also shown in Figure 5 is the three-point rolling ball transducer support system, which adjusts for desired couplant layer thickness. In addition, spring loading was provided to pressure couple the transducers to the cask. Figure 6 shows the mechanical fixture positioned for inspection of the cask.

ULTRASONIC TEST SYSTEM

Figure 7 shows a schematic representation of the demonstration ultrasonic test system for RSSF Cask inspection. The system consists of a Branson 303 ultrasonic pulser/receiver with a Hewlett-Packard two-channel strip chart recorder connected to the gated output of the Branson 303. A Tektronix 454 oscilloscope is connected to the Branson 303 for photographing specific ultrasonic reponse data. The two-channel

---

(a) Trademark of the Corning Glass Co.
(b) Trademark of the Aerotect Corporation.

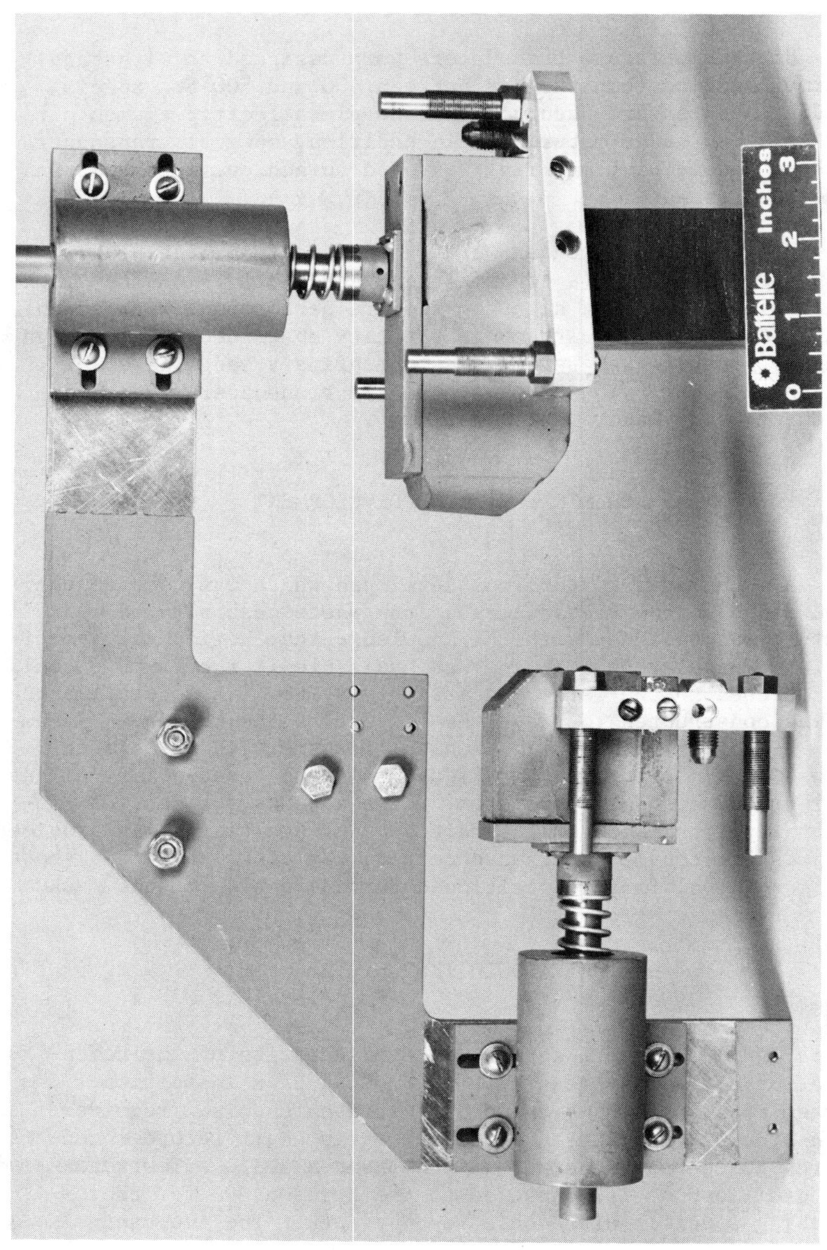

Fig. 5. Mechanical Fixture with High Temperature Transducers.

Fig. 6. Mechanical Fixture Positioned for Inspection.

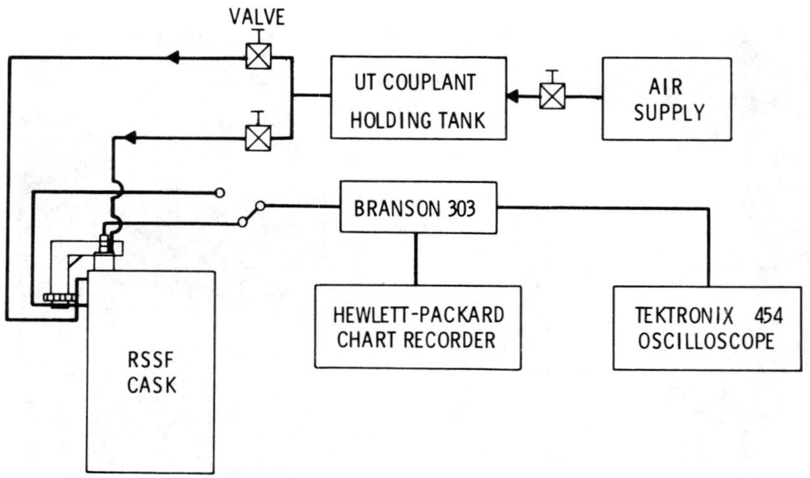

Fig. 7. Block Diagram of the Demonstration System Used for the Ultrasonic Test.

chart recorder provided an on-line record of the ultrasonic response during the remote scanning operation. The second channel indicated the response from defects which were greater than the 50% accept/reject threshold. The air supply provided 30 psi of pressure to the high temperature ultrasonic couplant (Pyrogel®[a]) holding tank. During operation, appropriate valves are sequentially opened and couplant is fed to the transducers and uniformly spread about the cask surface. Figure 8 shows the mockup hot cell test facility with the ultrasonic test system.

SYSTEM EVALUATION

The major problem encountered in the program was the failure to obtain a sufficient bond between the lead metaniobate piezoelectric element and the selected wedge material. Silver conductive epoxy, rated at 500°F, was used for this bond. Previous work of similar application at Westinghouse Hanford demonstrated that the silver epoxy performed satisfactorily at temperatures between 400 and 500°F. However, in this particular application, significant loss of ultrasound occurred after short exposures to temperatures ranging from 200 to 450°F. Because of this signal loss, inspection of the cask weld at

(a) Trademark of the Echo Laboratories.

Fig. 8. Demonstration Setup Showing the Ultrasonic Inspection System.

400°F was not completed. A scan of the cask weld was successfully performed at a temperature of 200°F. Due to the time constraints of the program, laboratory testing of the transducers at temperature did not take place before the mockup demonstration. Recently, PNL has tested transducers using Stycast 2762 FT ,[a] a high-temperature epoxy. The tests have shown that the present transducers perform satisfactorily at 350°F during a 72-hr test period. With an additional design alteration, the transducer will be capable of long-term operation at 400°F.

Another key problem that surfaced during the demonstration phase was that the attempt to calibrate the system using a reference target resulted in a totally unsatisfactory referencing procedure. A dynamic reference system is needed. The ideal system would consist of a reference block, a turntable for rotation, and a heating system which enables dynamic target referencing at temperature.

Several features of the inspection system performed successfully. The three-point ball system to support the transducers did an excellent job for a prototype system. The pressurized ultrasonic couplant and feed system performed satisfactorily and should not require further development. The couplant spreading concept was effective, producing a good signal to noise ratio at operating temperatures. The spring loading of the transducers also performed well, although the fixture did have a tendency to twist counterclockwise under pressure and thus cause the transducer on the surface of the cask to lift slightly. However, this was alleviated by reducing the spring load.

TEST RESULTS

The test results for the 2-in. full penetration weld, and the threaded plug and seal weld are presented in Figures 9 and 10, respectively. Figures 9 and 10 show the pre-drop and post-drop data with the impact area designated for each weld closure.

The 2-in. full penetration weld inspection results were produced using both the contact hand inspection and the remote inspection technique. The results obtained using the remote

(a) Trademark of Emerson and Cuming, Inc.

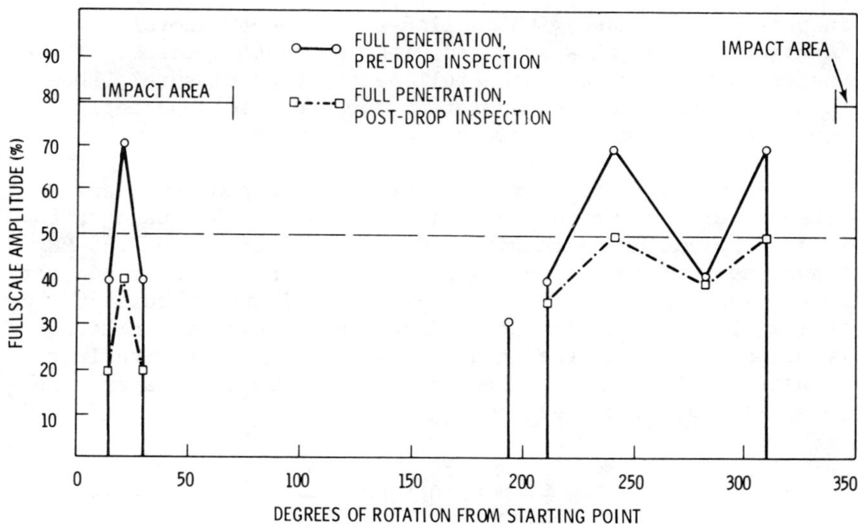

Fig. 9. Ultrasonic Examination Results for the Full Penetration Weld.

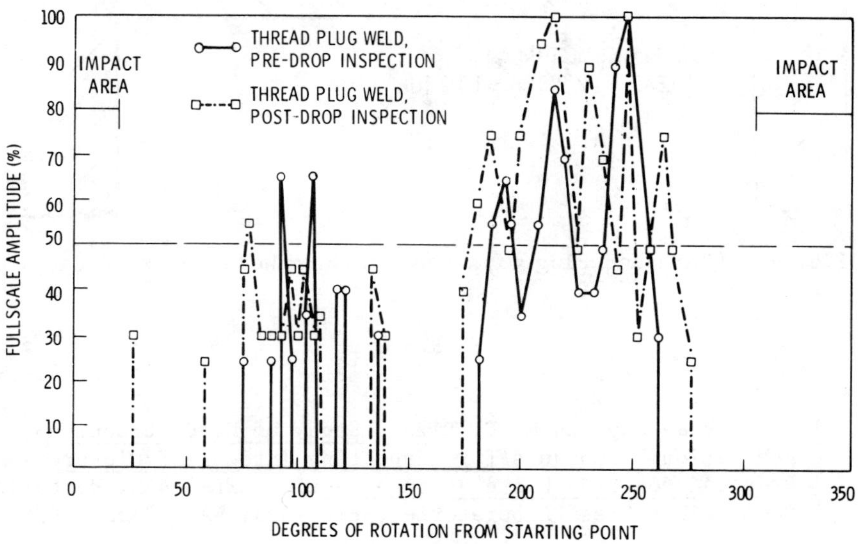

Fig. 10. Ultrasonic Examination Results for the Threaded Plug Closure.

inspection technique with on-line strip chart recording were identical to the data produced with the contact tests. The post-drop results (from the full penetration closure), show that the cracks detected prior to the drop had apparently closed slightly as a result of the drop test.

The ultrasonic testing of the threaded plug and seal closure weld (shown in Figure 11) was carried out using a 60° shear propagation angle with a 0.5-in. diameter and 2.25-MHz transducer. The testing was performed using the hand contact inspection technique. The post-drop results of Figure 10 show that a slight crack growth produced an increase in amplitude response. However, the changes shown are minor and reflect, in part, variations in transducers used, couplant layer thickness, and operator dependence.

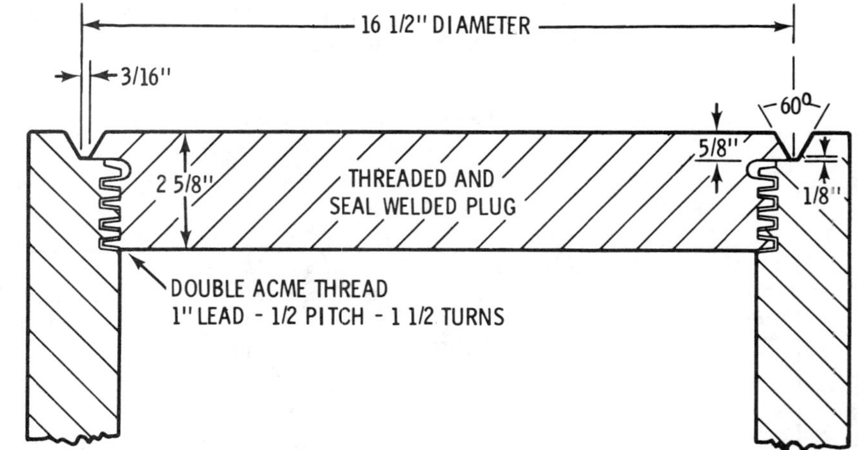

Fig. 11. Threaded Plug with Seal Closure Weld for RSSF Cask.

REFERENCE

(1) G. J. Posakony and N. E. Dixon, Study of Nondestructive Evaluation Techniques for Inspecting Three Weld Closures Proposed for RSSF Waste Storage Cask. BNWL-B-473, Battelle, Pacific Northwest Laboratories, Richland, WA 99352, 1976.